169

Advances in Polymer Science

Advances in Polymer Science

Recently Published and Forthcoming Volumes

Long-Term Properties of Polyolefins

Volume Editor: Ann-Christine Albertsson

With contributions by
A.-C. Albertsson · S. Al-Malaika · S. M. Desai · P. Eriksson
U. W. Gedde · M. Hakkarainen · K. Jacobson · S. Karlsson
H. Knuuttila · E. Kokko · A. Lehtinen · B. Löfgren · A. Mattozzi
A. Nummila-Pakarinen · C. J. G. Plummer · T. Reitberger
J. Seppälä · R. P. Singh · B. Stenberg

 Springer

The series presents critical reviews of the present and future trends in polymer and biopolymer science including chemistry, physical chemistry, physics and material science. It is addressed to all scientists at universities and in industry who wish to keep abreast of advances in the topics covered.

As a rule, contributions are specially commissioned. The editors and publishers will, however, always be pleased to receive suggestions and supplementary information. Papers are accepted for "Advances in Polymer Science" in English.

In references Advances in Polymer Science is abbreviated Adv Polym Sci and is cited as a journal.

The electronic content of APS may be found at http://www.springerLink.com

ISSN 0065-3195
ISBN 3-540-40769-3
DOI 10.1007/b13502
Springer-Verlag Berlin Heidelberg New York

Library of Congress Catalog Card Number 61642

Springer-Verlag is a part of Springer Science+Business Media

springeronline.com

© Springer-Verlag Berlin Heidelberg 2004
Printed in Germany

Typesetting: Stürtz AG, Würzburg
Cover: Künkellopka GmbH, Heidelberg; design&production GmbH, Heidelberg

Printed on acid-free paper   02/3020/kk – 5 4 3 2 1 0

Volume Editor

Prof. Dr. Martin Möller

Deutsches Wollforschungsinstitut
an der RWTH Aachen e.V.
Veltmanplatz 8
52062 Aachen, Germany
E-mail: moeller@dwi.rwth-aachen.de

Prof. Oskar Nuyken

Lehrstuhl für Makromolekulare Stoffe
TU München
Lichtenbergstr. 4
85747 Garching, Germany
E-mail: oskar.nuyken@ch.tum.de

Dr. E. M. Terentjev

Cavendish Laboratory
Madingley Road
Cambridge CB 3 OHE
United Kingdom
E-mail: emt1000@cam.ac.uk

Prof. Brigitte Voit

Institut für Polymerforschung Dresden
Hohe Straße 6
01069 Dresden, Germany
E-mail: voit@ipfdd.de

Prof. Gerhard Wegner

Max-Planck-Institut für Polymerforschung
Ackermannweg 10
Postfach 3148
55128 Mainz, Germany
E-mail: wegner@mpip-mainz.mpg.de

Preface

We dedicate the current volume entitled "Long-Term Properties of Polyolefins" to Professor Kausch on his 25th anniversary as editor of Advances in Polymer Science. Professor Kausch pioneered the work on molecular effects in the fracture of polymers. This is beautifully summarized in his books on polymer fracture. Professor Kausch is also the perfect gentleman – always eager to help newcomers to make their entrance into the scientific community and to assist his colleagues in their work and accomplishments. With his work, Professor Kausch has demonstrated the importance of "source science" – to present new data – and to present reviews of previously published material. This book is presented in the spirit of Professor Kausch, namely showing a good selection of data and explaining what they mean.

The main focus of this book is the relation between structure and properties and the trend towards better quality and reproducibility. The first chapter describes the metallocene polymerisation catalysts and their possibility not only of tailoring polymer properties but also of manufacturing entirely new materials. Due to improved control of microstructure, it will also be possible to produce specialty polyolefins which could compete with non-olefinic polymers. The next chapter shows how in each new development step catalyst and process innovations have gone hand in hand and how the control over polymer structure and the ability to tailor material properties has increased. For a better understanding of properties and behaviour, the basic of morphology is fundamental and is described in chapter three, followed by chapter four about fracture properties and microdeformation behaviour. Promising model systems for the investigations of the relations between crack-tip deformation, fracture and molecular structure are also presented. Chapter five gives an overview of stabilization of polyethylene crucial for long-term properties. Two main approaches have been used; the first advocates the use of biological antioxidants, and the second relies on the use of reactive antioxidants that are chemically attached onto the polymer backbone for greater performance and safety.

Chemiluminescence is presented as a tool for studying the initial stages in oxidative degradation and is explained in chapter six. However, for many years, tailor-made structures specially designed for environmental degradation have also been a reality. One of the key questions for successful development and use of environmentally degradable polymers is the interaction between the degradation products and nature and this is illustrated in chapter seven. The development of chromatographic methods and use of chromatographic fingerprinting gives not only degradation products but also information about degradation mechanisms as well as interaction between the polymer and different environments. The obstacles and possibilities for recycling of polyolefins are discussed in chapter eight with special emphasis on analytical methods useful in the quality concept. It is also shown how recycled material could be a valuable resource in the future together with renewable resources. Finally, chapter nine gives examples of existing as well as emerging techniques of surface modification of polyethylene.

These chapters together will hopefully inspire to a new generation of polyethylene by mimicking nature and use of new molecular architecture, new morphology and also "activated" additives in microdomains, with even more reproducible properties within narrow limits and with predetermined lifetimes.

January, 2004   Ann-Christine Albertsson

Advances in Polymer Science
Also Available Electronically

For all customers who have a standingorder to Advances in Polymer Science, we offer the electronic version via SpringerLink free of charge. Please contact your librarian who can receive a password for free access to the full articles by registering at:

http://www.springerlink.com

If you do not have a subscription, you can still view the tables of contents of the volumes and the abstract of each article by going to the SpringerLink Homepage, clicking on "Browse by Online Libraries", then "Chemical Sciences", and finally choose Advances in Polymer Science.

You will find information about the

– Editorial Board
– Aims and Scope
– Instructions for Authors
– Sample Contribution

at http://www.springeronline.com using the search function.

Contents

Adv Polym Sci (2004) 169:1–12
DOI: 10.1007/b13518

Specific Structures Enabled by Metallocene Catalysis in Polyethenes

Barbro Löfgren · Esa Kokko · Jukka Seppälä

Polymer Technology, Polymer Science Centre, Helsinki University of Technology,
P.O. Box 3500, 02015 HUT, Finland
E-mail: barbro.lofgren@hut.fi

Abstract This chapter briefly describes the history of metallocene polymerisation catalysts. The behaviour of metallocenes, particularly in copolymerisation of ethene with higher linear α-olefins is discussed. A class of olefin polymerisation catalysts based on siloxy-substituted bis(indenyl) metallocenes is reported. These ligand systems offer efficient opportunities for copolymerisation of ethene with heteroatoms containing monomers and traditional α-olefins. In addition, the formation of long-chain branching, attributed to copolymerisation of vinyl terminated polyethenes, is discussed. Examples of siloxy-substituted bis(indenyls) with high copolymerisation capability and high vinyl selectivity are highlighted.

Keywords Siloxy-substituted metallocenes · Long-chain branching · Rheological properties · Comonomer response · Vinyl selectivity

1
Introduction

The main benefits of metallocene catalysts in comparison to the conventional Ziegler-Natta catalysts are well-defined microstructures, high activity, narrow molar mass distribution and the possibility of tailor-made polyolefins. The advantages in metallocene catalyst chemistry offer a promising novel way not only to tailor polymer properties, but also to manufacture entirely new polymeric materials.

The development of metallocene catalysts, also called single-site catalysts, was stimulated by the discovery of highly effective activators. Metallocenes combined with aluminium alkyls for polyolefin polymerisation are not a new discovery. Natta et al. [1] with the homogeneous $Cp_2TiCl_2/AlMe_3$ system and Breslow and Newburg [2] with the $Cp_2TiCl_2/AlEt_2Cl$ catalyst system conducted almost simultaneously the first metallocene polymerisations in the 1950s. Due to their inferior activity, compared to the heterogeneous $TiCl_3$ and $TiCl_4$ Ziegler-Natta catalysts, the time was not right for the metallocenes. Nevertheless, the research on metallocenes continued focusing on polymerisation mechanisms, which was more straightforward in homogeneous catalyst systems.

2
Milestones in Metallocene Catalysis

Reichert and Meyer [3] were the first to demonstrate in the early 1970s an enhancement in the polymerisation activity upon addition of water to the $Cp_2TiEtCl/AlEtCl_2$ catalyst system. Soon thereafter, Long and Breslow [4] reported similar observations for the Cp_2TiCl_2/Me_2AlCl system, as did Kaminsky and Sinn [5] for the $Cp_2TiMe_2/AlMe_3$ catalyst and Cihlár et al. [6] for the $Cp_2TiEtCl/AlEtCl_2$ catalyst. In 1975 Kaminsky and Sinn [7] conducted ethene polymerisation with the catalyst Cp_2ZrCl_2/TEA, which can be considered the first effective metallocene catalyst.

The breakthrough in metallocene catalyst development occurred in the early 1980s when a metallocene catalyst, instead of an aluminium alkyl, was combined with methylaluminoxane (MAO) [8, 9, 10]. This catalyst system boosted the activity of metallocene-based catalyst and produced uniform polyethene with the narrow molar mass distribution typical for single-site catalysts. Efforts to polymerise propene failed, however: the product was found to be fully atactic, indicating complete lack of stereospecificity of the catalyst [10].

The synthesis of isotactic polypropene had to wait for sterically rigid catalysts: a group of catalysts in which the movement of the rings is hindered and the angle between the ligands is widened with a bridge. In 1982 Brintzinger et al. [11] synthesised a chiral *ansa*-bis(indenyl) titanocene with C_2-symmetry. This catalyst, together with MAO, was used by Ewen to first demonstrate the synthesis of isotactic polypropene [12]. This discovery initiated an extensive study in the field of stereospecific polymerisation by varying the composition, structure and the type of symmetry (C_2, C_1 and C_s) of sterically hindered *ansa* metallocenes. Soon several papers concerning the isospecific [13, 14, 15] and syndiospecific [16] polymers were published. At the same time Ishihara et al. [17] synthesised syndiotactic polystyrene with a half-sandwich titanocene catalyst. The most important characteristic of metallocenes, the type of symmetry, determines the stereospecificity. The *racemic* form of metallocenes with a C_2-symmetry gives isotactic polymers, while metallocenes with the C_s-symmetry are used to prepare syndiotactic polymers, and C_1-symmetry leads to the formation of isotactic, hemiisotactic, and stereoblock polyolefins.

However, the properties of the polypropene produced by the catalysts $Et(Ind)_2ZrCl_2$ or $Et(H_4Ind)_2ZrCl_2$ discovered by Kaminsky and Brintzinger were not satisfactory from the industrial point of view as the catalytic selectivity and molar mass of the polymer tended to stay only moderate. The change of the ethene-bridge to a silyl-bridge doubled the molar mass and increased the activity. Addition of a methyl group into the $2(\alpha)$ position enhanced the catalytic performance further, but the molar mass was still low [18]. After attachment of a phenyl ring to the 4-position, the obtained catalyst rac-Me$_2$Si(2-Me-Benzindenyl)$_2$ZrCl$_2$/MAO showed both high activity and polymer molar mass [19]. The optimisation of the bis(indenyl) catalyst systems lead Spaleck et al. [20] to the silylene-bridged 2-methyl-4-naphthyl-substituted bis(indenyl) zirconium dichloride catalyst, which was an answer to the request of improved isospecific metallocenes. Along the same lines was the first siloxy-substituted bis(indenyl), ethylene bis (2-tert-Butyl-dimethylsiloxyindenyl)zirconium dichloride reported by Leino et al. [21]. This siloxy-substituted catalyst showed high activity in copolymerisation of ethene with higher α-olefins.

In the early 1990s supported metallocenes were introduced to enable gas phase polymerisation. Also ethene/α-olefin copolymers with high comonomer content, cycloolefin copolymers and ethene-styrene interpolymers became available. In 1990 Stevens at Dow [22] discovered that titanium cyclopentadienyl amido compounds (constrained geometry catalysts) are very beneficial for the copolymerisation of ethene and long-chain α-olefins.

In the beginning, the major focus was on the early transition metals such as Ti, Zr and Hf, but today the potential of the late transition metals complexes of Ni, Pd, Co and Fe is well recognised. As the late transition metals are characteristically less oxophilic than the early metals, they are more tolerant towards polar groups. Therefore, it was assumed that with late transition metals catalysts one could produce a wide range of different polymers.

A nickel-based catalyst system, which produces, in the absence of comonomers, highly short-chain branched polyethene was developed by Brookhart et al. [23]. Independently, the groups of Brookhart [24, 25, 26] and Gibson [27, 28, 29, 30] developed efficient iron- and cobalt-based catalyst systems. Nickel or palladium is typically sandwiched between two α-diimine ligands, while iron and cobalt are tridentate complexed with imino and pyridyl ligands.

Grubbs' group [31, 32] developed another type of Ni-based catalyst. This neutral Ni-catalyst, based on salicylaldimine ligands, is active in ethene polymerisation without any co-activator and originated from the Shell higher olefin process (SHOP). Shortly thereafter another active neutral P,O-chelated nickel catalysts for polymerisation of ethene in emulsion was developed by Soula et al. [33, 34, 35]. The historical development of single site catalysts is represented in Fig. 1.

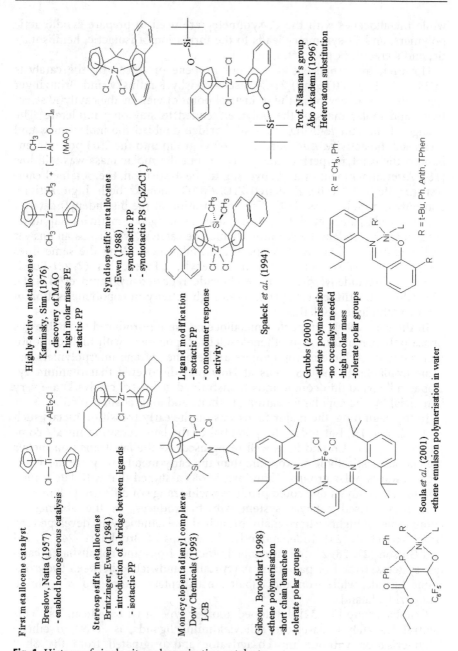

Fig. 1 History of single site polymerisation catalysts

The key to the extremely high activity of single-site catalysts in polymerisation is the activator. MAO is the most common one and it is well recognised that MAO forms cationic alkyl complexes containing a weakly or non-coordinating anion, which is complexed by the MAO molecule. Although MAO is used in industrial processes, a number of other activators have been developed lately. Some boron organic compounds, such as trisphenylmethyl-tetrakis(pentafluorophenyl)borate $[C(CH_3)_3]^+$ $[B(C_6F_5)_4]^-$, seem to especially fulfil a role as non-coordinating, non-nucleophilic counteranion to the active cationic species [36, 37, 38, 39, 40].

3
Metallocenes with High Activity for Ethene Polymerisation

Unbridged, bridged, substituted and half-sandwich complexes have been used as catalysts for ethene polymerisation. In the series of various *ansa* cyclopentadienyl, indenyl and fluorenyl derivatives of metallocene complexes, the bridged indenyl complexes show the highest activities as homogeneous ethene polymerisation catalysts. In general, zirconium catalysts are more active than the hafnium or titanium systems. These metallocene catalyst systems provided a unique opportunity for new tailor-made polymers and increased the range of comonomers that could be used. An interesting peculiarity of metallocene systems is that the difference in the reactivities of ethene, propene and higher α-olefins is markedly less compared to the conventional Ziegler-Natta catalysts [41]. This opens many new possibilities for the synthesis of copolymers.

3.1
Copolymerisation Behaviour

3.1.1
α-Olefin Comonomer Response

Ethene copolymers of higher α-olefins such as 1-butene, 1-hexene and 1-octene are industrially important materials. In contrast to Ziegler-Natta catalysts, metallocene catalyst systems offer several advantages. Metallocenes give, as already earlier has been mentioned, narrow molar mass distribution and high comonomer incorporation and, in addition, even compositional distribution. Copolymerisation of ethene with α-olefins, ranging from C_3-C_{18}, has been reported by us [42, 43, 44] and by several other groups [45, 46, 47, 48 ,49, 50, 51, 52, 53, 54]. We have studied incorporation of comonomers as a function of ligand substitution pattern and interannular bridge. Some general observations about the influence of the ligand structure on the polymer structure have been made [55, 56], although the details are not known of how both electronic and steric effects, in combination with each other, determine the reactivities of reactants.

Generally, *ansa*-metallocenes with bridges between the ligands incorporate α-olefin comonomers better than unbridged ones. It has also been shown that benzannelation of bridged indenyl ligands increases reactivity toward α-olefins.

The transition metal has, in addition to ligand substitution and interannular bridge, a distinctive effect on the copolymerisation behaviour. Hafnium-based metallocenes are considered to exhibit higher comonomer response than zirconium based ones [45].

Recent developments in catalyst design have offered more sophisticated catalyst structures with even better copolymerisation ability. $Me_2Si(Me-Benz(e)Ind)_2ZrCl_2$ is an example of such a catalyst [57, 58]. This catalyst, as mentioned earlier, was originally designed for polymerisation of isotactic polypropene, but it also shows a very high comonomer response in ethene polymerisation [51].

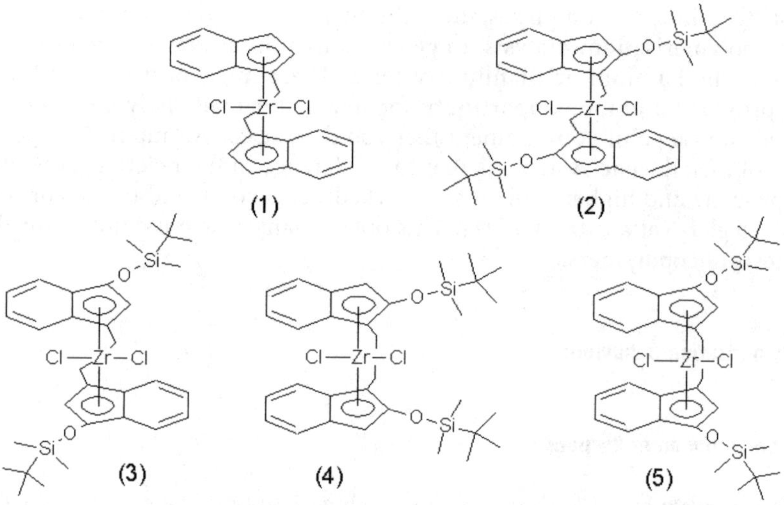

Catalyst	1	2	3	4	5	Comment
r_{Ethene}	48±4	55±3	19±4			at 80°C
$r_{1-Hexene}$	<0.02	0.005	0.006			
r_{Ethene}		36±3	16±1	11±1	10±1	at 40°C
$r_{1-Hexene}$		0.003	0.005	0.005	0.001	

Fig. 2 Comonomer response of siloxy-substituted metallocene catalyst in copolymerisation of ethene with 1-hexene. The catalysts are: (*1*) *rac*- Et(Ind)$_2$ZrCl$_2$; (*2*) *rac*- Et(2-*tert*-BuSiMe$_2$OInd)$_2$ ZrCl$_2$; (*3*) *rac*- Et(3-*tert*-BuSiMe$_2$OInd)$_2$ ZrCl$_2$; (*4*) *meso*- Et(2-*tert*-BuSiMe$_2$OInd)$_2$ ZrCl$_2$; (*5*) *meso*- Et(3-*tert*-BuSiMe$_2$OInd)$_2$ ZrCl$_2$

The use of heteroatoms, such as N, O, S and P, as ligand substituents has resulted in new catalyst families with improved comonomer response. For example, *rac*-Et(3-tert-BuSiMe$_2$O Ind)$_2$ZrCl$_2$ gives an almost threefold increase in comonomer response compared to *rac*-Et(Ind)$_2$ZrCl$_2$ catalyst (see Fig. 2). The siloxy-substituted single-site catalysts also have some other advantages over conventional metallocenes. Siloxy-substituted bridged bis(indenyl) metallocenes also exhibit, except for high 1-hexene response, high vinyl end group selectivity. In addition, the siloxy groups are beneficial when the metallocene is heterogenised to suit industrial purposes better.

Figure 2 presents examples of some siloxy-substituted metallocene catalysts (2–5) and, for comparison, an ordinary bis(indenyl) catalyst (1). The relative reactivity in ethene/1-hexene was investigated [44, 59, 60, 61, 62] in order to predict their copolymerisation ability compared with the conventional bis(indenyl) catalyst (1).

The comonomer response (compare *rac*-catalysts 2 and 3 with the corresponding *meso*-catalysts 4 and 5 in Fig. 2) can be improved by using *meso*-isomers of C$_2$ symmetric metallocenes [60]. In copolymerisation of ethene with 1-hexene the reactivity ratios of siloxy-substituted *meso*-isomers, catalysts 4 and 5, are comparable with those of Me$_2$Si(2-Me-Benz(e)Ind)$_2$ZrCl$_2$ [52], which is considered to be a highly efficient copolymerisation catalyst.

3.1.2
Copolymerisation of Functional Monomers with Ethene

Incorporation of functional groups into polyolefins has long been a scientifically interesting and technologically important subject. Metallocene catalysts have opened new perspectives in this area, through direct copolymerisation of ethenes with functional monomers [63, 64, 65, 66, 67, 68, 69, 70, 71, 72]. The direct copolymerisation of ethene with oxygen- or nitrogen-containing comonomers is, however, in general accompanied by a great loss in activity of the catalyst. Our recent studies [73] show that a siloxy-substituted catalyst is superior to the ethylene bis(indenyl) zirconocene catalyst and can successfully copolymerise ethene with 10-undecen-1-ol (see Table 1). The activity in copolymerisation of the siloxy-substituted catalyst is almost tenfold the ordinary bis(indenyl) catalyst.

Table 1 Effect of the catalyst on copolymerisation of ethene with 10-undecen-1-ol

Catalyst	Comonomer in feed	Activity	MW	Comonomer in polymer
	mmol L^{-1}	kg (mol Zr*h)$^{-1}$	g mol^{-1}	wt%
Et(Ind)$_2$ZrCl$_2$	-	90,000	130,000	-
Et(Ind)$_2$ZrCl$_2$	25	3,000	75,000	5.9
Et(3-tert-BuSiMe$_2$Oind)$_2$ZrCl$_2$	-	250,000	130,000	-
Et(3-tert-BuSiMe$_2$Oind)$_2$ZrCl$_2$	25	25,000	80,000	4.7

3.1.3
Chain Transfer Mechanisms

Vinyl end groups are dominant in polyethenes prepared by metallocene catalysts . The vinyl end groups are a result of chain transfer to monomer, β-H elimination or σ-bond metathesis. Vinyl selectivity was almost 100% with both the siloxy-substituted catalyst 2 and with indenyl zirconocene catalyst 1, while it was slightly less for the siloxy-substituted catalyst 3. The catalysts 1 and 3 showed some tendency towards isomerisation. The major chain-transfer mechanism in all these three catalysts is chain transfer to monomer [44, 59, 60]. In addition to vinyl end groups, Thorshaug et al. [74] have observed trans-vinylenes, verified by FT IR measurements. They proposed that an isomerisation reaction related to the chain transfer is the origin of the trans-vinylene end groups found in polyethene. In ethene-1-hexene copolymers, chain transfer after 2,1-insertion results in an internal trans-vinylene. Small amounts of vinylidene double bonds have also been found [44, 59, 61], which may originate from a side reaction where a vinyl-terminated polyethene macromonomer is reincorporated into the growing chain, followed by a subsequent termination [59, 75]. The vinylidene bonds can possibly result from the migration of an alkyl chain followed by a chain-transfer reaction [76].

3.2
Long Chain Branching

Commodity high pressure PE-LD grades have a broad polydispersity and they contain small amounts of LCB, which results in excellent melt flow properties. Due to the broad polydispersity (PD), the mechanical properties are inferior, even compared with conventional Ziegler-Natta resins. However, narrow PD metallocene resins have poor processing properties. Recent advances in catalyst technology, have resulted in metallocene polyethenes that contain small amounts of LCB to ease the processability.

The constrained geometry catalyst (CGC), developed at Dow, was the first single-site catalyst [77, 78] discovered to be capable of producing long-chain branched polyethene. Since the announcement of the ability of CGC catalyst to produce small amounts of LCB, several conventional bis (cyclopentadienyl)-based metallocene complexes have been reported to produce low levels of LCB [59, 79, 80, 81, 82, 83].

The LCB in metallocene-catalysed ethene polymerisation is considered to occur via a copolymerisation reaction where a vinyl-terminated polyethene chain is reinserted into a growing chain. Thus, the choice of the catalyst used will be extremely crucial. When the prerequisites of LCB are fulfilled, the process conditions will then be even more important [44, 60].

Accordingly, siloxy-substituted complexes have been found to give polymers with rheological behaviour indicative of long-chain branching [44, 60, 61, 83, 84], but the relations between molar mass, PD and the low shear-rate rheological behaviour are distinctive to each catalyst [83].

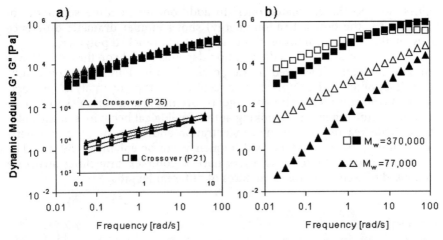

Fig. 3a, b Dynamic frequency sweeps of **a** long-chain branched and **b** linear polyethylenes. See Table 2 for details. The *filled symbols* denote G′ and *open symbols* G″ values

Figure 3a–b illustrate the LCB effect on the melt rheological properties. The response of the rheological behaviour to the copolymerisation ability and vinyl end group selectivity of the siloxy-substituted metallocenes has been investigated from their dynamic modulus curves. The frequency dependency of the dynamic modulus of the polyethenes produced with catalysts 2 is demonstrated in Fig. 3a. For comparison dynamic modulus for a linear polyethene, prepared by the catalyst n-BuCp$_2$ZrCl2, is shown in Fig. 3b.

The polyethenes prepared with catalyst 2 (Fig. 3a) have greatly elevated elastic modulus G′ values due to LCB compared to the linear polymers shown in Fig. 3b. LCB also shifts the crossover point to lower frequencies and modulus values. The measured complex viscosities of branched polymers (see also Table 2) are more than an order of magnitude higher than calculated zero shear viscosities of polymers having the same molecular weight but a linear structure. The linear polymers have, in turn, $\eta^*_{(0.02 \text{ rad/s})}$

Table 2 Characteristics of branched and linear homopolymers

Polymer	MW	$\eta^*_{(0.02 \text{ rad/s})}$ (measured)	η_0^a (calculated)	E_a
	g/mol	Pa s	Pa s	kJ/mol
Branched (P21)	110,000	110,000	5,000	48
Branched (P25)	110,000	220,000	5,000	57
Linear	77,000	1,400	1,300	27
Linear	370,000	307,000	380,000	26

^{a} The calculated η_0 is based on the equation $\eta_0=3.4\times10^{-15}$ MW$^{3.6}$ proposed by Raju et al. [85]

values close to the calculated ones. In addition, temperature sensitivity of viscosity (or the flow activation energy) shows similar dramatic difference between the rheological properties of linear and branched polymers.

The extent of LCB and its distribution depends mainly on the catalyst system and the conditions used in the polymerisation. Polymerisation conditions (monomer and comonomer concentration, type of catalyst, temperature and concentration of transfer agents) are important variables to be taken into account when one is looking at the rheological behaviour of the polymers. By decreasing the ethene concentration and increasing the polymerisation time in the reactor the LCB frequency can be enhanced [59, 81]. The polymers made with these catalysts have a complex branching structure composed of comb and tree structures of different lengths.

4
Perspectives

Materials produced with novel single-site catalysts will be superior to conventional Ziegler-Natta or PE-LD grades in applications where better mechanical properties are required. However, in many applications commodity Ziegler-Natta grades are fully competitive. Also, the ability of the new catalysts to introduce long-chain branching in polyethene structures will provide products challenging the materials produced by the high pressure route. Due to improved control of polymer microstructure achieved with single-site catalysts, it will also be possible to produce specialty polyolefins to compete with non-olefinic polymers and thus open completely new areas in higher performance applications.

The product design capability will expand to include polar comonomer incorporation. Copolymerisation of polar comonomers with α-olefins will alter the properties significantly and lead to materials with improved dyeability and adhesion properties, as well as better compatibility with non-olefinic polymers. In particular, the novel non-metallocene single-site catalysts, developed by Brookhart, Grubbs and others, are extremely tolerant to polar groups.

References

1. Natta G, Pino P, Mazzanti G, Giannini U (1957) J Am Chem Soc 79:2975
2. Breslow DS, Newburg NR (1957) J Am Chem Soc 79:5072
3. Reichert KH, Meyer KR (1973) Makromol Chem 169:163
4. Long WP, Breslow DS (1975) Liebigs Ann Chem 463
5. Andresen A, Cordes H-G, Herwig J, Kaminsky W, Merck A, Mottweiler R, Pein J, Sinn H, Vollmer H-J (1976) Angew Chem 88:689
6. Cihlár J, Meizlik J, Hamrik O (1978) Makromol Chem 179:2553
7. Kaminsky W, Sinn H (1975) Liebigs Ann Chem 424
8. Sinn H, Kaminsky W, Vollmer H-J, Woldt R (1980) Angew Chem 92:396
9. Kaminsky W (1983) In: Quirk RP (ed) Transition metal catalyzed polymerizations - alkenes and dienes, vol 4. Harwood, New York, p 225
10. Kaminsky W, Miri M, Sinn H, Woldt R (1983) Makromol Chem Rapid Commun 4:417

11. Wild FR, Zsolanai L, Huttner G, Brintzinger H-H (1982) J Organomet Chem 232:233
12. Ewen JA (1984) J Am Chem Soc 106:6355
13. Ewen JA, Haspeslagh L, Atwood JL, Zhang H (1987) J Am Chem Soc 109:6544
14. Herrmann WA, Rohrmann J, Herdtweck E, Spaleck W, Winter A (1989) Angew Chem 101:1536
15. Röll W, Brintzinger H-H, Rieger B, Zolk R (1990) Angew Chem 102:339
16. Ewen JA, Jones RL, Razavi A, Ferrara JD (1988) J Am Chem Soc 110:6255
17. Ishihara N, Seimiya T, Kuramoto M, Uoi M (1986) Macromolecules 19:2464
18. Rieger B, Reinmuth A, Röll W, Brintzinger H-H (1993) J Mol Catal 82:67
19. Jüngling S, Mülhaupt R, Brintzinger H-H, Fischer D, Langhauser FJ (1995) J Polym Sci Part A: Polym Chem 33:1305
20. Spaleck W, Küber F, Winter A, Rohrmann J, Bachmann B, Antberg M, Dolle V, Paulus EF (1994) Organometallics 13:954
21. Leino R, Luttikhedde H, Wilén C-E, Sillanpää R, Näsman JH (1996) Organometallics 15:2450
22. Stevens J (1993) Proceedings of MetCon. Houston, p 157
23. Johnson LK, Killian CM, Brookhart M (1995) J Am Chem Soc 117:6414
24. Small BL, Brookhart M, Bennett AMA (1998) J Am Chem Soc 120:4049
25. Small BL, Brookhart M (1998) J Am Chem Soc 120:7143
26. Ittel SD, Johnson LK, Brookhart M (2000) Chem Rev 100:1169
27. Britovsek GJP, Gibson VC, Kimberley BS, Maddox PJ, McTavish SJ, Solan GA, White AJP, Williams DJ (1998) J Chem Soc Chem Commun 849
28. Britovsek GJP, Bruce M, Gibson VC, Kimberley BS, Maddox PJ, Mastroianni S, McTavish SJ, Redshaw C, Solan GA, Strömberg S, White AJP, Williams DJ (1999) J Am Chem Soc 121:8728
29. Britovsek GJP, Mastroianni S, Solan GA, Baugh SPD, Redshaw C, Gibson VC, White AJP, Williams DJ, Elsegood MRJ (2000) Chem Eur J 6:2221
30. Britovsek GJP, Gibson VC, Wass DF (1999) Angew Chem Int Ed 38:428
31. Wang C, Friedrich SK, Younkin TR, Li RT, Grubbs RH, Bansleben DA, Day MW (1998) Organometallics 17:3149
32. Younkin TR, Connor EF, Henderson JI, Friedrich SK, Grubbs RH, Bansleben DA (2000) Science 287:460
33. Soula R, Novat C, Tomov A, Spitz R, Claverie J, Drujon X, Malinge J, Saudemont T (2001) Macromolecules 34:2022
34. Soula R Broyer JP, Llauro MF, Tomov A, Spitz R, Claverie J, Drujon X, Malinge J, Saudemont T (2001) Macromolecules 34:2438
35. Soula R, Saillard B, Spitz R, Claverie J, Llauro MF, Monnet C (2002) Macromolecules 35:1513
36. Bochmann M, Lancaster SJ (1993) Makromol Chem Rapid Commun 14:807
37. Bochmann M, Lancaster SJ (1992) J Organometallics 434
38. Jia L, Yang X, Ishihara A, Marks TJ (1995) Organometallics 14:3135
39. Chien JCW, Tsai WM, Rausch MD (1991) J Am Chem Soc 113:8570
40. Chien JCW, Tsai WM (1993) Makromol Chem Macromol Symp 66:141
41. Seppälä JV (1985) J Appl Polym Sci 30:3545
42. Härkki O, Lehmus P, Leino R, Luttikhedde HJG, Näsman JH, Seppälä JV (1999) Macromol Chem Phys 200:1561
43. Koivumäki J (1995) Polym Bull 34:413
44. Lehmus P, Kokko E, Härkki O, Leino R, Luttikhedde HJG, Näsman JH, Seppälä JV (1999) Macromolecules 32:3547
45. Heiland K, Kaminsky W (1992) Makromol Chem 193:601
46. Herfert N, Montag P, Fink G (1993) Makromol Chem 194:3167
47. Karol FJ, Kao SC, Wasserman EP, Brady RC (1997) New J Chem 21:797
48. Kim I, Kim SY, Lee MH, Won MS (1999) J Polym Sci Part A: Polym Chem 37:2763
49. Quijada R, Scipioni RB, Mauler RS, Galland GB, Miranda MS (1995) Polym Bull 35:299

50. Quijada R, Dupont J, Miranda MS, Scipioni RB, Galland GB (1995) Macromol Chem Phys 196:3991
51. Schneider MJ, Suhm J, Mülhaupt R, Prosenc MH, Brintzinger H-H (1997) Macromolecules 30:3164
52. Suhm J, Schneider MJ, Mülhaupt R (1998) J Mol Catal A: Chem 128:215
53. Uozumi T, Soga K (1992) Makromol Chem 193:823
54. Wigum H, Tangen L, Støvneng JA, Rytter E (2000) J Polym Sci Part A: Polym Chem 38:3161
55. Kaminsky W, Engelhausen R, Zoumis K, Spaleck W, Rohrmann J (1992) Makromol Chem 193:1643
56. Lehmus P (2001) Acta Polytechnica Scandinavica, Chemical Technology Series No 280. Espoo, p 56
57. Spaleck W, Küber F, Winter A, Rohrmann J, Bachmann B, Antberg M, Dolle V, Paulus EF (1994) Organometallics 13:954
58. Stehling U, Diebold J, Kirsten R, Röll W, Brintzinger H-H, Jüngling S, Mülhaupt R, Langhauser F (1994) Organometallics 13:964
59. Kokko E, Malmberg A, Lehmus P, Löfgren B, Seppälä J (2000) J Polym Sci Part A: Polym Chem 38:376
60. Kokko E, Lehmus P, Leino R, Luttikhedde HJG, Ekholm P, Näsman JH, Seppälä JV (2000) Macromolecules 33:9200
61. Kokko E, Lehmus P, Malmberg A, Löfgren B, Seppälä J (2001) In: Blom R, Follestad A, Rytter E, Tilset M, Ystenes M (eds) Organometallic catalyst and olefin polymerisation, catalysts for a new millennium. Springer, Berlin Heidelberg New York, p 335
62. Kokko E, Pietikäinen P, Koivunen J, Seppälä JV (2001) J Polym Sci Part A: Polym Chem 39:3805
63. Aaltonen P, Löfgren B (1995) Macromolecules 28:5353
64. Aaltonen P, Fink G, Löfgren B, Seppälä J (1996) Macromolecules 29:5255
65. Aaltonen P, Löfgren B (1997) Eur Polym J 33:1187
66. Hakala K, Löfgren B, Helaja T (1998) Eur Polym J 34:1093
67. Goretzki R, Fink G (1998) Macromol Rapid Commun 19:511
68. Wilén C-E, Luttikhedde H, Hjertberg T, Näsman JH (1996) Macromolecules 29:8569
69. Radhakrishnan K, Sivaram S (1998) Macromol Rapid Commun 19:581
70. Ahjopalo L, Löfgren B, Hakala K, Pietilä L-O (1999) Eur Polym J 35:1519
71. Hakala K, Helaja T, Löfgren B (2000) J Polym Sci Part A: Polym Chem 38:1966
72. Hakala K, Helaja T, Löfgren B (2001) Polym Bull 46:123
73. Hakala K (2001) unpublished results
74. a) Thorshaug K, Støvneng JA, Rytter E, Ystenes M (1997) Macromol Chem Rapid Commun 18:715, b) Thorshaug K, Rytter E, Ystenes M (1998) Macromolecules 31:7149
75. Dang VA, Yu L-C, Balboni D, Dall'Occo T, Resconi L, Mercandelli P, Moret M, Sironi A (1999) Organometallics 18:3781
76. Yano A, Hasegawa S, Kaneko T, Sone M, Akimoto A (1999) Macromol Chem Phys 200:1542
77. Lai Y, Wilson JR, Knight GW, Stevens JC, Chum PWS (1993) US Patent 5 272 236
78. Lai Y, Wilson JR, Knight GW, Stevens JC (1997) US Patent 5 665 800
79. Howard P, Maddox PJ, Partington SR, EP 0 676 421 A1
80. Harrison D, Coulter IM, Wang S, Nistala S, Kuntz BA, Pigeon M, Tian J, Collins S (1998) J Molecular Catalysis A 128:65
81. Malmberg A, Kokko E, Lehmus P, Löfgren B, Seppälä JVS (1998) Macromolecules 31:8448
82. Vega JF, Santamaría A, Muñoz-Escalona A, Lafuente P (1998) Macromolecules 31:3639
83. Malmberg A, Liimatta J, Lehtinen A, Löfgren B (1999) Macromolecules 32:6687
84. Seppälä J, Löfgren B, Lehmus P, Malmberg A (2001) Macromol Symp 173:227
85. Raju V, Smith GG, Marin G, Knox JR, Graessley WW (1979) J Polym Sci Polym Phys Ed 17:1183

Received: July 2003

Adv Polym Sci (2004) 169:13–27
DOI: 10.1007/b13519

Advanced Polyethylene Technologies—
Controlled Material Properties

Hilkka Knuuttila · Arja Lehtinen · Auli Nummila-Pakarinen

Borealis Polymers Oy, P.O. Box 330, 06101, Porvoo, Finland
E-mail: hilkka.knuuttila@borealisgroup.com

1
Introduction

Polyethylene (PE) is the most widely utilised thermoplastic polymer today. In the next five years its worldwide demand is forecast to grow about 5% annually and be about 66 million tonnes in 2005 [1]. With the growing demand, the material itself has become technically more sophisticated and more application specific. Utilisation of latest findings in polymer science have led to new catalyst and process innovations which have made the required improvements in material properties possible, and in many cases also resulted in better production economy. Examples of these new technologies are single-site catalysts [2], condensed mode technology in the gas phase process, which allows retrofittting a reactor and increasing the production capacity up to 60% [3], use of supercritical propane in a loop reactor, and several new multi-reactor processes (e.g. Unipol II, Spherilene, Advanced Sclairtech, Borstar) [3, 4].

The big revolutionary improvements in polyethylene technology have occurred in steps of 20 years (Fig. 1) [2]. In each new development step, catalyst and process innovations have gone hand-in-hand and the control over the polymer structure and the ability to tailor material properties have increased. Today ethylene can be polymerised under various conditions to yield polyethylenes having markedly different chain structures and physical

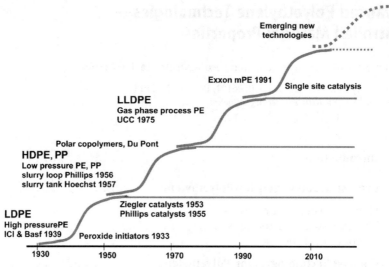

Fig. 1 Polyethylene technology S curve (modified from S curve by ChemSystem)

and mechanical properties. The structural parameters generally influencing the ultimate properties of polyethylenes are: type, amount and distribution of comonomer(s) and branching, average molecular weight and molecular weight distribution.

Polyethylenes are usually classified into groups according to their density, which can theoretically range from 855 to 1000 kg/m³ [5], but in commercial polyethylene grades it is normally in the range of 865–970 kg/m³. Each of these polyethylene groups has its own performance characteristics with different thermal, physical and rheological properties. Copolymerisation of ethylene with α-olefins has long been used to decrease the crystallinity and density of linear ethylene homopolymer to a desired level suitable for any specific application. If the polymer backbone is linear and contains no, or only few, short branches, the term high density polyethylene (HDPE) is used. These polymers are highly crystalline and have densities in the range 940–975 kg/m³. Other groupings of substantially linear polyethylenes are medium density (MDPE), linear low density (LLDPE) and very low density (VLDPE) polyethylenes, which have higher amounts of α-olefins incorporated in their chains, the amount increasing from MDPE to VLDPE. Sometimes MDPE is classified as LLDPE or HDPE depending on its density. The novel polyethylene plastomers and elastomers contain even larger concentrations of comonomer and therefore have the lowest densities of all polyethylenes (Fig. 2). All these linear polyethylenes are produced by coordination polymerisation using a catalyst (chromium, Ziegler-Natta or single site) at low or moderate pressures and temperatures.

A somewhat different type of polyethylene is low density polyethylene (LDPE). This oldest member of the polyethylene product family differs structurally from the linear polyethylenes by being highly branched and

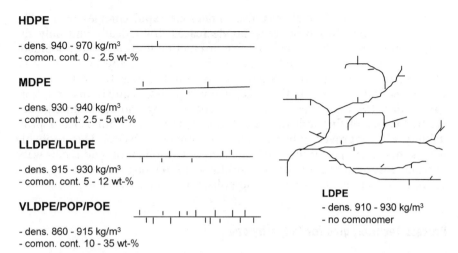

Fig. 2 Polyethylene types

containing both long-chain branching (LCB) and different types of short-chain branches (SCB). The long branches can be of the same length as the polymer backbone. This type of polymer microstructure is a product of many side reactions during a high temperature and high pressure free radical polymerisation. The density of LDPE can vary from 910–930 kg/m^3.

World largest polyolefins producers in 2002 are presented in Fig. 3. In addition to technology changes, in the last 10 years the polyethylene, and poly-

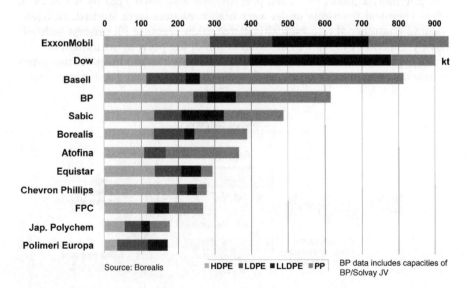

Fig. 3 World's largest polyolefin producers in 2002

olefins industry in general, has also undergone rapid changes [6, 7]. The number of polyolefins producers has decreased dramatically and only the larger companies with cost-competitive production units and more sophisticated products have survived in the highly competitive and cyclic markets. Many small national producers have vanished in the industry restructuring, which has included international company mergers, acquisitions and reorganisations. Even after many successive cost-cutting and streamlining programs the polyolefins industry is still under pressure to build better profitability and improved robustness against business cycles. This probably means that it will go through still one step of consolidation and a new wave of plant closures and decreases in workforce. In the long term this may have an influence on the development and utilisation of new technologies.

2
Process Technologies for Polyethylene

Today, several high pressure and low pressure process technologies have been developed for the production of the different polyethylene grades [4, 8, 9, 10, 11]. In 2001 about 20 million tonnes of the PE capacity used high pressure and about 45 million tonnes used low pressure technology. In 2000, 32% of the worldwide PE capacity was estimated to be for the production of LDPE, 31% for HDPE and 37% for LLDPE [6, 12, 13].

2.1
High Pressure PE Processes

The polymerisation of branched polyethylene was discovered by ICI in 1933, when chemical reactions under very high pressures were studied. Development of commercial free-radical-initiated, high pressure PE process technology by the end of 1930s was the foundation of the polyolefins industry.

High pressure low density polyethylene resins are produced in two types of reactors. One is a continuous-flow mechanically stirred autoclave (origi-

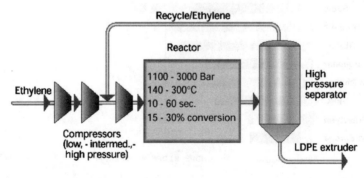

Fig. 4 Flowchart of high pressure polyethylene process

Autoclave process   **Tubular process**

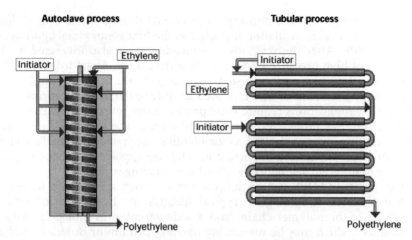

Fig. 5 Autoclave and tubular high pressure polyethylene reactors

nally developed by ICI) and the other is a tubular reactor (originally developed by BASF). Apart from differences in the reactor type, the process schemes for both reactor types are similar (Figs. 4 and 5). The high pressure process requires pressures of 1100–3000 bar and temperatures of 140–300 °C. Product properties may be controlled, e.g. by the pressure, temperature, temperature profile across the reactor and addition of chain transfer agent. Generally, pure ethylene is introduced into the reactor and no solvent is used. Free-radical initiators, such as oxygen and peroxides, are used to initiate the polymerisation that is a highly exothermic and fast reaction. The residence time in the reactor is in the range of 10–60 s, with about 15–25% conversion per pass for autoclave and 20–40% for a tubular reactor. After polymerisation, the reactor effluent is flashed in separator(s) and the unreacted monomer recycled. The molten polyethylene obtained is solidified by cooling and extruded into pellets.

Both reactor types have specific operating characteristics and product range capabilities. In addition to ethylene homopolymers, copolymers with polar comonomers such as vinyl acetate and different types of acrylates can be produced by high pressure processes. In the 1980s the death of LDPE and its replacement by linear low density polyethylene was widely predicted, especially in film applications, but still the LDPE markets have continued to grow and new production capacity has come on stream or is under construction, especially in the Middle East and China. All the new world-scale LDPE plants use tubular process technology. This technology is widely licensed by a number of companies, including Sabic EuroPetrochemicals/Stamicarbon, Basell (Lupotech T), ExxonMobil, EniChem and Atofina. Autoclave technology is available from Equistar, EniChem, ExxonMobil and Simon-Carves (ICI process).

High pressure polyethylene is a good electric insulation material (without any metal residues) and retains its flexibility at low temperatures. These properties, along with the military requirements during World War II, lead to the choice of cable insulation materials as the first commercial application area for LDPE. After the war, other application areas also increased as the properties of high pressure PE were further developed. Apart from wire and cable insulation, film and extrusion coating (EC) are the main application areas. Today over 70% of LDPE is used in different film applications. Both film and EC applications require good processability, meaning balanced melt strength and flow properties, to be able to extrude the material and form a film or coating layer as desired. Organoleptics, mechanical strength, sealing and optical properties are important for end-use applications that comprise a variety of different segments in a flexible packaging area.

High pressure LDPE has certain properties which are difficult to obtain by low pressure technologies. Typical features for LDPE are long-chain branching of the polymer chains and a wide variation in molecular weight distribution, which may be reasonably narrow, medium or broad, as well as symmetrical or asymmetrical. The molecular structure is usually tailored according to the specific application. For example, autoclave reactors produce homopolymers with a high degree of long-chain branching. A high amount of LCB allows easy processing at high molecular weights and makes the resin especially useful for extrusion coating. In contrast, tubular reactors produce resins that contain a much lower degree of LCB. These resins are well suited e.g. for wire and cable insulation and certain packaging film markets. In addition to long-chain branching, high pressure PE contains a reasonably high amount of short-chain branching. The degree of short-chain branching affects the crystallinity (density) and the stiffness of the material in solid state.

2.2
Low Pressure PE Processes

Ethylene polymerisation at low pressures and low temperatures took place for the first time in the early 1950s. Then Karl Ziegler's discovery offered a low pressure route to polyethylene through a catalyst consisting of an active transition metal compound on a support and a suitable aluminium alkyl co-catalyst, and that of Phillips Petroleum through supported metal oxides. The use of Ziegler catalysts in PE production grew at a much lower pace than the chromium oxide catalysts until commercial production of LLDPE started in the 1970s. Today, low pressure polyethylenes are produced by a variety of processes based on different catalyst systems. Numerous companies have developed their own proprietary technologies [9, 11].

The low pressure PE processes can be classified into three basic groups according to their reaction conditions:

- Slurry processes in which dissolved ethylene is polymerised to form solid polymer particles suspended in a hydrocarbon diluent
- Solution processes in which dissolved ethylene is polymerised to form a polymer that is dissolved in the reaction solvent

Slurry Loop reactor

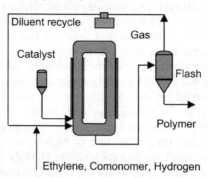

Fig. 6 Low pressure slurry loop process

– Gas phase processes in which ethylene is polymerised to a solid polymer in a fluidised bed of polymer particles.

In the low pressure PE processes comonomers such as α-olefins can be used, and all of the basic types of processes are economically competitive and in commercial use.

By definition, slurry processes are those in which polymerisation takes place at a temperature below the melting temperature of polyethylene in a liquid in which the polymer is essentially insoluble. A slurry process may use stirred tank or loop type reactors (Fig. 6). Stirred tank reactors are normally used in two-stage slurry processes. In a typical slurry process ethylene, a possible α-olefin comonomer (1-butene or 1-hexene) and a supported catalyst are fed under low pressure into the reactor containing isobutane, hexane or heptane diluent. Depending on the reactor and diluent type used, the polymerisation is carried out at a temperature of 70–110 °C and pressure of about 5–40 bar in the absence of oxygen, water or any other impurity which could reduce catalyst activity. The catalyst remains suspended and the polymer, as it is formed, precipitates from the diluent. In the case of controlled stirred tank reactors (CSTR) the polymer slurry is fed into a separator after which the wet polyethylene powder is dried and transferred to an extruder for homogenisation and pelletisation. In a slurry loop process, the polymer and diluent are separated in a flashing step. Diluent is recovered and fed back to the reactor, polymer is purged free of hydrocarbons and transferred to an extruder. The ethylene conversion in a slurry process is over 95%. The residence time in a slurry loop reactor is about 1 h.

The primary product in the slurry units is HDPE with MDPE as a secondary product. With single-site catalysts production of mLLDPE in slurry loop reactors is also possible. Today for example, Chevron Phillips, BP Solvay, Basell, Borealis and Sumitomo Mitsui have their own slurry process technology.

In solution processes, the polymerisation takes place in a hydrocarbon solvent above the melting temperature of polyethylene. These processes were developed in order to produce linear low and very low density polyethylenes without the limitations of unwanted solubility of the polymer. Typically, the polymerisation is carried out in a stirred tank reactor, the solvent used is cyclohexane or a paraffinic hydrocarbon solvent, reaction temperature is 160–300 °C, pressure is 25–100 bar and residence time is 1–5 min. The polymer formed is recovered by evaporating the solvent and unreacted monomers. The conversion in the solution process is over 95%. The main disadvantages of solution process are high energy consumption in the solvent evaporation step and that the production of high molecular weight polymers is limited because solution viscosity increases sharply with increasing molecular weight. Three main types of solution processes have been developed by Dow (Dowlex), DSM (Compact, now owned by Sabic EuroPetrochemicals) and DuPont (Sclairtech, now owned by Nova Chemicals). The key strengths of a solution process are short product transition times, ability to produce polyethylene of very low density (e.g. VLDPE, PE plastomers and elastomers) and ability to incorporate higher α-olefin comonomers such as 1-octene. LLDPE and other low density polyethylenes with higher α-olefins show good strength, toughness and sealing properties.

One of the drawbacks associated especially with slurry and solution CSTR processes is the necessity of removing the solvent or diluent in a post-production step. In a gas phase reactor the polymerisation takes place in a fluidised bed of polymer particles. Inert gas or gas mixture is used for fluidisation. The gas flow is circulated through the polymer bed and a heat-exchanger in order to remove the polymerisation heat. Gaseous ethylene and comonomer are fed into the fluidisation gas line of the reactor, and a supported catalyst is added directly to the fluidised bed (Fig. 7). Polymerisation occurs at a pressure of about 20–25 bar and a temperature of about 75–110 °C. The polymer is recovered as a solid powder which is, however, usually pelletised. Due to the limited cooling capacity of the fluidising gas, reactor

Gas Phase reactor

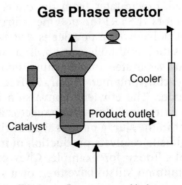

Fig. 7 Gas phase process

sizes and bed volumes in the reactors are quite high, resulting in long residence times (about 3–4 h). Conversion of ethylene is about the same as in slurry reactors. Condensed or super-condensed mode technologies have been developed to increase productivity in the gas phase reactors. All these technologies utilise, in one form or another, liquid evaporation to improve heat removal [3, 14].

In principle, PE gas phase plants are so-called swing units which are able to produce both HDPE, MDPE and LLDPE, even if cost-wise it is better to avoid frequent switches between the PE grades with different densities. The gas phase process is the most commonly used technology for the production of LLDPE with 1-butene or 1-hexene as a comonomer. Fluidised bed gas phase process technology is licensed by Univation (Unipol) and BP (Innovene), stirred bed gas phase technology by Basell (Lupotech G).

The low pressure polyethylene resins with higher densities, MDPE and HDPE, are used in applications where stiffness, temperature resistance and/or strength against mechanical and environmental stresses are important. Examples of applications are bottles, containers, pipes, wire and cable jacketing and stiff film. Processability of high and medium density polyethylenes is defined by molecular weight and molecular weight distribution (MWD). Mechanical properties are dependent on average molecular weight, and the amount and type of comonomer; higher density resulting in lower impact values. Optical properties of high and medium density polyethylenes are poor.

Commercial production of linear low density polyethylene started in the late 1970s. Typically LLDPE polymers are used in applications where high impact strength, low stiffness, good optical properties and/or low melting temperature are needed (e.g. packaging films, sealing layers in multilayer films). Linear low density polyethylenes produced with Ziegler-Natta or single site catalysts have relatively low molecular weight and narrow MWD and their processability is poor compared with LDPE. Therefore, they are often used as blends or in co-extrusion with LDPE. LLDPE with broader MWD can be produced (e.g. with chromium catalysts) but as the MWD increases, impact strength and optical properties decrease. Seal strength and mechanical properties in general are affected by the comonomer type used (improved properties with longer α-olefins such as 1-octene). The amount of comonomer used and how it is distributed in the polymer chains has an influence on the melting temperature and optical properties of LLDPE.

2.3
Bimodal PE Processes

In all the low pressure PE processes the polymer is formed through coordination polymerisation. Three basic catalyst types are used: chromium oxide, Ziegler-Natta and single-site catalysts. The catalyst type together with the process defines the basic structure and properties of the polyethylene produced. Apart from the MWD and comonomer distribution that a certain catalyst produces in polymerisation in one reactor, two or more cascaded reactors with different polymerisation conditions increase the freedom to tailor

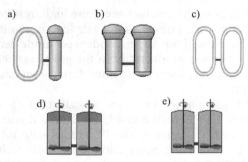

Fig. 8a–e Commercial bimodal processes: **a** slurry-loop process, **b** gas phase—gas phase process, **c** double slurry loop process, **d** dual/triple slurry tank process and **e** dual/triple solution tank process

polymer properties by in-situ blending of two or more structurally different polyethylenes. Mostly, the bimodal process technology has been used to improve processability and the mechanical properties of HDPE.

Bimodal polyethylene consists of two or more polymer fractions which usually differ from each other with respect to molecular weight and branching (comonomer) content [4, 15]. A two- or three-stage process gives the possibility to control the shape of the MWD and incorporation of comonomer as a function of molecular weight. Bimodal production technologies include different combinations of all the main low pressure polymerisation processes and reactor types [4]. In principle, five different two-stage process configurations are in commercial use: 2–3 cascaded stirred tank reactors (solution or slurry), two loop reactors, two gas phase reactors or a loop and a gas phase reactor (Fig. 8). The oldest bimodal process is the combination of two or more slurry tank reactors and the newest one the combination of a loop and a gas phase reactor.

In a bimodal process, the catalyst is fed into the first reactor where the first polymer fraction is produced. After that the polymer is transferred into the second reactor for the production of the second polymer fraction. Depending on the process type used, between the reactors there may be a separation unit to remove unreacted monomers, hydrogen and/or diluent from the reaction mixture before the polymer is passed into the second polymerisation stage. In principle, if there is a separation unit between the reactors, all the bimodal processes can be operated in two different modes. Either the low molecular weight polymer component is produced in the first reactor and the high molecular weight component in the second reactor (normal mode), or the high molecular weight component is produced first and then the low molecular weight component (reversed mode). The catalyst and reactor types used define which mode in practice yields a more stable process operation. In some of the bimodal processes the reactors may also be operated in parallel mode, not only in series.

Traditionally, bimodal polyethylene resins are produced with a Ziegler-Natta (Z-N) type catalyst, but for specialty grades (e.g. bimodal LLDPE) the use of metallocene catalysts is slowly increasing. As in the case of unimodal processes, all bimodal PE technologies are suitable for the production of high and medium density polyethylenes, but bimodal Z-N LLDPE can only be produced in bimodal solution, gas phase or loop-gas phase processes. However, single-site catalysts may change this picture. Several companies have developed, or acquired in company mergers, their own bimodal process technologies. Examples of these companies are Borealis (Borstar loop-gas phase process), Dow (Unipol II bimodal gas phase process, Dowlex bimodal solution process), Sumitomo Mitsui (Evolue bimodal gas phase process, CX bimodal slurry tank process), Basell (Hostalen triple slurry tank process, Spherilene bimodal gas phase process), BP Solvay (douple loop process), Atofina (double loop process), Equistar (bimodal slurry tank and triple solution processes), Nova Chemical (Advanced Sclairtech bimodal solution process). In 1995 the share of bimodal polyethylene of low pressure PE markets was about 10%, but the share is expected to grow to about 25% in 2005. Several new bimodal PE plants have come on-stream in the last 5 years and a big share of the new PE capacity under construction will be using bimodal technology.

The application areas of bimodal polyethylenes are the same as for corresponding uni-modal resins. However, improved product property combinations, such as stiffness-impact balance, result in products with higher performance. For example, without bimodal polyethylene and control over the MWD and comonomer incorporation, development of HDPE pipe materials with higher pressure classification would not have been possible [15, 16]. Another new opportunity is material reduction (source reduction) without compromising the properties of the ready-made plastic products. Depending on the end use, e.g. in film application a thickness reduction of 10–30% is possible.

3
Catalyst Technologies for Polyethylene

Today, polyethylenes are commercially produced using free-radical initiators, Ziegler-Natta catalysts, chromium oxide (Phillips type) catalysts, and more recently metallocene or single-site catalysts. Generally speaking, a Ziegler-Natta catalyst is a complex formed by reaction of a transition metal compound (halide, alkoxide, alkyl or aryl derivative) of group IV–VIII transition metals with a metal alkyl halide of group I–III base metals [17]. The former component is usually known as the catalyst and the latter as the co-catalyst or activator. For industrial use, most Ziegler-Natta catalysts are based on titanium salts and aluminium alkyls. Ziegler-Natta catalysts have improved considerably since K. Ziegler and G. Natta discovered them 50 years ago [18]. The fourth and fifth generation catalysts have e.g. significantly higher activity and better control of active sites and particle morphology that enable them to meet the increasing demands of high performance materials [19, 20]. Ziegler-Natta catalysts have been used in homogeneous, heterogeneous and colloidal forms. They are industrially very important since they can produce all basic linear polyethylene types from linear HDPE to very branched VLDPE and are able to polymerise higher α-olefins such as propylene, 1-butene and 4-methyl-1-pentene to stereoregular polymers. Hydrogen feed into the reactor is used as a chain transfer agent to control molecular weight of the forming polymer [21].

At about the same time that Ziegler published the titanium tetrachloride-triethylaluminium catalyst for ethylene polymerisation, P. Hogan and R. Banks at Phillips Petroleum Company discovered that inorganic chromium salts were able to polymerise olefins [22, 23]. A traditional Phillips catalyst is based on chromium (VI) oxide supported on silica or aluminosilicate. Unlike the Ziegler-Natta catalysts it does not necessarily require a co-catalyst to be activated in polymerisation. Activation is carried out by heat treatment (calcination). When the calcined hexavalent catalyst reacts with ethylene, it is reduced to the lower oxidation states, Cr(II) or Cr(III). The precise nature of the active site is still a subject for debate. Hydrogen response of chromium catalysts is poor, therefore the molecular weight of the polymer is not adjusted by hydrogen, but by temperature. There are numerous variations of the basic chromium catalyst in which either the chromium compound or its support is chemically modified before or during the catalyst preparation. For example, MWD and thus polymer properties may be influenced by changing the chemical surrounding of the chromium atom e.g. by using metal atoms like titanium or aluminium.

Due to their multi-sited nature, Ziegler-Natta and chromium catalysts produce structurally heterogeneous ethylene homo- and copolymers. This means that the polymers have broad MWD and broad composition (short-chain branching) distribution (Fig. 9). Catalyst active sites that produce lower molecular weights also have a tendency to incorporate more comonomer

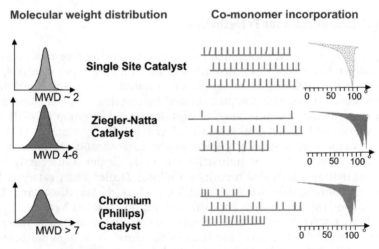

Fig. 9 Conventional Ziegler-Natta and chromium catalysts behave differently compared to a single-site catalyst producing more inhomogeneous polymer

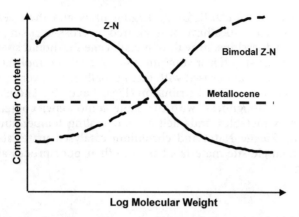

Fig. 10 Tailoring of polymer comonomer distribution gives greatest benefits in HDPE

and as a result there is an additional compositional drift, with shorter polymer chains containing most of the comonomer and the longer chains being more linear. In a two-stage process, the low and high molecular fractions are produced in different reactors and therefore their comonomer contents can be easily controlled by the amount of comonomer fed into each reactor. So-called reversed comonomer distribution, where the longer polymer chains contain more comonomer than the shorter ones, is beneficial for the mechanical properties of the materials (Fig. 10) [4, 15, 16].

Since most LLDPE resins are produced with Ziegler-Natta catalysts, much of the research carried out on these catalysts has focused on reducing their multisitedness and improving control over MWD and comonomer incorporation. The latest step in PE technology development is the use of metallocene and other single-site catalysts. Metallocene catalysts were originally discovered about the same time as the Ziegler-Natta and chromium oxide catalysts, but their commercial development did not start until alumoxane activation was discovered in the1970s. The first metallocene catalysts were metal (mainly Zr) complexes of unsubstituted or substituted bis-cyclopentadienyl ligands and they were activated by methylalumoxane (MAO). Today, numerous families of different single site catalysts have been developed for the production of polyethylenes and other polyolefins [2, 24, 25, 26].

Exxon Chemical introduced the first commercial polyethylene grades produced with metallocene catalysts in 1991 and Dow in 1992. They were polyolefin plastomers and elastomers with very low density and interesting new properties. The single-site catalysts were expected to revolutionalise the whole polyolefin industry and very fast cannibalise the conventional catalysts. However, the penetration of metallocene polymers to the markets has been much slower than expected due to many technical problems in full-scale production, unclear industrial rights and customer conservatism which has made launching of new types of materials slow and not very rewarding. Global demand for mPE was about 1.2 million tonnes in 2001 and the main application area is mLLDPE for packaging films and their sealing layers. Now that the patent issues have been cleared by cross-licensing agreements, alliances and partnerships, growth of metallocene PE should speed up.

Single-site catalysts offer a superior control over the molecular architecture. Polyethylenes polymerised with them usually have a very narrow MWD and a uniform composition distribution (Figs. 9 and 10). Therefore, in addition to improved toughness, with the same comonomer content they have clearly lower extractables and melting and sealing temperatures than the corresponding Ziegler-Natta and chromium catalysed materials. The main problem with single-site materials has been their poor processability e.g. in film blowing.

4
Conclusions

The big revolutionary improvements in polyethylene technology have occurred in steps of 20 years. In each new development step, catalyst and process innovations have gone hand-in-hand and the control over the polymer structure and the ability to tailor material properties have increased remarkably. Advances in coordination polymerisation catalysis have been the breakthroughs that have led to improved products. Today, ethylene can be polymerised under various conditions to yield polyethylenes having markedly different chain structures and physical and mechanical properties. Development of new generation bimodal PE technologies has made possible a better utilisation of the properties of enhanced Ziegler-Natta and well-defined single-site catalysts in tailoring polymer structure and in production of materials with dramatically improved and/or desirable combination of properties. These technologies will have a great impact on many sectors of the plastic industry, e.g. packaging, pipe applications and automotive industries. Today, bimodal polyethylene is already the global standard for high strength thin film and is becoming the global standard for PE80 and PE100 pipe materials.

New catalyst and process technologies play a major role in development of cleaner, easily tailorable and better synthetic materials which fulfil even the toughest application demands and can be produced cost-effectively saving raw materials and energy. Many new PE technologies developed during the past 15 years are now under commercialisation.

One of the driving forces in the search for new catalyst systems has been the desire to understand the structure-property relationships between catalyst and polymer structure in order to better control the material properties and to obtain novel property combinations (smart materials). New catalysts are needed if ethylene copolymers with functional groups (e.g. polar groups) are to be produced using low pressure processes. What will be the next technology step change in polyethylene is still unclear. However, there are some potential discoveries in late transition metal catalyst systems [27] and polymerisation processes (e.g. Basell's multi-zone reactor) which may be part of the next generation of polyethylene technology.

References

1. Kaiser UH, Müller WF (2001) Proc. PE 2001 World Congress, Maack Business Services, 13–15 February, Zürich
2. Montagna AA, Burkhart RM (1997) CHEMTECH 27:26
3. Shut JH (1995) Plastics World 53:12
4. Knuuttila H, Lehtinen A, Salminen H (2000) In: Scheirs J, Kaminsky W (eds) Metallocene-based polyolefins. Wiley, Chichester, pp 364–378
5. Zhu L, Chiu F-C, Fu Q, Quirk RP, Cheng SZO (1999) In: Brandrup J, Immercut EH, Grulke EA (eds) Polymer handbook, 4th edn. Wiley, New York, p V/9
6. Scheildl K (2001) Proc. PE 2001 World Congress, Maack Business Services, 13–15 February, Zürich
7. Zaby G (2001) Kunststoffe 91:238
8. Hogan JP, Norwood DD, Ayres CA (1981) J Appl Polym Sci: Appl Polym Symp 36:49
9. Hydrocarbon Proc. (2001) 80:119
10. Hästbacka K (ed) (1993) Neste—From oil to plastics, 3rd edn. Neste Oy, Espoo
11. Peacock AJ (2000) Handbook of polyethylene, structures, properties, and applications. Dekker, New York
12. Beer G (2001) Kunststoffe 91:248
13. Beulich I, Cook M (2001) Kunststoffe 91:254
14. Williams G (1995) ECN Chemscope, May 1995, pp 4–5
15. Scheirs J, Böhm LL, Boot JC, Leevers PS (1996) TRIP 4:408
16. Böhm LL, Enderle HF, Fleißner M (1992) Adv Mater 4:234
17. Huang J, Rempel GL (1995) Prog Polym Sci 20:459
18. Ziegler K (1953) Ger Patent 878 560
19. Böhm LL, Franke R, Thum G (1988) In: Kaminsky W, Sinn H (eds) Transition metals and organometallics as catalysts for olefin polymerization. Springer, Heidelberg Berlin New York, pp 391–404
20. Böhm LL, Enderle HF, Fleissner M (1994) In: Soga K, Terano M (eds) Catalyst design for tailor-made polyolefins. Elsevier, Amsterdam, pp 351–363
21. Boor J Jr (1979) Ziegler-Natta Catalysts and polymerization. Academic Press, New York
22. Hogan JP, Banks RL (1958) US Patent 2,825,721
23. Welch MB, Hsieh H (1993) In: Vasile C, Seymour RB (eds) Handbook of polyolefins. Dekker, New York, pp 21–38
24. Montagna AA (1995) CHEMTECH 25:44
25. Shut JH (1996) Plastics World 54:43
26. Shut JH (1997) Plastics World 55:27
27. Britovsek GJP, Gibson VC, Wass DF (1999) Angew Chem, Int Edn 38:428

Received: July 2003

Adv Polym Sci (2004) 169:29–73
DOI: 10.1007/b94176

Polyethylene Morphology

Ulf W. Gedde · Alessandro Mattozzi

Department of Fibre and Polymer Technology, Royal Institute of Technology,
100 44 Stockholm, Sweden
E-mail: gedde@polymer.kth.se

Abstract The morphology of polyethylene has been an important theme in polymer science for more than 50 years. This review provides an historical background and presents the important findings on five specialised topics: the crystal thickness, the nature of the fold surface, the lateral habit of the crystals, how the spherulite develops from the crystal lamellae, and multi-component crystallisation and segregation of low molar mass and branched species.

Keywords Polyethylene · Morphology · Review

1
Introduction

The modern polyethylene morphology era started in 1957 with the discovery of chain folding. In the following 50 years there were great discoveries, triggered by the development of new experimental techniques applied to well-defined polyethylene samples. Debate and disagreement have also occurred, sometimes induced by a lack of adequate experiments and insufficiently defined samples, but which have led to new strategies and findings and sometimes to a consensus among scientists. It is fair to state that polyethylene is the model semicrystalline polymer and that it has provided data and ideas valuable in the study of other semicrystalline polymers. The polymerisation methods available for decades have made it possible to make polyethylenes of different crystallinities and morphologies. The high-pressure process first developed in the 1930s yielded branched polyethylene with ~50% crystallinity. Low-pressure processes, developed in the 1950s utilising metal-organic chemistry, yielded linear polyethylene (~75% crystallinity). Low- and medium-density polyethylenes were made by the low-pressure technique, replacing part of the ethylene with higher 1-alkenes. The metallocene technology was put into commercial use in the 1990s and it provides polyethylenes with a narrow molar mass distribution and a uniform distribution of comonomer units. Preparation techniques using chromatography made it possible to obtain fractions with a narrow molar mass distribution, useful for scientific work. This availability of polyethylene samples of different molar masses and different degrees of branching is one of the reasons why polyethylene morphology has been the subject of so many papers. The recent availability of strictly monodisperse n-alkanes with several hundred carbon atoms using a preparation method developed by Paynter, Simmonds and Whiting [1] has provided new insight into several important aspects of polyethylene morphology. Polyethylene is used in large quantities and in many different applications. The properties of polyethylene are controlled by the morphology. This is not a topic covered by this paper, but rather a motivation for the reader to learn more about polyethylene morphology.

It is not intended to present a comprehensive review of the extensive literature on polyethylene morphology. Several themes have been selected on the basis of novelty and importance in the author's eyes. The historical review, which is part of Sect. 1, presents the old discoveries in a brief form and it also serves as an introduction to the more specialised themes presented in the subsequent sections. The link to the theory of polymer crystallisation—historically closely related to discoveries made within the polyethylene morphology field—is briefly discussed.

The crystal structure of polyethylene was first determined by Bunn [2]. The orthorhombic unit cell, identical with the crystal structure of straight-chain alkanes, is the most stable crystal structure. The unit cell with the space group notation $Pnam$-D_{2h} contains 4 CH_2 groups and the all-*trans* chains are parallel (Fig. 1a). The zigzag planes of the chains of the orthorhombic cell have different orientations (Fig. 1a). The dimensions at 23 °C of the orthorhombic cell of linear polyethylene are according to Busing [3]: a=0.74069 nm, b=0.49491 nm and c=0.25511 nm (chain axis) which gives a crystal density of 996.2 kg m^{-3}. The angle between the zigzag planes of the two chains and the **b** axis of the unit cell was first determined by Bunn [2] to be 41 °. A later study by Chatani et al. [4] on linear polyethylene gave a somewhat larger value, 45 °. The orthorhombic unit cell of branched polyethylene is expanded along **a** and to a smaller degree along **b** [5, 6, 7, 8, 9], and this causes an increase in the setting angle to 49–51 ° for this class of polyethylene. Martinez Salazar and Baltá Calleja [9] proposed that a fraction of the short branches, methyl and ethyl groups, are included in the

(a)

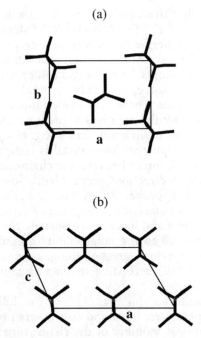

(b)

Fig. 1a, b Crystal structures of polyethylene showing views along the chain axis: **a** ortho-rhombic unit cell, **b** monoclinic unit cell

crystals in the vicinity of 2g1 kinks. According to this view, the extra 'room' provided by the kinks is not sufficient to house larger pendant groups, and the cell expansion along **a** and **b** is negligible in butyl-branched polyethylene [7].

The monoclinic cell with the space group notation $C2/m$-C_{2h}, first discovered by Teare and Holmes [10] is less stable and is found in samples subjected to mechanical stress (Fig. 1b). The dimensions at 23 °C of the monoclinic cell of linear polyethylene with 4 CH_2 groups are according to Seto at al. [11]: a=0.809 nm, b=0.253 nm (chain axis) and c=0.479 nm, which gives a crystal density of 998 kg m^{-3}. The zigzag planes of the chains of the monoclinic cell have a uniform orientation (Fig. 1b). The high-pressure hexagonal phase was discovered by Bassett et al. [12]. A triple point involving three phases—melt, orthorhombic and hexagonal phases—is positioned at 220 °C and 0.33 GPa [13]. The hexagonal phase has very special properties: a regular hexagonal packing of conformationally disordered chains with extreme translational chain mobility that allows rapid crystal thickening.

The orthorhombic crystal phase of polyethylene has, like most polymer crystals, pronounced anisotropic properties. The elastic moduli along the three orthogonal crystallographic directions of the orthorhombic unit cell are at 23 °C: 3.2 GPa along **a** and 3.9 GPa along **b** according to Sakaruda et al. [14] and 240–360 GPa along **c** [14, 15, 16]. The thermal expansion coefficients at $-70\ ^\circ$C in the three crystallographic directions are, according to Davis at al. [17]: 1.7×10^{-4} K^{-1} (a), 0.59×10^4 K^{-1} (b) and -0.12×10^4 K^{-1} (c). The difference in the properties parallel to and perpendicular to the chain axis c is a direct consequence of the pronounced difference in the attractive bond forces between the atoms. The all-*trans* conformation of the crystalline chain sequence makes it susceptible only to bond angle deformation and stretching of covalent bonds. These deformation modes are not very compliant, as is demonstrated by the high elastic modulus along c. The attractive forces acting on the atoms in different chains (e.g. along a and b) are due to weak London forces. The polyethylene crystal is composed of very stiff all-*trans* chains but the weak forces between the chains make the crystal compliant to stresses acting in directions perpendicular to c.

Optical and dielectric properties of the orthorhombic crystal phase are also anisotropic. The polarisability is greater along c than along a and b; the polarisabilities along a and b are approximately the same. The refractive index (n), which is related to the polarisability according to the Lorentz–Lorenz equation, is therefore greater along c (n=1.575) than along a (n=1.514) and b (n=1.519) according to data obtained by Bunn and de Daubeny [18].

The stems in the orthorhombic crystal possess mobility at room temperature and higher temperatures. Boyd and co-workers presented truly elegant proofs of the translational mobility of the chains through the polyethylene crystal. The work has been summarised by Boyd [19, 20]. Lightly oxidised polyethylene with a few carbonyl groups showed a dielectric relaxation process (α process) at temperatures well above the glass transition temperature. The dielectric α process is extremely sharp; the Cole–Cole width parameter is 0.7–0.8 for linear polyethylene [21]. The relaxation strength of the dielectric α process is proportional to the degree of crystallinity with an activation energy proportional to the crystal thickness up to a certain thickness level above which it remains constant [21]. The molecular interpretation of the dielectric α process presented by Mansfield and Boyd [22] is schematically shown in Fig. 2. The dielectric process is exclusively crystalline. It involves a 180 $^\circ$ rotation of the carbonyl dipole and a $c/2$ translation of the chain along its own axis to keep the chain in register with surrounding chains. The 180 $^\circ$ twist of the chain is accomplished through a smooth twist that propagates from one side of the crystal to the other. The motion of the twist from one site to the next ($c/2$ translation) involves the passage of a 17 kJ mol^{-1} energy barrier. The creation energy of the smooth twist with 12 main chain bonds was calculated to be 54 kJ mol^{-1}. There is no strain involved in any of the two terminal states. They are the same in that sense and the dielectric α process lacks mechanical activity. Polymers with small or no pendant groups, with short repeating units and with weak *inter*molecular forces are the polymers most likely to have an α process. Examples of polymers belonging to this category are polyethylene, isotactic polypropylene, polyoxymethylene and poly(ethylene oxide) [19, 20].

The efficient mechanism for translative motion of a stem through an orthorhombic crystal discovered by dielectric relaxation measurements has

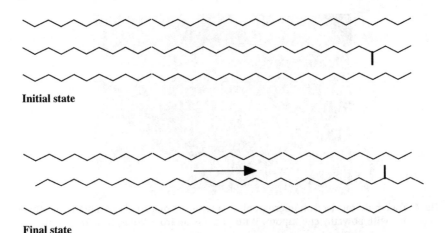

Initial state

Final state

Fig. 2 Dielectric α process in polyethylene showing initial and final states. The combined 180 ° rotation and $c/2$ translation of the central chain is accomplished by a smooth twist, consisting of 12 carbon atoms out of register, that propagates through the crystal

important implications for the mechanical behaviour of polyethylene. A mechanically stressed amorphous chain may 'pull out' a portion of the crystal chain by the α process (Fig. 3). Thus, the mechanical α process requires the presence of both crystalline and amorphous components [23, 24, 25, 26]. The α process has a low activation energy and, hence, it is a possible mechanism for the required translative motion of crystal stems to accomplish crystal thickening.

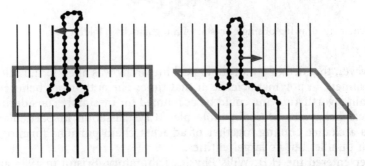

Fig. 3 Mechanical α relaxation process in polyethylene. Drawn after Boyd [20]

The lamellar shape of the polyethylene crystal was discovered in 1957 independently by Keller [27], Fischer [28] and Till [29]. The polymer chains in dilute solution were found to crystallise in 10 nm thick sheets; although the sheets were considerably larger, of the order of 10 μm, in the perpendicular

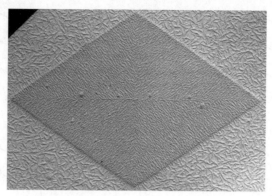

Fig. 4 Transmission electron micrograph of a replicate of a single crystal of polyethylene decorated with polyethylene vapour. With permission from Wiley, New York [30]

directions (Fig. 4 [30]). The fascinating discovery that the chain axis was aligned along the thin direction of the sheets led Keller to suggest chain folding (Fig. 5). The coining of the concept of chain folding was the birth of modem crystalline polymer morphology, well worth celebrating by citing Andrew Keller [31]:

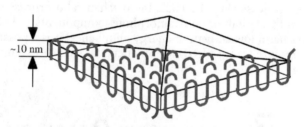

Fig. 5 Schematic drawing of single crystal with regular chain folding

However, the idea of a morphological hierarchy was alien to the scientific establishment in polymer science at that time. The authorities believed that everything worth knowing could be accounted for by simply considering the statistical behaviour of chain molecules. Crystallisation in particular was seen as a chance coming together of adjacent chain portions forming little micellar bundles but no larger entities.

Keller entered the H.H. Wills Physics Laboratory, Bristol in 1955 and he was stunned by what he saw:

The most positive aspect was the extraordinary intellectual ferment coupled with open-mindedness which permeated the whole place'. 'There was no distinction between high and low brow, it was all one intellectual adventure. That is how polymers eventually slotted in between quantum mechanics, dislocations, particle physics, liquid helium, design of new optical instruments and much else'. 'Amongst my negative experiences was first and foremost the nearly total lack of equipment'. 'I cannot deny that the above experimental conditions were frustrating to the extreme, yet they were inducive to make the best use of the little there was and always to concentrate on the essentials, lessons well worth learning'. 'Another negative experience was the total absence of anybody knowledgeable in polymers. As I was still unknown in the field nobody visited me and I had no funds to visit anybody else. Also my access to the polymer literature was highly limited. So I lived and worked for two full years in near complete isolation from the relevant scientific community'. 'Further, when I told him (Sir Charles Frank) that I cannot see how long chains, which I found to lie perpendicular to the basal surface (by combined electron microscopy and diffraction) of layers much thinner than the molecules are long (thickness assessed by electron microscopy shadowing and small-angle X-ray scattering), can do anything else but fold, he said "of course" and encouraged me to publish immediately.

Keller ends with: 'That is how in 1957 in an "office" filled with fumes, sparks and scattered X-rays, amidst total isolation from, in fact ignorance of the rest of polymer science, single crystals and chain folding were recognised'.

The single crystals of polyethylene grown from solution showed a number of other important features:

1. Sectorisation (Fig. 6); the four lateral surfaces of single crystals grown at low temperatures are crystallographically exact {110} planes. This nomenclature for single polymer crystals was introduced by Bassett et al. [32]. The notation {110} is for the group of the following four planes: (110), ($\bar{1}$10), (1$\bar{1}$0), ($\bar{1}\bar{1}$0). Crystallisation at higher temperatures led to the formation of crystals with six lateral surfaces; four {110} and two {100} faces. The latter is the group of the (100) and ($\bar{1}$00) planes. Bassett [33] discovered that the sub-cell in the different sectors was slightly distorted from the orthorhombic structure: the inter-plane spacing of the fold plane, e.g. (110), is shifted by a small quantity (0.001 nm) from the spacing of the 'equivalent' non-fold plane, e.g. (1$\bar{1}$0).

2. The single crystals grown from solution at moderate temperatures were, in their undistorted form, non-planar and shaped like tents (Fig. 7). The base of the tent with the {110} lateral faces is not planar. The regular chain folding model would suggest, on the basis of these results, that each chain along the periphery of the crystal is shifted a certain distance along the chain axis with respect to the nearby chain. Furthermore, the {110} strips must also be gradually shifted along the c-axis (Fig. 7). Sectorisation has also been reported for extended-chain crystals of n-alkanes [34]. However, the tilt angle (the angle between the chain axis and the normal to the crystal surface) is considerably larger in the fold-chain single crystals (~35 °) than in the extended-chain alkane crystals. Hence, sectorisation and the non-planar shape are characteristic features of the single crystals, suggesting the dominance of regular chain folding. Other non-planar shapes such as corrugated

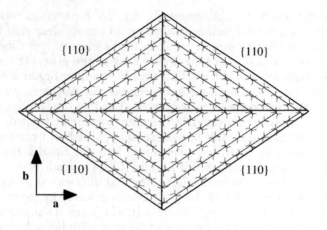

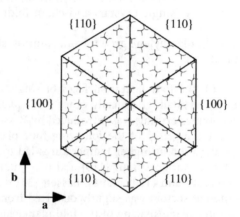

Fig. 6 Sectorisation of polyethylene single crystals. The *upper* crystal shows only {110} sectors whereas the *lower* also has {100} sectors. With permission from Kluwer, Doordrecht, Netherlands [91]

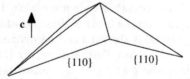

Fig. 7 Schematic drawing of tent-shaped polyethylene single crystals with only {110} sectors

layers and chair-shaped crystals have been reported [35, 36, 37]. Melt-grown crystals may adapt to other, more complex geometries (see Sect. 5).
3. Multilayer crystals with a central screw dislocation were commonly seen. A micrograph of a beautiful solution-grown multilayer crystal with regularly rotated terraces was presented by Keller [38]. This 'mechanism' to multiply a single crystal layer into many crystal layers is important for the crystal growth from the melt to form spherulites.

The existence of crystal lamellae in melt-crystallised polyethylene was independently shown by Fischer [28] and Kobayashi [39]. They observed stacks of almost parallel crystal lamellae with amorphous material sandwiched between adjacent crystals. At the time, another structure was well known, the spherulite (from Greek meaning 'small sphere'). Spherulites are readily observed by polarised light microscopy and they were first recognised for polymers in the study of Bunn and Alcock [40] on branched polyethylene. They found that the polyethylene spherulites had a lower refractive index along the spherulite radius than along the tangential direction. Polyethylene also shows other superstructures, e.g. structures which lack the full spherical symmetry referred to as axialites, a term coined by Basset et al. [41].

The structural hierarchy of melt-crystallised polyethylene is schematically displayed in Fig. 8. Crystal thickness and its controlling factors are discussed in Sect. 2. It includes one phenomenologically resolved issue, the initial crys-

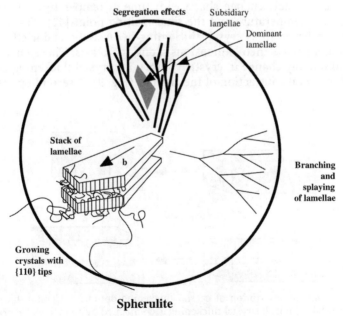

Spherulite

Fig. 8 Schematic drawing of the morphological hierarchy of a polyethylene spherulite

tal thickness, and one still being discussed, crystal thickening. The nature of the fold surface has been disputed and is the topic of Sect. 3. A closely related topic also treated in Sect. 3 is the interfacial component present in melt-crystallised samples. The lateral habit of the crystal lamellae is the topic of Sect. 4. It has been very important for the development of the theory of polymer crystallisation. Diffusion of small-molecule penetrants is confined to the amorphous component, and the extension of the diffusion path is affected by the lateral size and shape of the crystals. Section 5 deals with the link between the lamellar structure, planar or shaped like tents, and the superstructure, spherulites or axialites. The question asked is: how it is possible to obtain a spherical structure from the tent-shaped or planar crystallites? Finally, multi-component crystallisation is discussed in Sect. 6. This is the common case since commercial polyethylenes have a broad molar mass distribution and some materials also show heterogeneity in the degree of chain branching. Such polydisperse polymers show segregation of low molar mass and branched species. The general scheme is that the higher molar mass species or the less branched material crystallises first in so-called dominant lamellae (a term introduced by Bassett) leaving the low molar mass or branched material in pockets between the dominant lamellae. The segregated material then crystallises at a later stage in the so-called subsidiary lamellae.

2
Crystal Thickness

Polyethylene crystals change shape with time at temperatures between the crystallisation temperature and the final melting point [42]. The process is commonly referred to as crystal thickening because the major effect is that the crystals increase their dimensions along the c axis. These changes may occur without any change in crystal volume, i.e. crystal thickening is associated with a lateral contraction of the crystals (Fig. 9). A second option is that

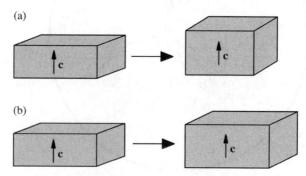

Fig. 9a, b Schematic description of crystal thickening options. **a** Crystal thickening at constant crystal volume. **b** Crystal thickening accompanied by an increase in crystal volume

crystal thickening is accompanied by a corresponding increase in crystal volume, i.e. amorphous material is converted into crystalline (Fig. 9). A third possibility proposed by Kawai [43] involves molecular fractionation as the mechanism for an 'apparent' crystal thickening: the higher molar mass species crystallise early in comparatively thin crystal lamellae, whereas the lower molar mass species, but still with a molar mass greater than M_{crit} at the prevailing temperature (for further explanation see Sect. 6), crystallise more slowly at a lower ΔT. The equilibrium melting point of these species is lower than that of the higher molar mass species and hence they give rise to thicker crystals. This mechanism thus causes crystal thickening without any restructuring of the crystal lamellae already formed.

What is the thermodynamic reason for crystal thickening at constant crystal volume? The surface free energy of the fold surface has been determined from melting point-crystal thickness data to be 90 ± 3 mJ m^{-2} [44, 45, 46], and the surface free energy of the lateral surfaces has been determined from linear growth rate data to be 12 mJ m^2 [45]. These data can be used to calculate the ratio of crystal thickness to crystal width of the equilibrium crystal (minimum free energy) to 90/12=73. If it is assumed that the equilibrium crystal is of the extended-chain type, then the non-lateral surfaces have approximately the same surface energy as the lateral surfaces. Hence, the crystal thickness:width ratio of such an equilibrium crystal would be 1. These numbers may be compared with the 'typical' value for the crystal thickness:width ratio for solution-grown crystals: $10\times10^{-9}/10\times10^{-6}=10^{-3}$. Melt-grown crystals show corresponding values in the range from 10^{-2} to 10^{-3}. Hence, there is always a thermodynamic driving force for crystal thickening. The thicker crystal has a lower free energy than the thinner analogue provided that the internal defects are kept at a constant level. Crystal thickening may also occur without any change in the lateral dimensions by converting amorphous material to crystalline. This also increases the stability of the crystals.

The 'old' work on crystal thickening was confined to medium to high molar mass polyethylenes with molar mass polydispersity, and the results obtained were in some respects different from the more recent findings obtained on strictly monodisperse n-alkanes.

Any structural rearrangement of the crystals resulting in a lower free energy requires translational mobility of the polymer chains in the crystals. Crystal thickening occurs only at temperatures higher than a certain minimum temperature necessary to overcome the kinetic barriers. The onset temperature (T_{ons}) for crystal thickening is related to factors such as initial crystal thickness and it has recently been shown that the molar mass distribution is an important factor. Early data by Mandelkern et al. [47] on polydisperse linear polyethylenes of intermediate molar mass showed that the onset temperature for crystal thickening increased with increasing initial crystal thickness (L^*_c): L^*_c=11.5 nm, T_{ons}=114 °C; L^*_c=17.2 nm, T_{ons}=127 °C. The occurrence of crystal thickening makes it imperative to distinguish between the thickness of the embryo crystal (L^*_c), i.e. the 'baby' crystal as it is immediately after its formation, and of the 'mature' crystal. Crystals grown

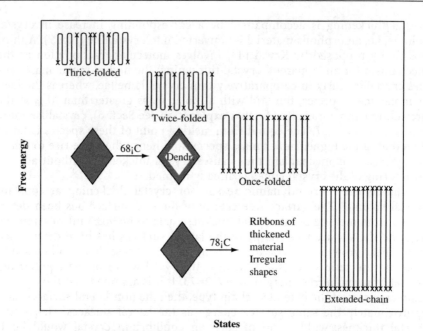

Fig. 10 Schematic representation of crystal thickening of $C_{294}H_{590}$. Drawn after findings of Hobbs et al. [51]

at very low absolute temperatures (e.g. 75 °C) from a solution of polydisperse medium to high molar mass polyethylene show no crystal thickening and the distinction between embryo and mature crystals is not relevant in these cases. Weaver and Harrison [48] crystallised linear polyethylene (Marlex 6001) from xylene solution (0.12% polymer in the solution) at 90 °C and showed (by small-angle X-ray scattering, differential scanning calorimetry and Raman spectroscopy) that no measurable crystal thickening occurred between 2 h and 168 h from the start of the crystallisation process.

Ultra-long strictly monodisperse n-alkanes behave differently from polydisperse polyethylenes of higher molar mass. Figure 10 shows the different crystalline states of $C_{294}H_{590}$. Solution-grown single crystals based on these compounds show crystal thickening even at low crystallisation temperatures (≥ 65 °C) from an initial integer folded crystal form to thicker integer folded or extended chain forms [49, 50, 51]. Electron microscopy of solution-grown single crystals of $C_{294}H_{590}$ (M=4126 g mol^{-1}; length of fully extended chain ≈ 37 nm) showed that thickening from a thrice-folded crystal (fold length ≈ 9 nm) to a twice-folded crystal (fold length ≈ 12 nm) occurred without loss of the original lozenge shape, but the interior was greatly changed from a uniform structure to a fine web of thickened material [51]. It should be noted that these experiments were conducted at 68 °C in dilute solution. The authors refer to this morphology as 'picture-frame crystals'. The frame refers to the relatively intact peripheral part of the crystal and the picture is the internal fine web of thickened material. Similar structures have been observed in dried single crystals based on higher molar mass and polydisperse linear polyethylene after annealing at temperatures well above the crystallisation temperature [52].

Thickening occurring at 78 °C from twice-folded (fold length ≈12 nm) to once-folded (fold length ≈18 nm) crystals of $C_{294}H_{590}$ and was accompanied by a complete loss of the initial crystal shape to a fine dendritic structure with 20–30 nm wide stripes of thickened material in a matrix of material with the original crystal thickness [51].

Synchrotron X-ray scattering has provided valuable information about the initial structure of the monodisperse n-alkane crystals. Ungar and Keller [53] showed that the initial crystals were imperfect without any match in fold length with any integer fraction of the chain length. Interestingly, the initial crystals with disordered fold surfaces either perfected into the nearest larger integer fold length (e.g. with two folds per molecule) or decreased in thickness and perfected to the nearest lower fold length (e.g. with three folds per molecule). This finding is very important for an understanding of polyethylene morphology. It suggests that the initial fold surface is disordered with loops, long cilia and some vacancies. Patel and Bassett [54] reported that the crystal stems are essentially parallel to the lamellar normal at this early stage. This interfacial structure is changed into a more regular fold surface with inclined (tilted) chains, e.g. into a {201} fold surface [54]. This finding is important in explaining the cause of S-shaped lamellae and the lamellar twisting in banded spherulites, as will be discussed further in Sect. 4 and Sect. 5.

Why do solution-grown single crystals of the strictly monodisperse n-alkanes show integer thickening at the crystallisation temperature and why do crystals based on polydisperse and higher molar mass linear polyethylene grown in exactly the same manner not show crystal thickening at the crystallisation temperature? The greater stability of the thicker fully developed crystal (with fewer folds) is not a matter of discussion. The crystal thickening cannot be a completely synchronised and instantaneous process; it must start at a 'point' somewhere on the crystal and then spread to neighbouring regions. In fact, the electron microscopy findings of Hobbs et al. [51] support this assumption. The first few stems involved form a nucleus with a higher free energy than the initial state; the free energy barrier should be proportional to the length of the outgrowth. Furthermore, when a chain with e.g. three folds converts into a chain with only two folds, one of the chain ends has to move from one fold surface to the other in order to reach the final state. In this process, a row vacancy will be created which will increase the free energy barrier. The experimental results for the strictly monochromatic n-alkanes clearly show that these free energy barriers are overcome at relatively low temperatures. A single crystal based on longer chain molecules with a distribution in molar mass can, in principle, undergo the same process, and the free energy barrier associated with nucleation should be of the same order of magnitude as for the strictly monodisperse n-alkanes. The longer chains will have a greater number of folds and thus require a higher degree of cooperation by the different stems of the molecule to accomplish a stepwise increase in crystal thickness. Only a few out of a great number of possible thermal processes would result in an integer (stepwise) increase in fold length. It is suggested that this is the main reason for the greater stability of solution-grown single crystals of higher molar mass, polydisperse polyethylene than of the corresponding single crystals based on strictly monodisperse n-alkanes.

The isothermal thickening (avoiding very short times) of 'mature' crystals of conventional polyethylenes follows the equation [55]:

$$L_c = L_c(t = t_0) + B(T) \log \left(\frac{t}{t_0} + 1 \right) \tag{1}$$

where L_c is the crystal thickness at time t, L_c $(t=t_0)$ is the crystal thickness at time t_0, and $B(T)$ is a constant for a given temperature, which increases with increasing temperature. Equation 1 with given parameter values is valid only within a limited time period. Hoel [56] showed that the crystal thickening as viewed in a $\lin(L_c)$–$\log(\text{time})$ diagram proceeded through two different stages: a period of rapid crystal thickening was followed by a period of considerably slower crystal thickening. The initial stages of crystal thickening of melt-grown crystals may be associated with discontinuous jumps in the crystal thickness [57, 58]. However, synchrotron X-ray scattering of a linear polyethylene fraction presented no evidence in favour of any early stepwise increase of the crystal thickness [59]. Many papers have reported that melt-crystallisation of linear polyethylene at $\geq 125\,°C$ involves a gradual increase with crystallisation time in both the degree of crystallinity and the crystal thickness as recorded after cooling to room temperature. These studies, which possibly present genuine facts, suffer from the deficiency that physical changes occur in the sample during cooling. Bark et al. [59] presented direct evidence—X-ray diffractograms were taken at the crystallisation temperature—of the occurrence of simultaneous isothermal crystal thickening and an increase in the degree of crystallinity. Crystal growth in this case occurs by a gradual ordering of the interfacial region into a crystalline register. Some differences of opinion exist concerning the magnitude of the possible effect of molecular fractionation on the recorded crystal thickness as a function of crystallisation time. Kawai [60] observed less crystal thickening in fractionated linear polyethylenes than in samples with a broad molar mass distribution. Hoffman and Weeks [61] also observed appreciable crystal thickening in fractionated samples. Later Raman spectroscopy studies on linear polyethylenes by Hoel [56] showed that crystal thickening induced by molecular fractionation was less important for the isothermal crystal thickening.

A remarkable experiment was carried out by Blackadder et al. [62]. They annealed a linear polyethylene grade (Rigidex 50) by gradually heating it from 120 °C to 135.8 °C. The thermal treatment was conducted in vacuum to avoid thermal oxidation. The experiment took 2 years and the mass crystallinity and the melting point after subsequent cooling to room temperature

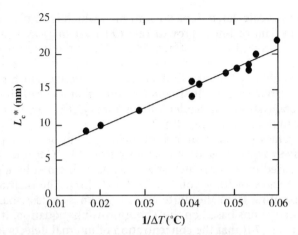

Fig. 11 Initial crystal thickness as a function of the reciprocal degree of super-cooling (ΔT). Drawn after data of Barham at al. [63]

were 92–94% and 136.6 °C. These values are in the same range as those of pressure-crystallised linear polyethylene. The samples showed also considerable brittleness, just as did the pressure-crystallised samples.

Crystals grown from the melt at high temperatures (>120 °C) show crystal thickening and it was a demanding task to assess the thickness of the initial crystals (L^*_c). This problem was solved by the Bristol group [63] in the 1980s; they showed that L^*_c depended only on the degree of super-cooling [$\Delta T = T^0_m - T_c$ where T^0_m is the equilibrium melting point and T_c is the crystallisation temperature] according to:

$$L^*_c(\text{in nm}) = \frac{276.4}{\Delta T} + 4.16 \tag{2}$$

where $T^0_m = 144.7$ °C. Figure 11 shows the initial crystal thickness as a function of the reciprocal of the degree of super-cooling for a series of melt-crystallised linear polyethylenes. The experimental data follow the straight line given by Eq. 2. Solution-grown crystals fall on the same line [63]. Thus, Eq. 2 is general for linear polyethylene and includes crystals grown from both solution and melt. The variation in L^*_c for a given ΔT according to the data presented by Barham et al. [63]—data for solution-grown single crystals from Organ and Keller [64] were included—was typically ±1 nm for crystals in the thickness range of 12–24 nm. Equation 2 has the same form as the average crystal thickness according to the theory of Lauritzen and Hoffman [65]:

$$L^*_c = \frac{2\sigma T^0_m}{\Delta T} \times \frac{1}{\Delta T} + \delta L \tag{3}$$

When appropriate values for the parameters are inserted in Eq. 3—$\Delta h^0 = 293$ kJ kg^{-1} [66], $\rho_c = 996$ kg m^{-3} [3] and $T^0_m = 144.7$ °C—the fold surface free energy (σ) is calculated to be 96.5 mJ m^{-2}, which is similar to the values obtained from data of linear growth rate (90.5 mJ m^{-2}) and melting point (93±8 mJ m^{-2}) according to Hoffman et at [46]. If the lower reported equilibrium melting point, 141.4 °C [44, 67], is used in these calculations, 74.5 mJ m^{-2} is obtained for the fold surface free energy, which seems low in view of other reported data. It should be noted that a competing theory for polymer crystallisation, the Sadler-Gilmer theory [68, 69], predicts the experimentally found relationship, Eq. 2.

Reported values for the equilibrium melting point of linear polyethylene cover a relatively wide range of temperature from 141 °C to 145.5 °C. Experimental data based on 10 μm thick single crystals obtained by high-pressure crystallisation show melting points of 141.4 °C [67]. The melting point depression originating from the finite crystal thickness is less than 0.1 °C according to calculations based on the Thomson-Gibbs equation. It was argued by Hoffman et al. [70] that the concentration of internal defects in the 10 μm thick crystals may be considerably higher than the equilibrium defect content at normal pressure in the melting temperature range and that a small decrease in the heat of fusion (1%) would suffice to explain the difference between the experimental and some extrapolated values. The 141.1 °C reported by Broadhurst [71, 72] was based on the extrapolation of melting point data of n-alkanes. The extrapolation method used by Broadhurst assumed that both the enthalpy and entropy of melting are linear functions of the number of repeating units of the n-alkanes. Flory and Vrij [73] included in their analysis of the melting of n-alkanes the effect on the melting entropy of unpairing of the chain ends that are confined to the crystal surface. This process adds a term, $R \ln n$ (n is the number of repeating units of the n-alkane), to the entropy of melting and the equilibrium melting point for a n-alkane of infinite molar mass was determined by Flory and Vrij [73] to be 145.5±1.0 °C. The melting point data used by Flory and Vrij for the extrapolation was taken from Broadhurst [71]. Grubb [74] found that the choice of data for extrapolation is also important for the extrapolated value obtained. The orthorhombic phase of n-alkanes with $n \leq 44$ transforms into the hexagonal phase before melting, and very long n-alkanes may not be strictly monodisperse and may even fold. The mean value of n used by Flory and Vrij was 44. Databases with mean values of n equal to 70-straight-chain alkanes of $n=160$ were included in the analysis and yielded extrapolated values for the equilibrium melting point of 143.2 °C [74]. Extrapolation of melting point–crystallisation temperature data gave an equilibrium melting point of 145 °C [75]. Determination of the equilibrium melting point by extrapolation of melting point–crystal thickness data and applying the Thomson-Gibbs equation yielded 141.5 °C [44] and 142.0±0.3C °C [46]. It may, however, be argued that the assumption of constant fold surface free energy independent of crystal thickness is not valid and that the extrapolation to $1/L_c = 0$ should not be linear.

3
Nature of the Fold Surface

The conclusion drawn by Andrew Keller [27], on the basis of the lamellar shape and orientation of the c-axis, that the chain must be folded was qualitative but, as it turned out later, essentially correct. The regularity or adjacency of the chain folding was impossible to assess at this stage. Keller realised this and continued searching for more definite answers. The tent-like structure of solution-grown single crystals, i.e. that the molecules are inclined to the lamellar normal, was explained by the dominance of a particular fold structure [32]. Other early evidence for regular chain folding was presented by the Bristol group using nitric acid etching followed by size exclusion chromatography. Etched solution-grown single crystals showed several distinct peaks in the chromatogram corresponding to the molar mass of extended, once-folded and possibly twice-folded chains [76, 77, 78]. These findings indicated that regular chain folding, at least involving a few stems, occurred in solution-grown crystals and, furthermore, that a transitional zone is present at the crystal surface with internal buried chain folds. Keller [79] suggested a tight fold conformation on the basis of the diamond lattice. This included only a few carbon–carbon bonds and the structure was consistent with the observation that the stems are inclined to the lamellar normal. Further refined analysis of the detailed regular fold structure came with Petraccone et al. [80] who calculated the minimum energy for adjacent folds in the {110} and {100} sectors. These folds are displayed in Fig. 12 with the following conformational sequences: ...g(56 °)g(57 °)tg(69 °)g(59 °)... (dihedral angles are shown within parentheses) for the {200} sector and ...g(75 °)tg(68 °)g(92 °)g′(−58 °)g′(64)... for the {110} sector. The bond angle distortion was within 4 ° from the unstrained value [80].

Figure 12 shows that the fold of the {110} sector yields stems inclined to the lamellar normal. Regular folds have been identified in solution-grown crystals by infrared spectroscopy from the detailed spectrum of the CH_2

(a)   (b)

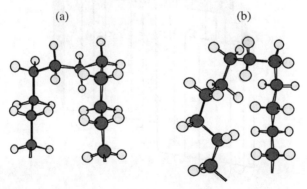

Fig. 12a, b Regular folds in **a** {100} and **b** {110} sectors according to Petraccone et al. [80]

46   Ulf W. Gedde · Alessandro Mattozzi

wagging region—1342 cm^{-1} peak assigned to ggtgg [81]. Wolf et al. [82] were able to assign the infrared absorption peaks associated with {110} and {100} folds by vibrational analysis but they also pointed out that quantitative analysis was not possible because 'standards' with known concentrations of the different fold types were not available and the observed intensities of the peaks were extremely weak. Earlier infrared studies of solution-grown single crystals based on mixtures of deuterated and hydrogenated polyethylene reported by Cheam and Krimm [83] showed incompatibility with the random switchboard model of Flory [84]. The data were instead consistent with a dominance of adjacent re-entry along the {110} and {100} planes. Neutron scattering studies of solution-grown single crystals based on mixtures of deuterated and hydrogenated polyethylene showed that the radius of gyration (s) decreased on crystallisation, with a very weak molar mass dependence: $s \propto M^{0.1}$ [85]. The 'super-folding' model displayed in Fig. 13 was obtained on the basis of these data and data from higher scattering angles [85]. The molecular trajectory is within the folded 'ribbon' and the folds with adjacent re-entry form clusters interrupted by 'statistical loops' bridging stems within the folded ribbon [85, 86]. The cluster aspects were based on both neutron scattering data and infrared spectroscopy of single crystals of mixtures of deuterated and hydrogenated polyethylene. This model has been

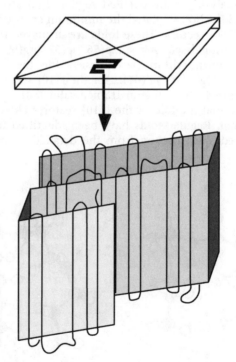

Fig. 13 The super-folding model according to Sadler and Keller [85] showing the conformation of a single molecule in a solution-grown single crystal

further substantiated by the studies of ultra-long n-alkanes. Ungar at al. [87] showed that the fold length of solution-grown crystals was an integer fraction of the chain length thus confirming the occurrence of both adjacent and regular chain folding. They found that alkanes as short as 150 carbon atoms are capable of folding. Decoration techniques (see Fig. 4) developed by Wittman and Lotz [30] and atomic force microscopy (frictional force measurements) by Nisman et al. [88] show that chain folding is dominantly directed along the lateral sector faces of the single crystals.

There are important differences between crystallisation from the melt and crystallisation from dilute solution especially for high molar mass polyethylene. The conformational changes necessary for crystallisation occur readily in dilute solution. The chain entanglements present in the high molar mass melt will obviously strongly retard the conformational changes necessary to obtain a more complete folded macro-conformation. The competition for crystallisation sites is also more severe in the crystallisation from the melt. It can thus be expected, on the basis of these general remarks that the semi-crystalline morphology will possess a certain degree of memory of the original molten chain structure. This picture is clear, viewing the fold structure of a single chain on a smaller scale including only a few stems. Studies by synchrotron X-ray scattering [53] and electron microscopy [54] show that the initial structure (on the scale of a few stems) is imperfect and non-inclined and that with time it perfects into the inclined {201} fold surface structure. These direct observations of fold structure were preceded by general arguments about regular chain folding with regard to amorphous density [89, 90]. The low amorphous density can be achieved only by having a large proportion of tight folds (Fig. 14). A simple calculation of the fraction of tight folds (f_{tf}) can be made, based on the equation:

$$\rho_a = \rho_c \times \frac{L_a}{Cl} \times \cos\theta \times (1 - f_{tf}) \tag{4}$$

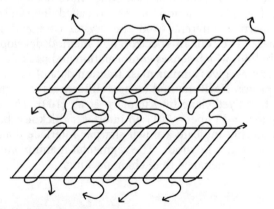

Fig. 14 Schematic description of how overcrowding of the amorphous phase is avoided by adjacent, regular folding

where ρ_a and ρ_c are the amorphous and crystalline densities, L_a is the thickness of the amorphous interlayer, C is the characteristic ratio of the Gaussian chain, l is the bond length and θ is the angle between c and the lamellar normal. Calculations based on realistic values for medium to high molar linear polyethylene yield f_{tf} values close to 80% [91]. Further theoretical arguments using the Gambler's ruin method in favour of regular chain folding was presented by Guttman et al. [92]. The result of their exercise is that at least 2/3 of the entries must be tight folds in a cubic lattice. Small-angle neutron scattering data obtained at higher angles provide information about the fold structure. The neutron scattering data taken by Schelten et al. [93] and Sadler and Keller [94] were analysed by several researchers. Yoon and Flory [95] reported that the random switchboard model was consistent with the experimental scattering profiles. A particularly unattractive feature of the random switchboard model is the overcrowding at the crystal interface, which makes the model physically unrealistic. Guttman et al. [96] came to a different result, which also had the attractive feature of providing a realistic density at the interface: 70% of the folds were to neighbour or next neighbour positions (adjacent folding) thus leaving 30% for random re-entry. Direct experimental evidence for adjacent re-entry by regular folding in melt-crystallised systems was presented by Ungar et al. [87] in a study on n-alkanes of strictly uniform molar mass.

The single molecule trajectory on a larger (global) scale in melt-crystallised polyethylene ($60,000<M<383,000$ g mol^{-1}) was revealed by small-angle neutron scattering, but only after rapid cooling conditions to avoid segregation of the labelled deuterated chains [97, 98]. The radius of gyration (s; in Å) was the same in the semi-crystalline state as in the melt prior to crystallisation: $s=0.46\times M^{1/2}$ [97]. The possible contraction of the chains on crystallisation has not yet been revealed by neutron scattering due to the segregation problem.

The crystal interface constitutes a third component of semicrystalline polyethylene. Several experimental techniques have been developed for the assessment of the mass fraction of the so-called interfacial component, which has properties different from the crystal core and the liquid-like amorphous phase. The Raman spectroscopy method developed by Mutter, Stille and Strobl [99], which senses conformation and packing different from the two other phases, gives similar mass fraction data of the interfacial component as the proton NMR method (senses differences in segmental mobility) as applied to polyethylene by Kitamaru et al. [100]. The extensive data presented by Vonk and Pijpers [101] on interlayer thickness based on small-angle X-ray scattering indicate a much sharper interface than has been reported by the two other methods. The segmental mobility constraint seems thus to have a longer 'memory' than chain packing.

4
Lateral Habit of Crystal Lamellae

The lateral habit of solution-grown crystals has been extensively studied. The first studies go back to the pioneering work of Keller [27], Fischer [28] and Till [29]. These early reports together with a later paper by Keller [52] established that lozenge-shaped crystals with only {110} were formed at low crystallisation temperatures (70 °C) in xylene. Crystallisation at higher temperatures (~85 °C) from the same solvent led to the formation of 'truncated lozenge-shaped' crystals with six lateral faces: four {110} and two {100}. The recognition of crystals showing curved {100} faces was due to Keith [102]. The lateral habit of solution-grown crystals was systematically studied by Organ and Keller [103] over a wide range of temperature from 70 °C to 115 °C using a range of different solvents. Figure 15 shows sketches of the crystals grown at different temperatures. The lateral habit depends primarily on the absolute crystallisation temperature and to a lesser extent on the degree of super-cooling [103]. The crystals are more elongated along the **b** axis at higher crystallisation temperatures. The two {100} faces show a progressively more pronounced curvature with increasing crystallisation temperature. The same observation was made with regard to the {110} faces, although this change became obvious only at the highest temperatures of the study. The ratio of the dimensions along **b** and **a** was 3.25 at the highest crystallisation temperature, 115 °C [103].

Labaig [104] presented the first comprehensive study of the lateral habit of melt-grown single crystals of linear polyethylene. This study was based on the development of a new preparation method revealing melt-grown single crystals. Important later studies using the technique developed by Labaig and other techniques (extraction of samples first isothermally crystallised and then rapidly quenched) were reported by Bassett et al. [105], Keith et al. [106], Toda [107] and Toda and Keller [37]. Crystallisation was performed at temperatures where spherulites (<127 °C) and axialites (>127 °C) are formed. A new shape was discovered in these studies, the lenticular single crystal (Fig. 15). The systematic study by Toda [107] of melt-grown single crystals of linear polyethylene with a molar mass of 14,000 g mol^{-1} crystallised at different temperatures between 125 °C and 131 °C showed the lateral habits displayed in Fig. 15. Crystals grown in the Regime II temperature range (<127 °C) showed the features of truncated lozenges with curved {200} and {110} faces. Regime I crystallisation (>127 °) yielded highly elongated (along b) lenticular crystals. The aspect ratios of these crystals were of the order of 10. Toda [107] argued that tiny {110} faces must surround the tips although no direct evidence for this supposition could be obtained from the micrographs. Hence, the {110} faces growing along the b axis—the radial direction in spherulites and axialites—determine the linear growth rate of the superstructure. This means that the substrate length, i.e. the lateral dimension of the unit of independent crystal growth, becomes very small at high crystallisation temperatures. This should have a retarding effect on the Regime I linear growth rate because the latter is proportional to the substrate

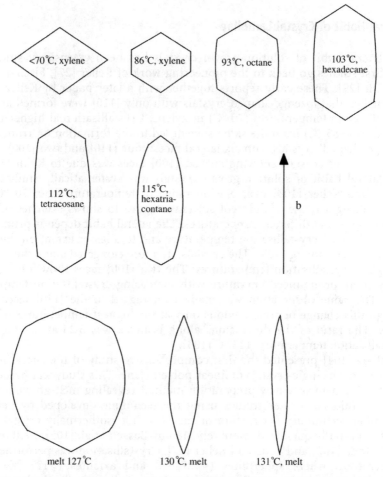

Fig. 15 Lateral habit of solution- and melt-grown polyethylene single crystals. The drawing are based on electron micrographs of Organ and Keller [64]—grown from solution—and of Toda [107]—grown from the melt

length [45]. The crystals that grow at temperatures above 127 °C have {201} fold surfaces, which means that the tilt angle (angle between the crystalline chain axis and the normal of the fold surface) is 35 ° [105, 106]. These single crystals are usually planar [54,104, 105, 106] but lower molar mass linear polyethylenes also show ridged lamellae with {201} fold surfaces [108]. Keith et al. [106] noted asymmetries in the lateral habit in the crystal grown at high temperatures. The shape of the lenticular crystals with {201} fold surfaces was asymmetric; the lateral growth was slower at the face with 'overhanging' folds.

5
Relation Between Growth of Crystal Lamellae and Superstructure

The scientific problem discussed in this section is: how it is possible that crystal lamellae (planar or tent-shaped) can fill the three-dimensional space of a spherulite? A stack of parallel lamellae obviously cannot be the answer. The number of radially growing crystal lamellae fronts must increase with increasing radial distance (r) from the centre of the spherulite in order to preserve the same degree of crystallinity throughout the spherulite. Provided that the size of the crystal lamellae front is the same at different radial positions, the number of growing crystal lamellae fronts must then be proportional to r^2. The only sensible solution to this problem is that there has to be one or several means by which (1) the number of growing crystal fronts increases during the course of spherulite crystallisation and (2) splaying or branching of crystal lamellae occurs. The analogy with the tree provides some insight into the problem, although important differences exist between the branched structure of a tree and the arrangement of crystal lamellae in a polyethylene spherulite.

It is important to point out that several different superstructures are formed in polyethylene depending on the molar mass of the polymer and the crystallisation temperature (Fig. 16). Linear polyethylenes (narrow molar mass distribution; $\overline{M}_w/\overline{M}_n \approx 1.1$) of molar mass between 18,000 and

Fig. 16a–c Polarised photomicrographs of different linear polyethylenes showing **a** non-banded spherulites, **b** banded spherulites (from [115] with permission from Elsevier, UK) and **c** axialites. *Scale bars* represent 20 µm

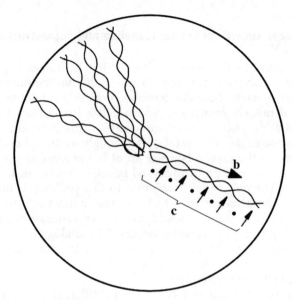

Fig. 17 Schematic (simplified) representation of twisting lamella radiating out from the centre of a banded spherulite

115,000 g mol^{-1} form negative spherulites on crystallisation at 17.5 °C or greater degrees of super-cooling [45]. Keller [109] and Point [110] showed independently that the crystallographic **b** axis was mainly along the spherulite radius, whereas the chain axis is preferentially oriented in the tangential plane of the spherulites. Banded spherulites—the optical texture is concentric rings overlapping the Maltese cross—are formed on crystallisation at low temperatures, typically at 20–25 °C or higher degree of super-cooling. Banded polyethylene spherulites were first reported by Keller [111]. Keith and Padden [112, 112] related this periodic structure to the periodic twisting of the crystal lamellae earlier observed by Fischer [28]. Hence, while the **b** axis is along the radius, the **c** axis shows a periodic twisting in the transverse plane along the spherulite radius (Fig. 17). Keller [114] showed that the band period decreases with decreasing crystallisation temperature. More recently, Rego Lopez and Gedde [115] reported for a narrow fraction of linear polyethylene ($\overline{M}_w$=66,000 g mol^{-1}; $\overline{M}_w/\overline{M}_n$=1.2) the following formation temperatures for the different spherulitic structures: non-banded spherulites: 121.5–127.5 °C; banded (5–10 μm band period) spherulites: 118–121.5 °C; banded (2–5 μm band period) spherulites: 114–118 °C; banded (1–2 μm band period) spherulites: 110.5–114 °C; banded (<1 μm band period) spherulites: <110.5 °C.

Crystallisation at higher temperatures (degree of super-cooling ≤17.5 °C) of linear polyethylene fractions of molar masses between 18,000 and 115,000 g mol^{-1} yields axialites [45]. Low molar mass linear polyethylenes (M<17,000 g mol^{-1}) form axialites at all crystallisation temperatures [45]. It seems that the mechanism(s) leading to the multiplication and splaying of the growing crystal lamellae occurs less frequently at higher crystallisation temperatures.

Linear polyethylene fractions ($\overline{M}_w/\overline{M}_n$=1.26–1.81) in the molar mass range 119,000 to 800,000 g mol^{-1} also crystallise at high temperatures in spherulites; Hoffman et al. [45] refer to these as 'irregular' spherulites.

Very high molar mass polyethylene ($M \geq 2,000,000$ g mol^{-1}) crystallises without the, formation of a clear superstructure, sometimes referred to as the random lamellar structure [116]. The great many chain entanglements present in high molar mass polymers obstruct crystallisation and the crystals become small and their orientation less correlated with surrounding crystal lamellae.

Our understanding of the lamellar structure underlying the different superstructures is based on studies by transmission electron microscopy of replicates or stained thin sections. The early studies of Anderson [117] and Mandelkern et al. [118] on replicates of fracture surfaces provided the view along the **a** axis, with the **b** axis (i.e. the spherulite radius) in the fracture surface plane [119]. Complementary information was obtained from studies of sections stained with chlorosulphonic acid according to the method developed by Kanig [120]. This technique permits a clear view only of crystal lamellae with their lamella normal perpendicular to the electron beam (i.e. 'edge on'), see Fig. 18. Tilted lamellae appear less sharp and areas without lamellar contrast are found over large areas in the electron micrographs [91]. The location of the sections should be random with respect to the centres of the spherulites. From simple geometrical considerations, it can be shown that the average distance between the section and the spherulite centre should be $R/\sqrt{3}$ where R is the radius of the spherulite. It is well established that the **b** axis is parallel to the radius of a mature spherulite. Hence, sharply appearing crystal lamellae are dominantly viewed along **b**, which must be only a limited

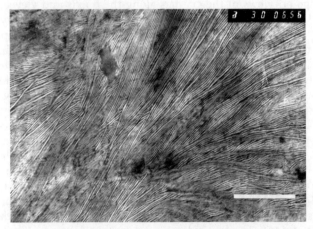

Fig. 18 Transmission electron micrograph of a chlorosulphonated section of polyethylene. *Scale bar* represents 0.5 μm. Courtesy of M.T. Conde Braña

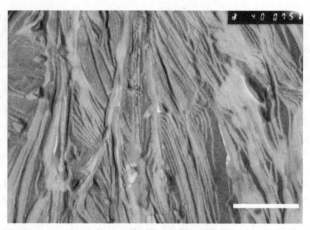

Fig. 19 Transmission electron micrograph of a replicate of permanganic-etched polyethylene. *Scale bar* represents 0.5 μm. Courtesy of M. T. Conde Braña

area of the section: 40% of the surface is within a 20 ° angle from the spherulite radius and 60% within a 30 ° angle from the spherulite radius. The permanganic etching technique using solutions prepared from H_2SO_4 and $KMnO_4$ was introduced by Olley at al. [121]. This paper was submitted in 1977 but the paper by Bassett and Hodge [119] showing electron micrographs of replicates of samples treated with the permanganic etching technique was published first. The permanganic etching technique has been developed further by reducing the concentration of KMnO4 from 7% to between 0.7 and 2% and by introducing orthophosphoric acid in order to avoid the formation of artificial structures, observed primarily at low magnification. The acid etches to a depth of 1–2 μm and the crystal lamellae can be observed at all the angles with respect to their crystallographic axes [119]. Hence, the method provides edge-on-views of the crystals along the **b** axis and views along **c** revealing the lateral habit (Fig. 19). The morphology of individual crystal lamellae, mainly views along the lamellar normal, were obtained by depositing dilute solutions of polyethylene on freshly cleaved mica, evaporating the solvent, crystallising the melt and shadowing [106]. It should be noted that the crystals revealed in this case are not from a true three-dimensional spherulite.

Let us start by looking at the lamellar morphology viewed along the radius of the spherulite (axialite). The general picture of the dominant crystal lamellae morphology was provided in 1981 by Bassett et al. [108] who studied a wide range of linear polyethylenes including fractions after crystallisation at 120–131 °C. Crystallisation at 130 °C of the low molar mass samples (20,000–50,000 g mol^{-1}) yielded both planar sheets and ridged sheets, both 4 μm wide as viewed along the **b** axis (Fig. 20). Later studies by Keith et al. [106], Toda [107] and Toda and Keller [37] of melt-grown single-crystal linear polyethylenes in this molar range were, in principle, in accordance with the findings of Bassett et al. [46]. The lenticular single crystals shown in Fig. 21 were reported to be flat, with a 35 ° chain tilt angle corresponding to a fold surface

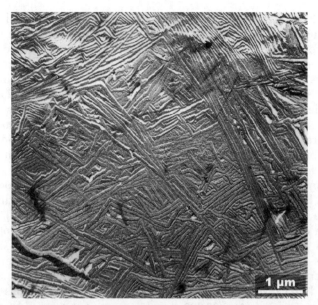

Fig. 20 Transmission electron micrograph of permanganic-etched linear polyethylene fraction crystallised at 130.4 °C for 27 days. Courtesy of D.C. Bassett. From [46] with permission from the Royal Society of London, UK

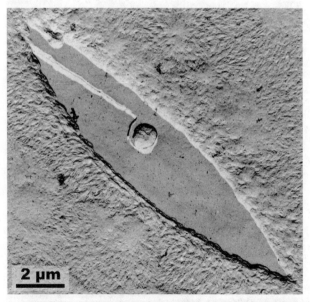

Fig. 21 Transmission electron micrograph of permanganic-etched linear polyethylene ($\overline{M}_n$=17,000 g mol^{-1}, $\overline{M}_w$=54,000 g mol^{-1}) crystallised at 130 °C for 13 h showing a lenticular crystal lamella. Courtesy of D.C. Bassett. From [54] with permission from Elsevier, UK

close to {201}. One of the studies [106] reported an asymmetric lateral habit that was suggested to be related to the fact that the chains were inclined so that the growth rate was slower on the side with overhanging folds and cilia. The fraction of planar sheets increased with increasing molar mass in this molar mass range. Higher molar mass polyethylenes (>100,000 g mol^{-1}) crystallising at 130 °C showed curved lamellae and occasional ridged and planar lamellae viewed along **b**. The maximum width of these structures from the b view was 4 μm. Bassett et al. [46] reported further that the structures at 128 °C were similar to those found at 130 °C, except that the ridged sheets were not found in the low molar mass samples. Crystallisation at 125 °C resulted in dominantly curved and S-shaped lamellae except for the sample with the lowest molar mass (M≈20,000 g mol^{-1}), which still showed planar lamellae. The latter sample showed mostly S-shaped dominant lamellae, but still with a few planar sheets, after crystallisation at 120 °C [46]. The main conclusion drawn by Bassett et al. [46] was that the lamellar profile viewed along **b** is planar or ridged for axialites and curved (C- or S-shaped) for spherulites (Fig. 22). The angle between the lamellar normal and **c** of the S-shaped crystals changed continuously along the crystal with a maximum of 35 °. Toda [107] revealed the lateral habit for melt-grown crystals of a linear polyethylene fraction ($\overline{M}_w$=32,000 g mol^{-1}) formed at temperatures between 125 and 130 °C (see also Sect. 4). The crystals grown within the Regime II

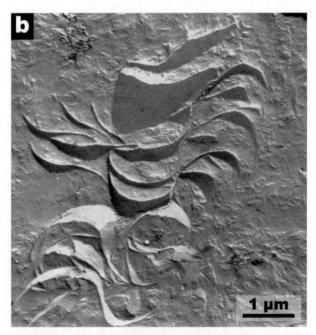

Fig. 22 Transmission electron micrograph of permanganic-etched branched polyethylene crystallised at 120 °C for 1 min showing dominant S-shaped lamellae. From Patel and Bassett [54] with permission from Elsevier, UK

temperature region (<128 °C) were basically truncated lozenges with rounded {100} and {110} faces. These crystals were also less extended along **b**. In a subsequent study, Toda and Keller [37] reported that these crystals were chair-like. These crystals appear curved in a view along **b**. It is pointed out by the authors that the chair-like crystals are curved in the opposite way to Bassett's S-shaped lamellae, i.e. they constitute a new class of curved lamellae. It is suggested that the restructuring of the fold surface is the fundamental reason for the different observed shapes of the crystal lamellae as viewed along **b**. The idea of an initial irregular and short-lived fold surface is relatively old [122, 123], but direct evidence came in the paper of Ungar and Keller [53]. Abo el Maaty and Bassett [124] found by electron microscopy that the initial (disordered) structure was non-inclined, i.e. the crystal stems were parallel to the lamellar normal. This structure is only temporary and it rearranges into a more ordered and inclined (tilt angle 35 °) fold structure. Abo el Maaty and Bassett [124] proposed that the different crystal shapes as viewed along b are a consequence of the relative rates of fold surface ordering and the advance of crystal front. The rate of advance of the crystal front should be proportional to $\exp[-K_g/(T_c\Delta T)]$, i.e. it decreases strongly with increasing temperature [45], whereas the rate of fold surface ordering may even be faster at higher temperatures. The most 'efficient' mechanism for a longitudinal shift of stems—a process that must be important for the fold surface ordering—is by the α process, which shows an Arrhenius temperature dependence, i.e. the rate is proportional to $\exp[-\Delta E/(RT_c)]$ [21]. The discovery by Abo el Maaty and Bassett [124] that high temperature crystallisation (>127 °C) led to immediate fold surface ordering and the formation of planar sheets with an inclined (35 ° tilt angle) fold surface is thus logical. Crystallisation at lower temperatures led, according to electron microscopy findings, to the formation of a more long-lived planar lamellar structure with non-inclined fold surfaces. The transformation to the more energetically favourable inclined {201} fold surface occurred at a later stage behind the crystal front [124]. This means that the fold surface ordering (chain tilting) occurs within a lamella with a non-inclined 'outer' part which provides some constraint on the process. The stresses built up in such lamellae trigger the deformation of the initially plane sheets to the S-shape. Abo el Maaty and Bassett [124] suggested that the S-shape is more easily adapted in more open textures with a low degree of crystallinity.

The length of the crystals along the spherulite (axialite) radius is not easily determined from electron micrographs because of lamellar branching and because the view is seldom perpendicular to the spherulite radius. Bassett and Hodge [119] reported, however, that they traced individual crystals over 20 µm, which was a very significant part of the spherulite radius in the particular case. There seems to be a consensus that the crystals are continuous from the centre to the periphery of the spherulites. The implication of the lamellar branching, as will be discussed in the next paragraph, is that the individual crystal lamella is a highly branched structure. It starts as a single sheet but soon branches into many growth faces that are continuous at the periphery of the spherulite.

Let us now consider how more crystals are created during the course of crystallisation from the centre towards the periphery of the spherulite. Non-crystallographic branching and self-epitaxy must be mentioned. The first mechanism means that the crystallographic regularity is broken between the mother and daughter crystals and that the latter can grow at any angle with respect to the mother crystal. Self-epitaxy, most prominent in the cross-hatched morphology of isotactic polypropylene with the α-crystal phase, means that the chains are nucleating in a particular, well-defined manner on the surface of the mother crystal. Self-epitaxy occurs at a specific angle with respect to the mother crystal. A third possibility, a crystallographic branching that not only provides the required multiplication effect but also causes splaying of mother and daughter crystals, has been demonstrated by Bassett [125], Bassett and Olley [126] and Bassett and Vaughan [127], who observed splaying dominant lamellae with a certain angle of divergence, ~20 °. Bassett [125] argued, on the basis of these results from studies of strictly monodisperse n-alkanes, that the reason for the splaying of the crystals is 'ciliation'. During the process of attachment of stems to the growth surface of a growing crystal, the remaining un-crystallised part of a single chain is in the form of a cilium. This phenomenon is referred to as transient ciliation. The ensemble of cilia in the vicinity of the contact point should generate a positive internal pressure that makes the crystal arms to diverge by an angle of approximately 20 °. Direct evidence of lamellar branching was reported for linear polyethylene by Bassett et al. [105]. They observed spiral terraces around screw dislocations, providing in each branch two growing crystal 'arms' with splaying of these two successive layers (Fig. 23). An important discovery reported by these authors was that the upper and lower layers of a three-layer

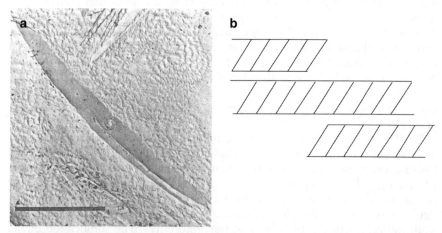

Fig. 23 a Transmission electron micrograph of permanganic-etched linear polyethylene. The micrograph shows the spiral development around a screw dislocation in a sample crystallised at 130 °C. *Scale bar* represents 10 μm. Courtesy of D.C. Bassett. From [105] with permission from Elsevier, UK. **b** Sketch showing c with respect to the fold surface of the crystal layers around the screw dislocation

terrace stopped their growth at the side where their lateral surfaces make an acute angle (angle <90 °) to the fold surface of the adjacent layer (Fig. 23). This gives the spiral a certain chirality. Further confirmation was delivered by Bassett et al. [128, 129] who showed that the crystallisation of strictly monochromatic n-alkanes in extended-chain crystals (thus without ciliation) leads to the formation of axialites, which confirms the important role of ciliation in branching/splaying and spherulite formation. The importance of a permanent cilium (i.e. a portion of the polymer chain that preserves its amorphous character even after a long time) on lamellar splaying was recognised by Techoe and Bassett [130] and Hosier and Bassett [131, 132]. Toda and Keller [37] found that the spiral terraces were mainly in the {101} sectors of the truncated lozenges in the Regime II, i.e. at crystallisation temperatures below 127 °C (Fig. 24). They also noted that the spiral terraces were much less frequent in the lenticular crystals formed in Regime I, i.e. at crystallisation temperatures above 127 °C. The relatively few lamellar branches created during Regime I growth are thus sufficient to achieve spherical symmetry and, hence, axialites are formed.

The twisting lamellar structure of banded spherulites has been debated for decades without obtaining any satisfactory answer until recently. The nature of the isochiral (certain uniform handedness) lamellar twisting and the synchronic character of the twisting of a group of adjacent dominant lamellae both require an explanation. The permanganic etching technique provided

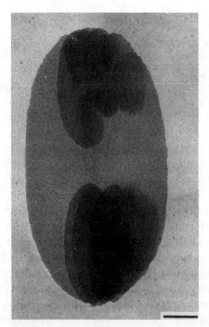

Fig. 24 Transmission electron micrograph of linear polyethylene crystallised at 127 °C for 2.5 min showing spiral terraces in the {110} sectors. *Scale bar* represents 1 μm. From Toda and Keller [37] with permission from Springer, Berlin Heidelberg New York

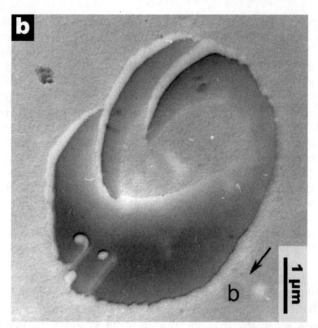

Fig. 25 Transmission electron micrograph of permanganic-etched branched polyethylene crystallised at 123 °C for 35 min. From Patel and Bassett [54] with permission from Elsevier, UK

the three-dimensional perspective of the structure that was needed to reveal the morphological facts necessary to understand the lamellar structure of banded spherulites. Let us first repeat that the dominant lamellae in banded spherulites are S- or C-shaped as viewed along the **b** axis and that they twist along **b** in a regular (on average) and coordinated fashion. Electron microscopy has identified two different lamellar regions where the twist occurs [108]:

1. The spiral terraces generated by screw dislocations are responsible for almost the entire twist. Figure 25 shows a crystal lamella with screw dislocations. Important features are the splaying and the twisting of the branching crystals and the redirection of **b** of the branching crystals with respect to that of the mother crystal. The branching crystals thus form diagonal links in the structure, a feature already found by Bassett and Hodge [108].

2. A much smaller contribution to the twist is that associated with the isochiral twist of the lamellae between the lamellar branches. In fact, the original view of Bassett and Hodge [108] was that these parts of the lamellae were untwisted. However, a more recent paper by Patel and Bassett [54] identifies a small twisting of the lamellae about **b**; 5 ° over the band period of 1 µm, which means that the angle of 175 ° originates from the regions near the lamellar branch regions. The circumferential coordination of the lamellar twisting is very obvious from many electron micrographs presented in the literature.

What is the fundamental reason for the lamellar twisting? According to Patel and Bassett [54], the answer is the asymmetry generated by the uniformly inclined fold surface. The delayed fold surface ordering occurring at lower temperatures leads to the deformation of the planar to the S-shaped sheet. The axis of the 'S', according to Patel and Bassett [54], is inclined with respect to the **b** axis. This will cause the lamella to make a twist about the **b** axis. This twist will be amplified at the screw dislocations because the torsional rigidity is lower at these sites due to the reduced crystal width. Patel and Bassett [54] explained the well-known decrease in band spacing with decreasing crystallisation temperature in terms of the lower torsional rigidity and the lower yield stress of thinner crystals formed at lower crystallisation temperatures.

6
Multi-Component Crystallisation and Structure Development in Melt-Crystallised Systems

The morphology of a polyethylene blend (a homopolymer prepared from ethylene is a blend of species with different molar mass) after crystallisation is dependent on the blend morphology of the molten system before crystallisation and on the relative tendencies for the different molecular species to crystallise at different temperatures. The latter may lead to phase separation (segregation) of low molar mass species at a relatively fine scale within spherulites; this is typical of linear polyethylene. Highly branched polyethylene may show segregation on a larger scale, so-called cellulation. Phase separation in the melt results in spherical domain structures on a large scale.

Hill and Barham [133] showed by transmission electron microscopy that blends of high and low molar mass polyethylene melts were homogeneous with no detectable phase separation. The blends were prepared by solution mixing to obtain an initially homogeneous blend before the thermal treatment in the melt. It should be realised that the mechanical mixing of high and low molar mass linear polyethylenes to obtain a homogeneous melt may require considerable work and time.

The melt morphology of blends of linear and branched polyethylene is another and more complicated story. Small-angle neutron scattering with one of the components being labelled (^1H being replaced with ^2H) provides direct information about the blend morphology of the molten systems: blends of linear polyethylene and branched polyethylene (low-pressure process using heterogeneous catalysts) are homogeneous provided that the branched polymer has less than 8 mol% of branches [134, 135]. The same authors reported phase separation in blends where the branched polyethylene component had 16 mol% branches. Barham et al. [136] developed an indirect method for the assessment of the blend morphology of the melt. This technique includes solution blending of the components, equilibration of the molten blend, rapid cooling to room temperature in order to minimise further phase separation and determination of the morphology of the semicrystalline polymer by transmission electron microscopy (linear and branched polyethylenes were expected to show different lamellar morphologies) and differential scanning calorimetry (bimodal melting was assumed to indicate phase separation in the molten state). The Bristol group has reported a great many studies suggesting that phase separation of linear and branched polyethylenes occurs in the molten state. Using this technique, phase separation was detected in binary blends with branched polyethylene with significantly lower degree of chain branching than was detected by the direct small-angle neutron scattering technique. Hill et al. [137] showed that the indirect methods could detect phase separation in blends with a branched polymer having less than 1 branch per 100 main chain carbon atoms. Similar results have been reported by Tanem and Stori [138, 139] using the indirect methods.

The main conclusions drawn from the studies of a great many binary (in some cases ternary) blends were: phase separation was insensitive to the molar mass of the linear polyethylene [140, 141] and to the branch type of the branched polyethylene [137, 142] but dependent on the branch content [137, 142, 143, 144]. The typical diameter of the minority phase (supposedly enriched in linear polyethylene) was reported to be ~1 μm and this minority phase showed a coarsening with holding time (diameter $\propto t^{1/3}$), which suggested the occurrence of Ostwald ripening [145]. The typical phase diagram constructed on the basis of results obtained by the indirect methods is a closed 'loop' (with both upper critical and lower critical solution temperatures) defining the two-phase region. The two-phase loop in the phase diagram is located near the 100% branched polyethylene, which typically extends in composition from 50–80% to 100% branched polyethylene and in temperature from ~120 °C to ~170 °C. Good examples of such phase diagrams are presented by Hill and Barham [136].

A recent paper from the Bristol group [146] provides a new perspective on the earlier findings obtained by the indirect methods. Micro-Raman imaging showed that phase-separated blends of linear and branched polyethylene remixed in the two-phase region of the phase diagram; the latter being mapped by the indirect methods. The authors concluded that the phase separation revealed by transmission electron microscopy does not occur on the basis of branch content. Morgan et al. [146] found evidence for regions of the same size as the domains observed by electron microscopy with either of the following two combinations: low crystallinity of the linear component and high crystallinity of the branched component, or vice versa. The origin of this heterogeneity is not yet clear. Hence, the current view is that linear and branched polyethylenes do not phase-separate in the melt unless the branched polymer is very highly branched (>10 mol% branching).

Crystallisation of most polymers is accompanied by the separation of different molecular species, a process referred to as molecular fractionation. Bank and Krimm [147] provided the first direct evidence of molecular fractionation in polyethylene. The first extensive study performed by Wunderlich and Mehta [148] indicated that, at each crystallisation temperature, there exists a critical molar mass (M_{crit}) such that the molecules of molar mass greater than M_{crit}, are able to crystallise at this temperature, whereas

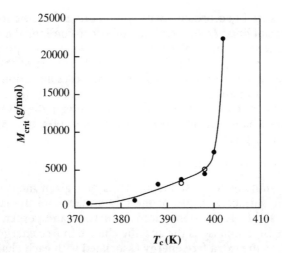

Fig. 26 Critical molar mass of melt-crystallised linear polyethylene as a function of crystallisation temperature. *Filled circles* data for a broad molar mass sample with $\overline{M}_n$=8500 g mol^{-1}, $\overline{M}_w$=153,000 g mol^{-1} of Mehta and Wunderlich [149]. *Open circles* data for a sample with $\overline{M}_n$=12,900 g mol^{-1}, $\overline{M}_w$=108,000 g mol^{-1} from Gedde et al. [152]. From [91] with permission from Kluwer, Doordrecht, Netherlands

molecules of molar mass less than M_{crit} are unable to crystallise. Fractionation was found to be relatively sharp in terms of molar mass. Figure 26 shows that M_{crit}, increases with increasing crystallisation temperature. The lower limit of segregation is set by the hypothetical equilibrium of crystallisation. It is assumed that dynamic equilibrium is achieved between fully extended-chain crystals and the surrounding melt. At equilibrium, the molecular length of the crystallisable species corresponds sharply to the lamellar thickness, and molecules that are shorter or longer than the fold length increase the free energy and are rejected from the crystal. The equilibrium melting point of a given molecular species is dependent not only on its molar mass but also on the molar masses of the other species present in the blended melt:

$$\frac{1}{T_m} - \frac{1}{T_m^0(M)} = \frac{R}{\Delta H}\left[-\ln v_p + (\bar{x}-1)(1-v_p) - \bar{x}\chi(1-v_p)^2\right] \tag{5}$$

where T_m is the melting temperature of the crystallising species in the mixture of different species, $T_m^0(M)$ is the equilibrium melting-crystallisation temperature of the pure species of the molar mass considered v_p is the volume fraction in the melt of the crystallising species. ΔH is the molar heat of fusion, χ is the Flory-Huggins interaction parameter and $\bar{x}$ is the volume fraction of crystallising species with respect to all species in the blend. It is thus possible using Eq. 5) to calculate an equilibrium critical molar mass for each temperature of crystallisation considering the molar mass distribution data of the polymer. Wunderlich and Mehta [149] showed that the experimental values were in accordance with the theoretical prediction at high degrees of super-cooling. At low degrees of super-cooling, the experimental data was significantly higher than the critical molar mass predicted by the equilibrium theory (Eq. 5). This led Wunderlich to suggest that each molecule undergoes a *molecutar nucleation* before crystallisation. Wunderlich claimed that fractionation is governed not by equilibrium considerations but rather by the size of the molecular nucleus under the given conditions. The free energy change on folded-chain crystallisation of a molecule on a crystal substrate is given by:

$$\Delta G = vabL_c\Delta g + 2bL_c\sigma_L + 2vab\sigma + 2vab\sigma_{ce} \tag{6}$$

where v is the number of crystallising stems of a given molecule, σ and σ_L are the specific surface energies of the fold surface and the lateral surface, respectively, a and b are the stem widths transverse and perpendicular to the crystal growth direction, Δg is the specific change in free energy on crystallization, and σ_{ce} is the extra free energy associated with each chain end. Zachmann [150, 151] suggested that the major part of σ_{ce}, is due to the entropy reduction of the non-crystallised cilia. The size of the critical nucleus (L_{crit}) can be calculated from Eq. 6 to be:

$$L_{crit} = \frac{4\sigma\sigma_L b\left(T_m^o\right)^2}{(\Delta h^o)^2\Delta T^2} + \frac{2\sigma_{ce}T_m^0}{\Delta h^0\Delta T} + \frac{2kT_c T_m^0}{ab\Delta h^0\Delta T} \tag{7}$$

The first term of Eq. 7 dominates at low degrees of super-cooling (ΔT), whereas the second and third terms predominate at higher ΔT. Equation 7 was fitted to experimental data of Wunderlich and Mehta [148], where the adjustable parameter σ_{ce} was given a value of 100 mJ m^{-2}.

Linear polyethylene shows fractionation of different molar mass species [148, 149, 152]. The low molar mass material crystallises at low temperatures in subsidiary lamellae located between the dominant lamellae and in the spherulite boundaries [46, 119, 153, 154]. Direct evidence for crystal continuity between dominant and subsidiary lamellae was presented by Bassett et al. [46]. Figure 27 shows dominant ridged sheets that are growing further and converting into much thinner and S-shaped subsidiary lamellae. Figure 28 displays isothermally crystallised high-density polyethylene after solvent extraction to remove the segregated low molar mass species. This particular sample demonstrated a certain preference for segregation towards the spherulite boundaries.

Most of the early studies concerned with molecular fractionation dealt with samples having a broad molar mass distribution. The crystallisation of binary mixtures of sharp fractions was studied to a lesser degree. The crystallisation of binary mixtures of linear polyethylene sharp fractions in the molar mass range from 1000 to 20,000 g mol^{-1} depended upon the cooling rate, and two types of crystallisation were observed [155]:

1. Separate crystallisation of the components occurred at low degrees of super-cooling.

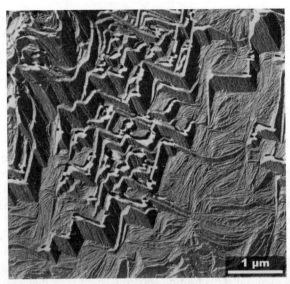

Fig. 27 Transmission electron micrograph of etched cut surface of a linear polyethylene after crystallisation at 130.4 °C for 27 days followed by quenching. Etching was performed with permanganic acid. Note the continuity between dominant ridges and thinner S-shaped lamellae. From Bassett et al. [46] with permission from the Royal Society of London, UK

2. Water-quenched mixtures crystallising at very extensive degrees of supercooling displayed only one melting peak and one small-angle X-ray scattering peak, which was taken as evidence of co-crystallisation of the components.

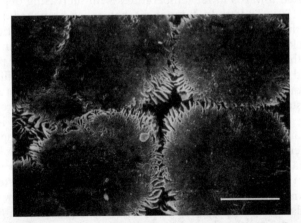

Fig. 28 Scanning electron micrograph of high density polyethylene first isothermally crystallised at 128 °C and then rapidly cooled to room temperature. The sample was etched with hot *p*-xylene to remove the material crystallising in the cooling phase. *Scale bar* represents 20 μm. From Gedde and Jansson [154] with permission from Elsevier, UK

Later work on binary linear polyethylene blends reported by Rego Lopez and Gedde [115], Rego Lopez et al. [156] and Conde Braña et al. [157] provided a somewhat different view. The blends studied were based on the combination of a low molar mass linear polyethylene ($\overline{M}_w$=2500 g mol^{-1}; $\overline{M}_w/\overline{M}_n$=1.1) with one of a series of higher molar mass linear polyethylenes (11,000<$\overline{M}_w$<66,000 g mol^{-1}; $\overline{M}_w/\overline{M}_n$=1.1). Different types of crystallisation were observed in the binary linear polyethylene blends [156]:

1. At high crystallisation temperatures, the high molar mass polymer crystallised alone. Data for the fold surface free energy obtained from linear growth rate data supported the view that the nature of the fold surface of the dominant lamellae was related only to the molar mass of the crystallising component and was not affected by the composition of the melt.
2. At intermediate temperatures, i.e. at temperatures below the temperature corresponding to M_{crit}=2,500 g mol^{-1}, both components crystallised but in separate crystal lamellae. Crystallisation of the low molar mass component in the blend was promoted by the presence of crystals consisting of the high molar mass material. This finding was consistent with the crystal continuity between dominant and subsidiary crystals reported by Bassett et al. [46].
3. At low temperatures, partial co-crystallisation was indicated by transmission electron microscopy and differential scanning calorimetry [156, 157]. Both electron microscopy of stained sections and optical microscopy showed that the segregated low molar mass material was present as small domains between the stacks of dominant lamellae within the spherulites/axialites [115, 157, 158].

Branched polyethylene exhibits not only molar mass segregation but also fractionation due to structural irregularity. The crystallisation temperature range is shifted towards lower temperatures with increasing degree of chain branching [159]. The multi-component nature of branched polyethylene arises from the fact that the chain branches are randomly positioned on the polymer backbone chain. Segregation is thus never sharp, as in the case of linear polyethylene with differences in molar mass only. Linear polyethylene with a broad molar mass distribution, which was melt-crystallised at constant temperature and then rapidly quenched to room temperature, exhibited two crystal populations; one melting at high temperatures from the isothermally crystallised fraction and the second showing a low melting point associated with the material crystallised during quenching [148, 160]. In this particular case, it was possible to selectively remove the low melting point material by p-xylene extraction. When the fraction dissolved at a given extraction temperature was plotted as a function of extraction temperature, the curve almost exactly resembled the cumulative melting curve shifted by 31 °C towards lower temperatures [152]. This finding, shown in Fig. 29, suggests that the different molar mass species of the linear polyethylenes crystallise in different crystal lamellae. It is not possible to remove the low melt-

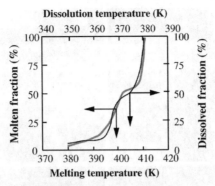

Fig. 29 Cumulative melting and dissolution (in *p*-xylene) curves of a linear polyethylene crystallised at 401 K to completeness and then rapidly cooled to room temperature. Drawn after data of Gedde et al. [152]

ing species to the same extent from branched polyethylenes, probably because of the statistical distribution of the branch points [152].

Blends of linear and branched polyethylene have received considerable attention [161, 162, 163, 164, 165, 166]. The two components in binary mixtures of linear polyethylene and branched polyethylene produced by the high-pressure process are unquestionably segregated in the solid state [161, 162, 163]. The conclusions drawn from studies of blends of linear polyethylene and branched polyethylene produced by a low-pressure process are diverse, although the studies were concerned with similar polymers of relatively high molar mass with medium to high polydispersity and with the branched polyethylene containing 1.4–1.8 mol% of ethyl groups [164, 165, 166]. Hu et al. [164] and Edwards [165] presented evidence obtained by differential scanning calorimetry, X-ray diffraction and Raman spectroscopy supporting the hypothesis that co-crystallisation of the components occurs in slowly cooled samples. In contrast to this view, Norton and Keller [166] reported data obtained by differential scanning calorimetry, polarised light microscopy and transmission electron microscopy.The results established, predominantly, segregation of linear and branched polyethylene (1.4 mol% of ethyl groups) components in a 50/50 blend of commercial HDPE and LLDPE crystallised at different constant temperatures between 394 and 403 K. The linear polymer crystallised first under isothermal conditions to form thicker and less curved dominant lamellae, whereas the branched polymer crystallised at a later stage during the rapid cooling in finer, S-shaped lamellae located between the stacks of dominant lamellae. Some limited co-crystallisation was, however, indicated in samples crystallised close to 394 K. According to differential scanning calorimetry, the quenched samples exhibited less pronounced segregation.

The morphology and crystallisation behaviour of a series of binary blends based on a low molar mass linear polyethylene ($\overline{M}_w$=2500 g mol^{-1}; $\overline{M}_w/\overline{M}_n$=1.1) and two higher molar mass branched polyethylenes

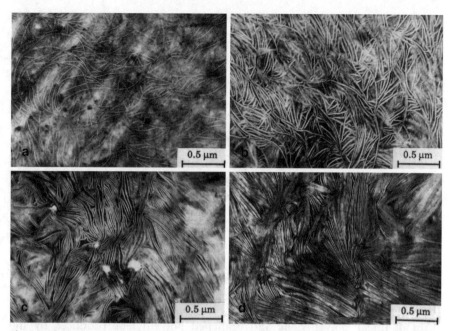

Fig. 30a–d Transmission electron micrographs of chlorosulphonated sections of L2.5/
BE1.5 crystallised at 114 °C for 0.7 h and then cooled at a rate of 80 °C min^{-1} to room
temperature: **a** 0% L2.5, **b** 20% L2.5, **c** 60% L2.5, **d** 80% L2.5. From Conde-Braña and
Gedde [170] with permission from Elsevier, UK

[166,000<$\overline{M}_w$<290,000 g mol^{-1}; $\overline{M}_w/\overline{M}_n$=6–15; 1.5 mol% ethyl branches
(BE1.5) and 0.5 mol% butyl branches (BB0.5)] were reported by Gedde and
co-workers [58, 167, 168, 169, 170].

In these blends, transmission electron microscopy (Fig. 30 shows a series
of binary blends with BE1.5 and the low molar mass linear polyethylene)
showed a progressive change from curved to straight and occasionally roof-
ridged lamellae and a strong decrease in average amorphous layer thickness
with increasing content of the linear polyethylene [168, 170]. Data obtained
by model calculations of the average amorphous thickness assuming complete co-crystallisation of the linear and branched polymers show good cor-
respondence with the experimental data obtained for the BE1.5 blends, except for the blend consisting of 80% of linear polyethylene, but a pronounced deviation for BB0.5 blends. For most of the BE1.5 blends there was
good agreement between the calorimetric crystallinity and the crystallinity
determined by transmission electron microscopy, indicating co-crystallisation of the components. There was, however, a significant deviation between
the two crystallinity values for all the BB0.5 blends, and for the BE1.5 blend
containing 80% linear polyethylene. This can be explained by partial segregation of the low molar mass linear polyethylene in these blends. The linear
growth rate and the super-molecular structure were found to be highly sen-

sitive to composition [169]. The pronounced increase in linear growth rate with increasing content of the linear fraction may be explained by an increase in the rate of diffusion of crystallisable segments due to a reduction in chain entanglement. The introduction of the linear polyethylene fraction changed the originally spherulitic structure into a predominantly axialitic superstructure. When segregation of low molar mass component occurred in these blends, it was confined to domains within spherulites between stacks of dominant lamellae [168, 170].

More recently, the Reading group discovered segregation of highly branched species in finger-like cells within spherulites on a length scale visible in the optical microscope [171, 172]. The descriptive term for this phenomenon is 'cellulation', which means 'separation of fingers of crystalline polymer by regions containing poorly and non-crystalline material' [172]. A series of branched polyethylenes (13.8 to 37.4 branches per 1000 carbon atoms) showed cellulation at the later stages of spherulite growth together with a continuous decrease in the spherulite radius growth rate [171, 172]. The radial distance to the onset of cellulation and the width of the cells were independent of spherulite growth rate but they both decreased with increasing degree of chain branching [172]. These parameters showed no scaling with the $\delta = D/G$ ratio (D=segregant diffusion coefficient, G=spherulite growth rate). It is important to point out that segregation of low molar mass species in linear polyethylene occurs without any continuous decrease in spherulite growth rate and without cellulation [156, 173].

7
Conclusions and Final Comments

Chain folding in a particular way leading to an inclined fold surface determines the lateral habit with growth sectors and the shape of the crystals as viewed along the crystallographic b axis. The lamellar branching through screw dislocation, which leads to lamellar twisting and a new growth direction of daughter lamellae, is another consequence of the fold structure. The continuity of crystal lamellae in polyethylene spherulites is well established but the detailed morphology is not easily described for the modelling of transport and mechanical properties. It is a demanding task to describe the morphology in sufficient detail to be able to predict the geometrical impedance factor for diffusion. Adjacent regular chain folding is a dominant feature of solution-grown single crystals and it is also very important in melt-crystallised polyethylene. The memory of the chaotic molten state persists to some extent in the semi-crystalline polymer. The nature of the crystal interface with constrained chains leaving the crystal is less well understood, although it is very important for certain properties, e.g. diffusivity [174].

Acknowledgements The financial support from the Swedish Research Council (grant 621–2001–1621) is gratefully acknowledged.

References

1. Paynter OI, Simmonds DJ, Whiting MC (1982) Chem Commun 1982:1165
2. Bunn CW (1939) Trans Faraday Soc 35:428
3. Busing WR (1990) Macromolecules 23:4608
4. Chatani Y, Ueda Y, Tadokoro H (1977) Annual Meeting of the Society of Polymer Science, Japan, Tokyo, Preprint, p 1326
5. Swan PR (1962) J Polym Sci 56:409
6. Holdsworth PJ, Keller A (1967) J Polym Sci, Polym Lett 5:605
7. Preedy JE (1973) Br Polym J 5:13
8. Balta Calleja FJ, Gonzales Ortega JC, Martinez Salazar J (1978) Polymer 19:1094
9. Martinez Salazar FJ, Baltá Calleja FJ (1979) J Cryst Growth 48:282
10. Teare PW, Holmes DR (1957) J Polym Sci 24:496
11. Seto T, Hara T, Tanaka K (1968) Japanese J Phys 7:31
12. Bassett DC, Block S, Piermarini GJ (1974) J Appl Phys 45:4146
13. Bassett DC, Turner B (1974) Phil Mag 29:925
14. Sakaruda l, Ito T, Nakamae K (1966) J Polym Sci C15:236
15. Mizushima R, Shimanouchi T (1949) J Am Chem Soc 71:1320
16. Shauffele RF, Shimanouchi T (1967) J Chem Phys 47:3605
17. Davis GT, Eby RK, Colson JP (1970) J Appl Phys 41:4316
18. Bunn CW, de Daubeny R (1954) Trans Faraday Soc 50:1173
19. Boyd RH (1985) Polymer 26:323
20. Boyd RH (1985) Polymer 26:1123
21. Ashcraft CR, Boyd RH (1976) J Polym Sci, Polym Phys 14:2153
22. Mansfield M, Boyd RH (1978) J Polym Sci, Polym Phys 16:1227
23. Boyd RH, Biliyar K (1973) Am Chem Soc Div Polym Chem Polym Prepr 14:329
24. Boyd RH (1984) Macromolecules 7:903
25. Crissman JM, Passaglia E (1971) J Appl Phys 42:4636
26. Crissman JM (1975) J Polym Sci, Polym Phys 13:1407
27. Keller A (1957) Phil Mag 2:1171
28. Fischer EW (1957) Z Naturf 12a:753
29. Till PH (1957) J Polym Sci 24:301
30. Wittmann JC, Lotz B (1985) J Polym Sci, Polym Phys 23:205
31. Keller A (1991) Chain-folded crystallisation of polymers from discovery to present day: a personalised journey. In: Chambers CG, Enderby JE, Keller A, Lang AR, Steeds JW (eds) Sir Charles Frank, OBE, FRS: an eightieth birthday tribute. Adam Hilger, Bristol, p 265
32. Bassett DC, Frank FC, Keller A (1963) Phil Mag 8:1739
33. Bassett DC (1968) Phil Mag 17:145
34. Dorset DL, Alamo RG, Mandelkern L (1993) Macromolecules 26:3 143
35. Vaughan AS, Bassett DC (1989) In: Allen G, Bevington JC (eds) Comprehensive polymer science, vol 2. Pergamon, Oxford, p 415
36. Keller A (1964) Kolloid Z Z Polym 197:98
37. Toda A, Keller A (1993) Colloid Polym Sci 271:328
38. Keller A (1967) Kolloid Z Z Polym 219:118
39. Kobayashi K (1962) Kagaku Chem 8:203
40. Bunn CW, Alcock TC (1945) Trans Faraday Soc 41:317
41. Bassett DC, Keller A, Mitsuhashi S (1963) J Polym Sci A 1:763
42. Wunderlich B (1976) Macromolecular physics, vol 2: Crystal nucleation, growth, annealing. Academic Press, New York
43. Kawai T (1965) Makromol Chem 84:290
44. Illers K-H, Hendus H (1968) Makromol Chem 113:1
45. Hoffman JD, Frolen LJ, Ross GS, Lauritzen JI (1975) J Res Nat Bur Std-A Phys Chem 79A:671

46. Bassett DC, Hodge AM, Olley RH (1981) Proc R Soc London A377:39
47. Mandelkern L, Sharma RK, Jackson JF (1969) Macromolecules 2:644
48. Weaver TJ, Harrison IR (1981) Polymer 22:1590
49. Ungar G, Organ SJ (1990) J Polym Sci, Polym Phys 28:2353
50. Organ SJ, Ungar G, Keller A (1990) J Polym Sci, Polym Phys 28:2365
51. Hobbs JK, Hill MJ, Barham PJ (2000) Polymer 41:8761
52. Keller A (1968) Rep Prog Phys 31:623
53. Ungar G, Keller A (1987) Polymer 27:1835
54. Patel D, Bassett DC (2002) Polymer 43:3795
55. Fischer EW, Schmidt GF (1962) Angew Chem 74:551
56. Hoel RH (1976) PhD Thesis, Department of Physics, University of Bristol, UK
57. Dreyfuss P, Keller A (1970) Polym Lett 8:253
58. Keller A, Priest DJ (1970) Polym Lett 8:13
59. Bark M, Zachmann HG, Alamo R, Mandelkern L (1992) Makromol Chem 193:2363
60. Kawai T (1967) Kolloid Z 229–2:116
61. Hoffman JD, Weeks JJ (1965) J Chem Phys 42:4301
62. Blackadder DA, Keniry JS, Richardson MJ (1972) Polymer 13:584
63. Barham PJ, Chivers RA, Keller A, Martinez-Salazar J, Organ SJ (1985) J Mater Sci 20:1625
64. Organ SJ, Keller A (1985) J Mater Sci 20:1602
65. Lauritzen JI, Hoffman JD (1960) J Res Nat Bur Std 64A:73
66. Wunderlich B (1980) Macromolecular physics, vol 3: Crystal melting. Academic Press, New York
67. Arakawa T, Wunderlich B (1967) J Polym Sci C16:653
68. Sadler DM (1983) Polymer 24:1401
69. Sadler DM, Gilmer GH (1984) Polymer 25:1446
70. Hoffman JD, Lauritzen JI, Passaglia E, Ross GS, Frolen LJ, Weeks JJ (1968) Kolloid Z Z Polym 231:564
71. Broadhurst MG (1962) J Chem Phys 36:2578
72. Broadhurst MG (1966) J Res Natl Bur Std A Phys Chem 70A:481
73. Flory PJ, Vrij A (1963) J Am Chem Soc 85:3548
74. Grubb DT (1985) Macromolecules 18:2282
75. Hoffman JD, Weeks JJ (1962) J Res Natl Bur Std—A Phys Chem 66A:13
76. Blundell DJ, Keller A, Connor T (1967) J Polym Sci A-2 5:991
77. Williams T, Blundell DJ, Keller A, Ward IM (1968) J Polym Sci A-2 6:1613
78. Keller A, Martuscelli E, Priest DJ, Udagawa Y (1971) J Polym Sci A-2 9:1807
79. Keller A (1962) Polymer 3:393
80. Petraccone V, Allegra G, Corradini P (1972) J Polym Sci C 38:419
81. Spells SJ, Organ SJ, Keller A, Zerbi G (1987) Polymer 28:697
82. Wolf S, Schmid C, Hägele PC (1990) Polymer 31:1222.
83. Cheam TC, Krimm S (1981) J Polym Sci, Polym Phys 19:423
84. Flory PJ (1962) J Am Chem Soc 84:2857
85. Sadler DM, Keller A (1979) Science 203:263
86. Spells SJ, Keller A, Sadler DM (1984) Polymer 25:749
87. Ungar G, Stejny J, Keller A, Bidd I, Whiting MC (1985) Science 229:386
88. Nisman R, Smith P, Vancso GJ (1994) Langmuir 10:1667
89. Frank FC (1979) General introduction. Faraday Soc General Discussion 68:7
90. DiMarzio EA, Guttman CM (1980) Polymer 21:733
91. Gedde UW (1995) Polymer physics. Kluwer, Dordrecht
92. Guttman CM, DiMarzio EA, Hoffman JD (1981) Polymer 22:1466
93. Schelten J, Ballard DGH, Wignall GD, Longman G, Schmaltz W (1976) Polymer 17:751
94. Sadler DM, Keller A (1977) Macromolecules 19:1128
95. Yoon DY, Flory PJ (1979) Faraday Soc General Discussion 68:452
96. Guttman CM, DiMarzio EA, Hoffman JD (1981) Polymer 22:597
97. Schelten J, Wignall GD, Ballard DGH (1974) Polymer 15:682

98. Schelten J, Wignall GD, Ballard DGH, Longman GW (1977) Polymer 18:1111
99. Mutter R, Stille W, Strobl GR (1993) J Polym Sci, Polym Phys 3 1:99
100. Kitamaru R, Horii F, Hyon S-H (1977) J Polym Sci, Polym Phys 15:821
101. Vonk CG, Pijpers AP (1985) J Polym Sci, Polym Phys 23:25 17
102. Keith HD (1964) J Appl Phys 35:3115
103. Organ SJ, Keller A (1985) J Mater Sci 20:1571
104. Labaig JJ (1978) PhD Thesis, University of Strasbourg
105. Bassett DC, Olley RH, al Rehail IAM (1988) Polymer 29:1539
106. Keith HD, Padden FJ, Lotz B, Wittman JC (1989) Macromolecules 22:2230
107. Toda A (1992) Colloid Polym Sci 270:667
108. Bassett DC, Hodge AM (1981) Proc R Soc London A377:61
109. Keller A (1955) J Polym Sci 17:351
110. Point J-J (1955) Bull Acad R Belg 41:982
111. Keller A (1955) J Polym Sci 17:291
112. Keith HD, Padden FJ (1963) J Polym Sci 39:101
113. Keith HD, Padden FJ (1963) J Polym Sci 39:123
114. Keller A (1959) J Polym Sci 39:151
115. Rego Lopez JM, Gedde UW (1988) Polymer 29:1037
116. Maxfield J, Mandelkern L(1977) Macromolecules 10:1141
117. Anderson FR (1964) J Appl Phys 35:64
118. Mandelkern L, Price JM, Gopalan M, Fatou JG (1966) J Polym Sci, Polym Phys 4:385
119. Bassett DC, Hodge AM (1978) Proc R Soc London A359:121
120. Kanig G (1973) Kolloid Z Z Polym 25 1:782
121. Olley RH, Hodge AM, Bassett DC (1979) J Polym Sci, Polym Phys 17:627
122. Frank FC, Tosi MP (1961) Proc R Soc London A263:323
123. Lauritzen JI, Passaglia E (1967) J Res Nat Bur Std A Phys Chem 71A:261
124. Abo el Maaty MI, Bassett DC (2001) Polymer 42:4957
125. Bassett DC (1984) CRC Crit Rev 12:97
126. Bassett DC, Olley RM (1984) Polymer 25:935
127. Bassett DC, Vaughan AS (1985) Polymer 26:7 17
128. Bassett DC, Olley RH, Sutton SJ, Vaughan AS (1996) Macromolecules 29:1852
129. Bassett DC, Olley RH, Sutton SJ, Vaughan AS (1996) Polymer 37:4993
130. Teckoe J, Bassett DC (2000) Polymer 41:1953
131. Hosier IL, Bassett DC (2000) Polymer 41:8801
132. Hosier IL, Bassett DC (2002) Polymer 43:307
133. Hill MJ, Barham PJ (1995) Polymer 36:1523
134. Wignall GD, Alamo RG, Londono JD, Mandelkern L, Stehling FC (1996) Macro-
     molecules 29:5332
135. Alamo RG, Graessley WW, Krishnamoorti R, Lohse DI, Londono JD, Mandelkern L,
     Stehling FC, Wignall GD (1997) Macromolecules 30:561
136. Barham PJ, Hill MJ, Keller A, Rosney CCA (1988) J Mater Sci Lett 7:1271
137. Morgan RL, Hill MJ, Barham PJ, Frye C-J (1997) Polymer 38:1903
138. Tanem BS, Stori A (2001) Polymer 42:4309
139. Tanem BS, Stori A (2001) Polymer 42:5689
140. Hill MJ, Barham PJ, Keller A (1992) Polymer 33:2530
141. Hill MJ (1994) Polymer 35:1991
142. Hill MJ, Barham PJ (1994) Polymer 35:1802
143. Hill MJ, Barham PJ, van Ruiten J (1993) Polymer 34:2975
144. Thomas D, Williamson J, Hill MJ, Barham PJ (1993) Polymer 34:4919
145. Hill MJ, Barham PJ (1995) Polymer 36:3369
146. Morgan RL, Hill MJ, Barham PJ, van der Pol A, Kip BJ, Ottjes R, van Ruiten J (2001)
     Polymer 42:2121
147. Bank MI, Krimm S (1970) J Polym Sci Lett 8:143
148. Wunderlich B, Mehta A (1974) J Polym Sci, Polym Phys 12:255
149. Mehta A, Wunderlich B (1975) Colloid Polym Sci 253:193

150. Zachmann HG (1967) Kolloid Z Z Polym 216–217:180
151. Zachmann HG (1969) Kolloid Z Z Polym 231:504
152. Gedde UW, Eklund S, Jansson J-F (1983) Polymer 24:1532
153. Dlugosz J, Fraser GV, Grubb DT, Keller A, Odell JA, Goggin PL (1976) Polymer 17:471
154. Gedde UW, Jansson J-F (1984) Polymer 25:1263
155. Smith P, St. John Manley R (1979) Macromolecules 12:483
156. Rego Lopez JM, Conde Braña MT, Terselius B, Gedde UW (1988) Polymer 29:1045
157. Conde Braña MT, Iragorri Sainz JI, Gedde UW (1989) Polym Bull 22:277
158. Gustafsson A, Conde Braña MT, Gedde UW (1991) Polymer 32:426
159. Gedde UW, Jansson J-F, Liljenström G, Eklund S, Wang P-L, Holding S, Werner P-E (1988) Polym Eng Sci 28:1289
160. Gedde UW, Jansson J-F (1983) Polymer 24:1521
161. Clampitt BH (1965) J Polym Sci 3:671
162. Datta NK, Birley AW (1982) Plast Rubber Proc Appl 2:237
163. Kyu T, Hu S-R, Stein RS (1987) J Polym Sci, Polym Phys 25:89
164. Hu S-R, Kyu T, Stein RS (1987) J Polym Sci, Polym Phys 25:71
165. Edwards GH (1986) Brit Polym 118:88
166. Norton DR, Keller A (1984) J Mater Sci 19:447
167. Rego Lopez JM, Gedde UW (1989) Polymer 30:22
168. Conde Braña MT, Iragorri Sainz JI, Terselius B, Gedde UW (1989) Polymer 30:410
169. Iragorri Sainz JI, Rego Lopez JM, Katime I, Conde Braña MT, Gedde UW (1992) Polymer 33:461
170. Conde Braña MT, Gedde UW (1992) Polymer 33:3123
171. Abo el Maaty MI, Hosier IL, Bassett DC (1998) Macromolecules 31:153
172. Abo el Maaty Ml, Bassett DC, Olley RH, Jääskeläinen P (1998) Macromolecules 31:7800
173. Hosier IL, Bassett DC (1999) Polymer J 31:772
174. Neway B, Hedenqvist MS, Mathot VBF, Gedde UW (2001) Polymer 42:5307

Received: August 2003

Adv Polym Sci (2004) 169:75–119
DOI: 10.1007/b13520

Microdeformation and Fracture in Bulk Polyolefins

Christopher J. G. Plummer

Laboratoire de Technologie de Composites et Polymères (LTC), Institut des Matériaux,
Ecole Polytechnique Fédérale de Lausanne (EPFL), 1015, Lausanne, Switzerland
E-mail: christopher.plummer@epfl.ch

Abstract The fracture properties and microdeformation behaviour and their correlation with structure in commercial bulk polyolefins are reviewed. Emphasis is on crack-tip deformation mechanisms and on regimes of direct practical interest, namely slow crack growth in polyethylene and high-speed ductile–brittle transitions in isotactic polypropylene. Recent fracture studies of reaction-bonded interfaces are also briefly considered, these representing promising model systems for the investigation of the relationship between the fundamental mechanisms of crack-tip deformation and fracture and molecular structure.

Keywords Polyethylene · Polypropylene · Microdeformation · Fracture · Slow crack growth

Abbreviations

a	Crack length (m)
$\dot{a}$	Crack speed (m s^{-1})
$\dot{a}_o$	Scaling constant (m s^{-1})
AFM	Atomic force microscopy
α	Craze anisotropy factor
BSE	Backscattered electron
CT	Compact tension
χ	Flory-Huggins interaction parameter
D	Craze fibril spacing (m)
d_e	Entanglement spacing (m)
δ	Crack opening displacement (m)
δ_c	Critical crack opening displacement (m)
DENT	Double edge-notched tension
E	Young's modulus (GPa)
EP	Ethylene-propylene copolymer
EPR	Ethylene-propylene rubber
EPDM	Ethylene-propylene-diene monomer
ε_b	Elongation at break
$\dot{\varepsilon}$	Strain rate (s^{-1})
$\dot{\varepsilon}_o$	Scaling constant (s^{-1})
ESC	Environmental stress cracking
ESR	Electron resonance spectroscopy
EWF	Essential work of fracture (J m^{-2})
FEG-SEM	Field emission gun scanning electron microscopy
FNCT	Full notch creep test
G_c	Critical strain energy release rate (J m^{-2})
G_{Ic}	Mode I plane strain critical strain energy release rate (J m^{-2})
HDPE	High density polyethylene
iPP	Isotactic polypropylene
J_c	Critical non-linear strain energy release rate (J m^{-2})
K	Stress intensity factor (MPa m$^{1/2}$)
K_c	Critical stress intensity factor (MPa m$^{1/2}$)
K_{co}	Scaling constant (MPa m$^{1/2}$)
K_{Ic}	Mode I plane strain critical stress intensity factor (MPa m$^{1/2}$)
K_{max}	Maximum stress intensity factor (MPa m$^{1/2}$) (cyclic tests)
l	Fibrillar zone length (m)

l_o	Statistical step length (m)
LDPE	Low density polyethylene
LLDPE	Linear low density polyethylene
LEFM	Linear elastic fracture mechanics
MDPE	Medium density polyethylene
M	Molar mass (g mol^{-1})
M_e	Entanglement molar mass (g mol^{-1})
M_w	Weight average molar mass (g mol^{-1})
MWD	Molecular weight distribution
MAH	Maleic anhydride
μ	Poisson's ratio
n	Power law exponent
ν_e	Entanglement density (m^{-3})
OM	Optical microscopy
PA6	Polyamide 6
PE	Polyethylene
PP	Polypropylene
R	Ratio of maximum to minimum K (cyclic tests)
RCP	Rapid crack propagation
SAXS	Small angle X-ray scattering
SCG	Slow crack growth
SEM	Scanning electron microscopy
SSY	Small scale yielding
Σ	Interfacial crossing density (m^{-2})
σ	Stress (MPa)
σ_c	Craze widening stress (MPa)
σ_f	Fibril breakdown stress (MPa)
σ_y	Tensile yield stress (MPa)
σ_{yo}	Scaling constant (MPa)
T	Temperature (°C)
t	Time (s)
t_b	Time to break (s)
t_c	Fibril breakdown time (s)
τ	Time constant (s)
T_1, T_2	Temperatures in non-isothermal welding (°C)
TEM	Transmission electron microscopy
T_g	Glass transition temperature (°C)
T_i	Interface temperature during welding (°C)
T_m	Melting temperature (°C)
UHMWPE	Ultra-high molecular weight polyethylene
v	Test speed (m s^{-1})
v_f	Fibril volume fraction
w	Interface width (m)
WAXS	Wide angle X-ray scattering
ξ	Creep rate coefficient (m s^{-1})

1
Introduction

In terms of tonnage, polyolefins are by far the most important polymeric materials for structural applications, and there is consequently enormous interest in optimising their fracture properties. A rational approach to this requires detailed understanding of the relationships between macroscopic fracture and molecular parameters such as the molar mass, M, and external variables such as temperature, T, and test speed, v. Considerable effort is therefore also devoted to characterising the irreversible processes (crazing and shear deformation) that accompany crack initiation and propagation in these polymers, some examples of which will given.

Although certain polyolefin rubbers, such as ethylene-propylene rubber (EPR), are commercially important, the discussion will essentially be limited to isotropic, unfilled polyethylene (PE) and isotactic polypropylene (iPP). These may loosely be described as stiff, chemically inert semicrystalline solids. Even so, the corresponding range of mechanical properties is very broad, as illustrated in Table 1, and although the values shown are fairly representative, they are highly sensitive to extrinsic factors such as solidification conditions and additive packages, as well as intrinsic materials parameters. Moreover, the palette of properties is continuing to expand with the development of new catalysts and processes, leading, for example, to novel ethylene polymers combining narrow molecular weight distributions (MWD) with homogeneous comonomer/branch distributions and an increasing diversity of ethylene/propylene (EP) copolymers. Other semicrystalline polyolefins are also commercially available, including polybutene and poly(4-methyl-1-pentene) (similar properties to PE and iPP respectively) and syndiotactic polypropylene (significantly lower melting point than iPP), but are less widely used and will not be discussed in detail here.

As may be inferred from Table 1, a high elongation to break, ε_b, which is an indication of ductility, also implies a high fracture resistance (the energy absorbed during a standard Izod impact test, say [1]). The ductility typically increases with T, particularly in the vicinity of the 'ductile–brittle transition temperature', often associated with the glass transition temperature, T_g, in un-notched specimens (although certain polymers, such as polycarbonate, remain ductile well below T_g). This implies not only a relatively low fracture resistance for $T<T_g$, but may also be reflected by embrittlement at high deformation rates for $T>T_g$, consistent with the notion of time–temperature equivalence in viscoelastic materials. Such embrittlement is of particular practical importance in iPP, whose T_g is just below room temperature. In PE, on the other hand, the impact behaviour is relatively good, high density polyethylene (HDPE) generally undergoing ductile failure for $M>10^6$ g mol^{-1} at room temperature. However, its behaviour under long-term low level loading, especially in the presence of a hostile environment, is of concern, owing to the phenomenon of slow sub-critical crack growth (SCG). SCG can lead to catastrophic failure after service times greatly exceeding the time-scales usually associated with laboratory experiments, whence current ef-

Table 1 Typical physical and (room temperature) mechanical properties (melting point T_m, glass transition temperature T_g, Young's modulus E, Izod toughness, tensile yield stress σ_y, elongation at break ε_b) and applications of commodity polyolefins

Polymer	T_m	T_g	E	Izod	σ_y	ε_b
	(°C)	(°C)	(GPa)	(Jm^{-2})	(MPa)	(%)
Low density polyethylene (LDPE)	80–100	−100	0.2	>100	10–30	800
	Injection mouldings (kitchenware, toys, packaging, linings, sealants etc.); blow mouldings (bottles, containers); film (food, clothes packaging); wire and cable (high frequency insulation, jacketing); pipe (chemical handling, irrigation systems, natural gas transmission).					
Linear low density polyethylene (LLDPE)	80–100	−100	0.2	>100	10	800
	Film (markets as for LDPE); mouldings (dustbins, kitchen ware etc.).					
High density polyethylene (HDPE)	120–135	−100	1.0	5–100	30	500
	Refrigerator parts, packaging, structural housing panels, pipe, defroster and heater ducts, sterilisable household items and hospital equipment, hoops, battery parts, blow moulded containers, including automotive petrol tanks, film wrapping materials, wire cable and insulation and chemical-resistant pipe.					
Isotactic polypropylene (iPP) (typically 90% isotactic)	165	−20	1–1.4	7	25–40	50–300
	Fibres, film appliance parts, industrial parts for fluid processing, automotive and electrical hardware, tank and pipe liner, carpeting, packaging materials, stadium seats, battery cases, air conditioning and heating ducts (injection moulding, sheet stamping).					

forts to develop adequate short-term test methods for the reliable prediction of long-term fracture resistance.

In this article, after a short review of issues relating to common test methods and phenomenology, long- and short-term fracture in PE and iPP, respectively, will be discussed within the framework of fracture mechanics with passing reference to results from more qualitative methods. Emphasis will be placed on microdeformation mechanisms accompanying crack propagation and their dependence on molecular parameters such as M and test conditions. The fracture resistance of polyolefins containing internal interfaces, including rubber-modified iPP and welded or reaction-bonded joints, will also briefly be discussed.

2
Experimental Methods

2.1
Fracture Testing

The fracture behaviour of polymers, usually under conditions of mode I opening, considered the severest test of a material's resistance to crack initiation and propagation, is widely characterised using linear elastic fracture mechanics (LEFM) parameters, such as the plane strain critical stress intensity factor, K_{Ic}, or the critical strain energy release rate, G_{Ic}, for crack initiation (determined using standard geometries such as those in Fig. 1). LEFM

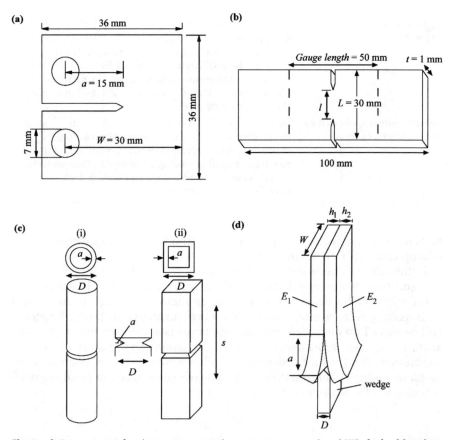

Fig. 1a–d Fracture mechanics test geometries: **a** compact tension (CT), **b** double edge-notched tension (DENT), **c** full-notched creep test (FNCT) and **d** double cantilever beam (DCB), widely used in fundamental studies of interfacial fracture. In this case, crack propagation is usually maintained by inserting a wedge at constant speed between the faces of a pre-crack aligned with the plane of the interface (the "wedge test"), and G_c can be determined from the crack length ahead of the wedge

may nevertheless be inappropriate for ductile polymers, including most polyolefins, for which the usual range of test conditions result in excessive non-linear deformation and crack-tip plasticity for the (generally processing limited) maximum specimen thickness. Indeed, even the small scale yielding (SSY) criteria for the validity of more elaborate elastic-plastic fracture-mechanics-based methods, such as J_{Ic} testing, may be difficult to meet [1]. This has prompted a search for improved test protocols involving geometries with relatively high plastic constraint factors, such as the FNCT geometry (Fig. 1c), for example, or alternative approaches to fracture testing. These latter are typified by essential work of fracture testing (EWF), based on extrapolation of data from double edge-notched tensile (DENT) (Fig. 1b) or single edge-notched tensile specimens with different ligament widths, whose physical interpretation is the subject of some debate, but for which full ligament yielding is a requirement rather than a restriction [2].

In many cases, the viscoelastic character of polymeric materials also needs to be considered explicitly. The resulting modifications to static schemes, collectively known as viscoelastic fracture mechanics, have been extensively applied to time-dependent SCG [3, 4]. Rate effects also constitute a difficulty with traditional impact test methods such as Izod or Charpy, where fracture performance is associated with the energy absorbed per unit energy of the broken cross-section, since the effective range of test speeds is limited (2–3 m s^{-1} for Izod and Charpy; 1–4 m s^{-1} for falling weight tests). Materials rankings based on such tests alone may therefore be misleading in terms of global performance. Moreover, although impact tests are cheap and easy to perform, and are adequate for demonstrating the existence of ductile brittle transitions, say, the impact energy is not an intrinsic property, but depends on the geometry and exact test method (fracture mechanics analyses are possible, but are complicated by dynamic effects and multiple energy terms [1]). There is now, therefore, considerable interest in quasi-static high rate testing of fracture mechanics specimens using hydraulic high-speed tensile test apparatus, both because it is straightforward to derive intrinsic fracture toughness parameters from such tests, and because essentially the same test can be performed over a wide range of test speeds, including speeds characteristic of impact testing [5, 6, 7]. Results from high-speed testing of compact tension (CT) specimens of iPP (Fig. 1a) will be discussed in Sects. 3.4 and 4.1.

2.2
Morphological Characterisation

Transmission electron microscopy (TEM) has traditionally been the mainstay of morphological investigations of polyolefins [8], but recent developments in low voltage high-resolution field emission gun scanning electron microscopy (FEG-SEM) [9] and the advent of atomic force microscopy (AFM) and related near-field techniques [10] have challenged its dominance at the length scales of the order of 10 nm, characteristic of both microdeformation (cavitation, fibrils) and structural components of semicrystalline polymers (crystalline lamellar thicknesses). Moreover, SEM and FEG-SEM can give rapid access to much a wider range of length scales than TEM, often with relatively cursory specimen preparation. Optical microscopy (OM) is also an important complement to TEM in this respect.

Near-field scanning techniques are relative newcomers, and the basis for their interpretation is less well established. However, AFM has opened up new perspectives for morphological studies, particularly given that excessive surface damage in soft specimens can be avoided by use of non-contact or intermittent contact modes. Its sensitivity to surface topography neverthe-less makes AFM prone to artefacts when used to observe surfaces prepared by microtoming, and its effective depth of field is limited compared with SEM. On the other hand, if lamellar surfaces can be prepared such that the surface relief (or hardness, friction variations) is representative of the bulk texture, very striking detail can be recorded at the nanometre scale in de-formed polyolefins [11].

Indirect techniques, such as small angle X-ray scattering (SAXS) and wide angle X-ray scattering (WAXS), are also extensively used to study changes in lamellar and crystallographic textures during homogeneous deformation of polyolefins and other semicrystalline polymers [12, 13, 14, 15]. Although these are not always well adapted to fracture specimens, in which the defor-mation gradients are large and scattering owing to cavitation may mask oth-er signals, this situation may well evolve with the increasing availability of fine, high-intensity probes [16].

2.2.1
Specimen Preparation

Conventional bright-field TEM observations of polyolefins often require contrast enhancement, usually by staining with RuO_4 or other suitable markers [17]. These accumulate in the amorphous phase, at lamellar sur-faces and in cavities, and differential staining can reveal the phase distribu-tion in blends. Staining also hardens the specimens, facilitating preparation of thin sections at room temperature (cryo-sectioning is required for un-fixed polyolefins).

In highly deformed specimens, not only is the lamellar structure substan-tially modified, but fibrillar structure may also be present. Since such struc-tures undergo relaxation and collapse on unloading, they may need to be embedded in a suitable resin (a low viscosity epoxy or acetate resin, for ex-ample) under load, prior to sectioning [18]. In partly cracked notched speci-mens, this is conveniently achieved by using a wedge to maintain the crack faces under load after the specimen has been removed from the tensile test apparatus, as shown in Fig. 2. Where this proves difficult (at test speeds ap-proaching impact speeds, for example), sub-critical crack-tip deformation can be induced by testing two specimens in series (the 'twin' method), so that fracture of one specimen instantaneously unloads the other [6, 19]. Sim-ilar techniques are also suitable for SEM, including observation of internal surfaces intersecting crack-tips [20, 21] in specimens that have been wedged

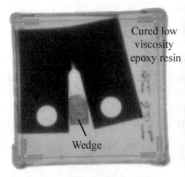 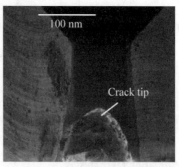

Fig. 2 Schematic of embedding and sectioning procedure used to limit relaxation of crack-tip microdeformation during specimen preparation for TEM (here for a CT specimen deformed under static loading)

open and either etched [22], or embedded and stained as described above [23]. In stained samples, chemical contrast provided by the backscattered electron detector (BSE-SEM) is also useful for examining phase or lamellar morphologies (Fig. 3) [24].

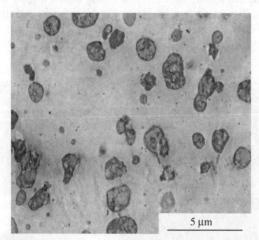

Fig. 3 EPR inclusions in an impact-modified iPP, sectioned with an ultra-microtome, stained with RuO_4 and observed in backscattered mode with a FEG-SEM; the lighter regions in the particles are ethylene rich, whereas the darker peripheral regions are rich in EPR [24]

3
Fracture of Bulk Polyolefins

Before discussing specific aspects of microdeformation and fracture in bulk polyolefins, some basic notions of microdeformation and the micromechanics of fracture mediated by generation and breakdown of cavitated or fibrillar deformation zones or 'crazes' are introduced. SCG in PE and rate-dependent fracture in iPP are then considered in more detail.

3.1
Microdeformation Mechanisms

Above T_g, that is, in the usual range of service temperatures for semicrystalline polyolefins, the crystalline lamellae undergo simple rigid displacements in the initial stages of deformation, the intervening rubbery amorphous regions deforming either by shear or lamellar separation, depending on their orientation with respect to the local tensile axis (the strains in the amorphous regions can approach 100% for a global deformation of 20% [25]). As deformation proceeds, stress relaxation occurs either by cavitation in the interlamellar material or by co-operative deformation of the lamellae, this latter leading ultimately to ductile necking and a highly drawn, non-cavitated texture locally, as shown in Fig. 4 for HDPE (the texture of iPP necks is very similar [26]). In HDPE, Hookean behaviour persists up to few percent strain, followed by crystal slip, twinning, martensitic phase transforma-

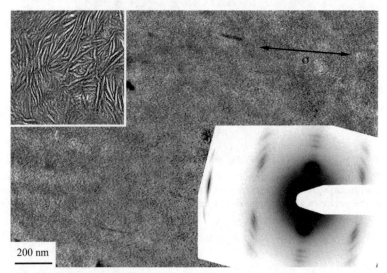

Fig. 4 TEM micrograph of part of a macroscopic neck in HDPE deformed at room temperature, stained in RuO_4 and microtomed; also shown are the corresponding electron diffraction pattern and (*top left*) the structure prior to deformation (deformation roughly horizontal as indicated)

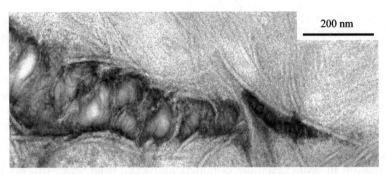

Fig. 5 TEM micrograph of a craze in bulk HDPE deformed in tension at room temperature (embedded in epoxy and stained in RuO_4)

tion, rotation of lamellar stacks, fibril formation and necking [27]. Increased branch content (and the resulting decrease in crystallinity and yield stress (σ_y)) accentuates the role of interlamellar deformation, and double yield points are sometimes observed, accounted for in terms of activation of fine chain slip (initial yield point) and coarse lamellar fragmentation (second yield point) [13].

Cavitation is often a precursor to craze formation [20], an example of which is shown in Fig. 5 for bulk HDPE deformed at room temperature. It may be inferred from the micrograph that interlamellar cavitation occurs ahead of the craze tip, followed by simultaneous breakdown of the interlamellar material and separation and stretching of fibrils emanating from the dominant lamellae visible in the undeformed regions. The result is an interconnected network of cavities and craze fibrils with diameters of the order of 10 nm. This is at odds with the notion that craze fibrils in semicrystalline polymers deformed above T_g are coarser than in glassy polymers [20, 28], as well as with models for craze formation in which lamellar fragmentation constitutes an intermediate step [20, 29] but, as will be seen, it is difficult to generalise and a variety of mechanisms and structures is possible.

Stress transfer between the lamellae is often attributed to 'tie-molecules', which are well-anchored molecular strands (or interconnected loops) bridging adjacent lamellae [30, 31]. These have been assimilated with the concept of entanglement, that is, topological constraints present in the melt and anchored by the crystalline lamellae on solidification [32]. The entanglement network model provides a convenient physical picture for this, namely a density (ν_e) of notional entanglement points that behave analogously to chemical crosslinks. In high ν_e polymers such as the polyolefins, the resulting high density of strong interlamellar connections will hinder breakdown of the interlamellar material, so that non-cavitational shear (as in Fig. 4) might be expected to be favoured over crazing [33]. However, the competition between the two mechanisms also depends on external conditions; crazing tends to dominate (i) at high strain rates and/or low T, where the yield stress is high, and (ii) at low strain rates and/or T approaching T_m, where the chain mobility is sufficient for entanglement loss to occur via chain slip or 'disentanglement' [32, 34]. Moreover, the presence of a notch in standard fracture tests leads to severe constraints on bulk plastic deformation, so that crazing occurs over a wider range of conditions than in simple tension. Similar behaviour is already well documented for high v_e amorphous glassy polymers, where (scission) crazing is reported to be favoured by high strain rates at $T \ll T_g$ and disentanglement crazing by low strain rates at T close to T_g [33, 35, 36, 37 38]. Morphological features such as lamellae and spherulites, by definition absent in amorphous polymers, nevertheless play in important role in determining the characteristic length scales associated with deformation in semicrystalline polymers above T_g, where the stiffness mismatch between the amorphous and crystalline regions is large [32].

3.2
Craze Micromechanics

The existence of a wedge-shaped cavitated or fibrillar deformation zone or craze, ahead of the crack-tip in mode I crack opening, has led to widespread use of models based on a planar cohesive zone in the crack plane [39, 40, 41, 42]. The applicability of such models to time-dependent failure in PE is the focus of considerable attention at present [43, 44, 45, 46, 47]. However, given the parallels with glassy polymers, a recent static model for craze breakdown developed for these latter, but which may to some extent be generalised to polyolefins [19, 48, 49], will first be introduced. This helps establish important links between microscopic quantities and macroscopic fracture, to be referred to later.

3.2.1
Craze Breakdown

Based on a model for crack propagation in a single crack-tip craze [50], verified by more detailed treatments [51, 52, 53], it is shown that under conditions of SSY, where the crack propagation is determined by the far field stress intensity K:

$$K_{\mathrm{Ic}} \approx \alpha^{-1/4} \left(\frac{1-v_f}{1-\mu^2} \frac{E}{\sigma_c} \right)^{1/2} \sqrt{\pi D} \sigma_f \qquad (1)$$

or, equivalently:

$$G_{\mathrm{Ic}} = \sigma_c \delta_c \approx \alpha^{-1/2} (1-v_f) \pi D \frac{\sigma_f^2}{\sigma_c} \qquad (2)$$

where δ_c is the critical crack opening displacement, μ is Poisson's ratio, α is related to the craze anisotropy, σ_c is the draw stress normal to the craze-bulk interface, D is the centre-to-centre fibril spacing, v_f is the fibril volume fraction in the craze and σ_f is the stress to break a craze fibril. This approach differs from other cohesive zone models in that it incorporates the stress-concentrating effect of cross-tie fibrils, widely observed in crazes in glassy polymers [35, 54]. Without this stress-concentrating effect, that is for $\alpha \to 0$, a time-independent failure criterion σ_f implies $\delta_c \to \infty$, since the stress in a given fibril can never exceed σ_c.

For a crossing density of covalently or topologically anchored polymer strands, Σ, such that $\sigma_f \propto \Sigma$, it follows from Eq. 2 that $G_c \propto \Sigma^2$. In an entangled polymer $\Sigma \approx d_e v_e$, where d_e is the rms spacing of topologically linked entanglement points in the network model, demonstrating the importance of v_e for toughness [50]. It should be borne in mind that the model assumes a single crack-tip craze, whereas multiple crazing, mixed shear and crazing or other types of deformation zones are often encountered in practice. The underlying principles remain the same, however, and recent results from reaction-bonded interfaces, discussed in more detail in Sect. 4.2.2, suggest the above scaling to have much wider generality for both amorphous and semicrystalline polymers [49, 55].

3.2.2
Time-Dependent Failure

In the model leading to Eq. 1 and Eq. 2, not only the elastic and plastic flow properties, but also σ_f must also be assumed to be time-dependent for the results to be consistent with experimental data obtained at different crack growth rates [52]. To illustrate the consequences of this, the time-dependence is grouped together in a single power law term, so that Eq. 1 becomes:

$$K \approx \alpha^{-1/4} \left(\frac{1 - v_f}{1 - \mu^2} \frac{E}{\sigma_c} \right)^{1/2} \sqrt{\pi D} \sigma_f \left(\frac{t}{\tau(M)} \right)^{-1/n} = K_{co} \left(\frac{\tau(M)\dot{a}}{D} \right)^{1/n} \qquad (3)$$

where $\dot{a}$ is the crack advance rate, τ is a characteristic time constant and the scaling constant K_{co} is equivalent to the critical stress intensity of a time-independent material ($1/n=0$). Although criteria for LEFM testing are usually met for large effective values of n, in strongly time-dependent systems (small n), significant crack growth may occur below K_{co}. Thus, in a standard K_{Ic} test carried out at a constant loading rate, the onset of crack propagation is not necessarily well defined, even when SSY conditions are satisfied. This raises questions as to the precise interpretation of fracture mechanics parameters determined under non-standard conditions, including most of the data to be presented in Sect. 3.4 and 4.1, whence the use of K_c and G_c, rather than K_{Ic} and G_{Ic} (more detailed discussion of the validity of these parameters in specific cases is given in the original references). Although K_c and G_c remain very useful for comparing the crack initation resistance of different materials under different conditions, a single parameter of this type provides little indication of the nature of crack propagation, which is also of considerable practical interest. In constant strain rate tests, it is therefore useful to distinguish between ductile behaviour (significant energy dissipation beyond the maximum in the force-displacement curve) and brittle behaviour (negligible energy dissipation beyond the maximum in the force-displacement curve), providing a convenient, if not always clear-cut, definition of the ductile–brittle transition(s).

3.3
Slow Crack Growth in PE

One advantage of PE piping is that it often flexible enough for one to be able to seal a length of pipe temporarily simply by squeezing it. However, enquiries following a serious gas explosion showed that this practice of 'pinch clamping' is sufficient to introduce small flaws. Although K generally remains very much less than K_{co} as defined by Eq. 3, under service loading conditions, sub-critical SCG is possible, increasing the effective flaw size and hence the local K, until catastrophic failure occurs at some relatively long time after the introduction of the initial damage [56].

In industry, hydrostatic pressure testing of pipes is still widely used to assess their resistance to this type of failure. Typical results are shown in Fig. 6 [57, 58], where failure times of HDPE pipes are given as a function of the circumferential (or hoop) stress. At relatively high stresses, ductile failure is observed (stage I); although deformation is initially homogeneous, small local variations (arising from variations in the specimen thickness, for exam-

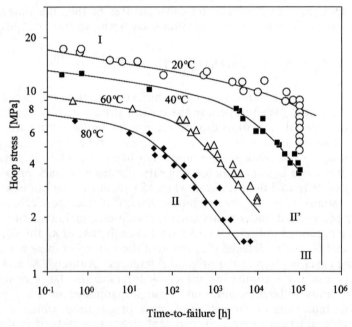

Fig. 6 Durability diagram for HDPE pipes under internal pressure, showing hoop stress vs. the time to fail at different temperatures, along with the regimes of ductile failure (*I*) and brittle failure (*II*) [57, 58]

ple) are amplified with time, leading to a plastic instability in the form of a blister [57, 59]. In this regime, σ vs. t_b is relatively flat, reflecting the weak dependence of σ_y on the strain rate, usually approximated to by $\sigma_y \sim \sigma_{yo}(\varepsilon/\varepsilon_o)^{1/n}$ with n between 10 and 20 for a wide variety of polymers, including PE. At lower stresses, however, SCG becomes dominant (stage II). Failure in this regime is reflected by a much stronger dependence of σ on t_b and brittle failure, in so far as SCG is often a precursor to unstable rapid crack propagation (RCP) [60].

For many modern PE grades, failure times in pipe tests are extremely long under service conditions. As suggested by Fig. 6, SCG can be accelerated by testing at high temperatures. The observed shifts are often fitted with an Arhennius expression, giving activation energies of the order of 100 kJ mol^{-1}. Room temperature lifetimes can then be inferred by extrapolation, although caution should be exercised since deviations from Arhennius behaviour are often significant [61, 62]. Testing may be accelerated further, if necessary, using cyclic loading conditions (Sect. 3.3.5) and/or a surfactant [63]. In this latter case, it is usually assumed that the surfactant penetrates into the specimen during crack-tip deformation, facilitating molecular level chain slip and disentanglement [64]. Non-ionic surfactants such as Igepal (nonyl phenol ether glycol) are particularly effective for reducing the failure times of tougher materials under low level loading without affecting their ranking with respect to other grades tested under the same conditions, although it is not clear to what extent the failure mechanisms are affected [58].

Environmental stress cracking (ESC), usually associated with SCG accelerated by surfactants or other aggressive media, is also of considerable interest in its own right, since pipe materials are often intended for service in contact with water or other industrial fluids. Oxidative degradation is another important concern, particularly at high temperatures, and stabilisers are often added specifically to retard SCG and catastrophic failure [65, 66, 67, 68], the stabiliser type and concentration being critical at very low SCG rates, even where they have little effect on rapid fracture [69].

3.3.1
Fracture Mechanics-Based SCG Testing

Pre-screening of different grades of PE for pipe applications typically involves accelerated testing of notched specimens either taken from a pipe or moulded ad hoc [70]. Fracture-mechanics-based SCG testing increasingly makes use of FNCT specimens (Fig. 1c), either under constant loading or under conditions of constant K, where the load is adjusted to take into account crack-tip advance, using video feedback techniques, for example [71, 72]. Results from constant load testing are given in Fig. 7 for different grades of HDPE in air [73], illustrating the sharp drop in the effective K_c as a function of failure time, which is associated with the onset of SCG (an overshoot is sometimes associated with this transition owing to crack-tip blunting [74]). The data points in Fig. 7 were obtained using two different geometries, con-

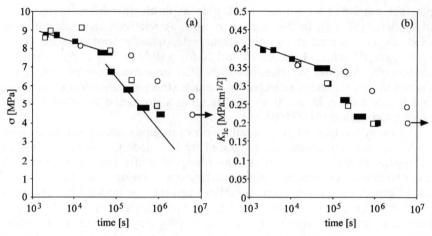

Fig. 7 a σ plotted against failure time in full edge notched cylindrical specimens of a first generation HDPE (*squares*) and a third generation HDPE (*circles*), subjected to constant tensile loading. **b** K plotted against failure time for the same specimens. The *open* and *filled* symbols are for notch depths of approximately 2 and 4 mm respectively [73]

firming failure to be controlled by the initial value of K rather than the overall ligament stress, σ [75].

3.3.2
Crack-tip Microdeformation Mechanisms

Short-term failure in notched creep specimens at high K is generally associated with yielding across the whole of the load-bearing ligament (sometimes accompanied by coarse cavitation in the specimen interior, where the triaxial stresses are high). As K decreases, the transition to SCG is marked by the formation of a fibrillar crack-tip damage zone or craze at the notch tip, and fibrillar debris on the fracture surfaces, although given that K will increase with a for fixed σ, necking generally persists in the final stages of fracture [59, 74, 75, 76]. For K immediately below its value at the transition, the deformed ligaments in the damage zones are relatively wide and, as shown in Fig. 8, their internal structure closely resembles that of macroscopic neck [77]. Indeed, 'fibrillar' is somewhat of a misnomer here, since the cavities are isolated in the early stages of deformation and coarse ribbon-like structures are visible on the fracture surface. The structure nevertheless become progressively finer as K decreases [73], fibril diameters becoming comparable with the lamellar thickness, as illustrated in Fig. 9 and Fig. 10. The draw ratio is roughly constant within the craze under these conditions, indicating a surface drawing mechanism, and there is significant interconnectivity between the craze fibrils.

At still lower K, the mode of crack propagation has been inferred from accelerated testing of SCG-resistant third generation (see Sect. 3.3.4) HDPE to

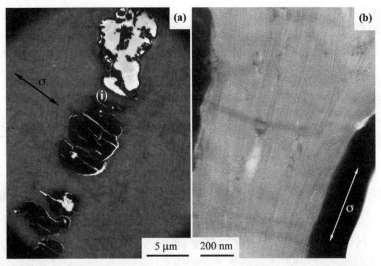

Fig. 8a, b The crack-tip deformation zone in a first generation HDPE tested with an initial K of 0.3 MPa m$^{1/2}$ in air. **a** Low magnification TEM image of the deformation zone tip (the dark areas of over-stained epoxy correspond to cavities). **b** Detail of the internal structure of the region marked (*i*) in **a** (embedded in epoxy and stained in RuO$_4$, tensile axis as indicated by the *arrows*) [73]

change to one of localised interlamellar cavitation and crack propagation via breakdown of interlamellar ligaments (Fig. 11), leading to a mirror-like fracture surface [73, 78]. The existence of zones of stable interlamellar cavitation behind the main crack front is assumed to reflect the relative stability of the interlamellar fibrils during the early stages of deformation, breakdown inter-

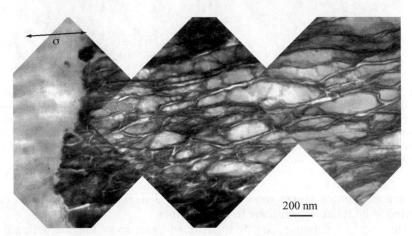

Fig. 9 Fibrillar deformation in a third generation HDPE specimen tested with an initial K of 0.26 MPa m$^{1/2}$ in air (embedded in epoxy and stained in RuO$_4$) [73]

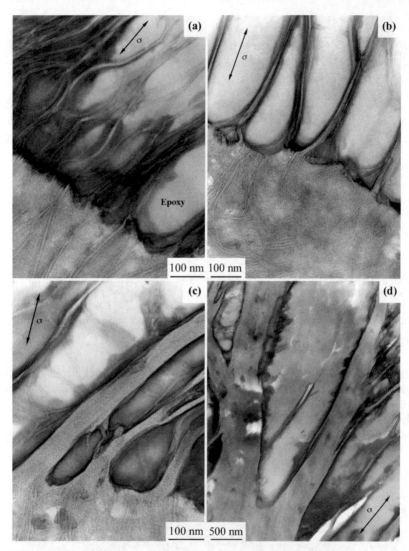

Fig. 10a–d TEM images from a first generation HDPE tested to failure with an initial K of 0.22 MPa m$^{1/2}$ in air showing the effect of increasing K as the crack-tip advances. **a** Section from just in front of the notch tip. **b** Section from about 2 mm from the notch tip. **c** Section from close to the centre of the specimen. **d** Section from the centre of the specimen (embedded in epoxy and stained in RuO$_4$, tensile axis as indicated by the *arrows*) [73]

vening after much longer times as a result of disentanglement, which is expected to be relatively sensitive to loading time.

The overall behaviour may be interpreted in terms of the competing deformation mechanisms: in the presence of tensile stresses approaching σ_y,

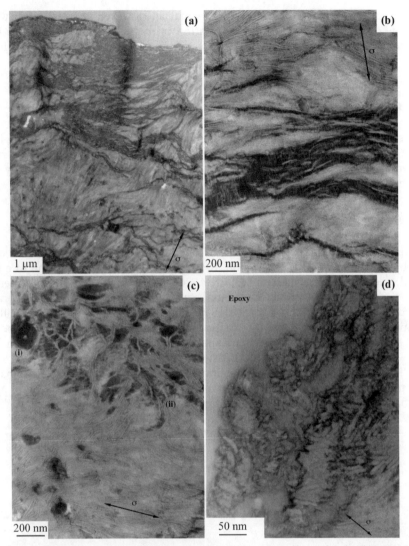

Fig. 11a–d TEM of a section through part of the fracture surface of a third generation HDPE tested with an initial K of 0.12 MPa m$^{1/2}$ in Igepol. **a** Overview, **b** localised interlamellar cavitation behind the fracture surface, **c** region of relatively diffuse deformation behind the fracture surface, showing both relatively coarse cavitation (*i*) and finer interlamellar cavitation (*ii*) and **d** detail of deformation at the fracture surface (embedded in epoxy and stained in RuO$_4$, tensile axis as indicated by the *arrows*) [73]

regions in which interlamellar cavitation first occurs (i.e. regions in which the lamellar trajectories are at high angles to the tensile axis) are assumed to be unstable with respect to further breakdown of the interlamellar material, leading to cavities that dilate rapidly owing to the local plastic constraint release, and necking of the surrounding material. However, at stress levels well below σ_y (σ_y is insensitive to the duration of the test), interlamellar cavitation becomes the dominant stress relaxation mechanism, and lamellar deformation occurs at a more local scale (if at all) [73].

3.3.3
Models

The data in Fig. 7 are broadly consistent with the widely reported scaling for the regime of steady state SCG, with $\gamma \approx 4$ for a wide range of PEs [75], assuming the failure time to be dominated by the initial value of K (as may be demonstrated from Eq. 4).

$$\dot{a} \approx \dot{a}_o \left(\frac{K}{K_{co}} \right)^{\gamma}$$ (4)

The ubiquity of this power-law behaviour in SCG tests on PE has been the subject of considerable discussion, usually based on the assumption of a fibril creep failure mechanism [43, 45, 46, 47, 76, 79]. At high and intermediate K, after a certain induction period, steady-state crack advance is generally observed to occur by a stick-slip mechanism; all or part of the fibrillar zone breaks down rapidly after an incubation time during which fibril creep takes place. The crack-tip then advances rapidly over a short distance and a new fibrillar zone stabilises, as sketched in Fig. 12.

The following simplified explanation for the observed power-law behaviour during stick-slip crack growth is a limiting case of a more general approach to SCG [43] based on crack layer theory [40, 41]. The 'crack layer' in

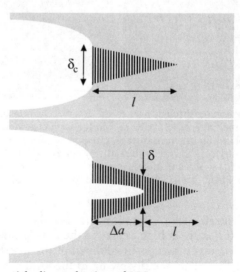

Fig. 12 Sketch of the stick-slip mechanism of SCG

this case consists of a crack of length a, and a fibrillar zone of length l. If the traction on the boundary of the fibrillar region is uniform and independent of K, the Dugdale model [39] applies, so that $l \propto K^2$ for $a >> l$. Similarly, the initial crack opening displacement is $\delta \propto K^2$. Assuming simple Newtonian viscous creep in the fibrils and a critical extension ratio for failure, the critical crack opening displacement is given by:

$$\delta_c = \delta(1 + \xi t_c) \tag{5}$$

where the failure time is:

$$t_c \propto \frac{\delta_c - \delta}{\delta} \approx \frac{\delta_c}{\delta} \tag{6}$$

Taking $\dot{a} \propto l/t_c$, it follows that $\dot{a} \propto K^4$.

This result may nevertheless be reached in other ways [76], and the assumptions regarding the fibril breakdown kinetics are in any case difficult to justify a priori. Test methods have therefore been developed for the direct measurement of the constitutive behaviour of the craze, using geometries with a high plastic constraint factors [45, 46, 80]. The results can then be incorporated in cohesive elements in finite element simulations, allowing discrimination between the performance of different materials [47]. Damage zone morphologies such as shown in Fig. 8, also imply a strong link between the creep behaviour of the damage zone and that of macroscopic necks, and reasonable agreement has been obtained between experimental SCG rates and predictions made on this basis [79]. Part of the interest in such approaches is that they can be used to infer long-term SCG performance directly from relatively rapid measurements. On the other hand, the sensitivity of the damage zone morphology to the applied K discussed in Sect. 3.3.2 implies the generality of any given model of this type to be limited (which makes the ubiquity of the scaling in Eq. 4 all the more remarkable).

3.3.4
Influence of Molecular Architecture

The sensitivity of SCG in PE to M is well established and has been widely cited in support of disentanglement mediated fibril breakdown during SCG under a wide range of conditions [22, 31, 77, 81, 82, 83, 84, 85]; the SCG resistance in certain grades of PE has been reported to improve dramatically for weight average molar masses, $M_w > 150$ kg mol^{-1}, for example, whereas it becomes negligible for $M_w < 18$ kg mol^{-1} [86]. Another important factor is the density and nature of chain branching; the presence of lateral groups (typically from methyl to octyl) strongly influences crystalline order and decreases the lamellar thickness, resulting in increased molecular connectivity between the lamellae for sufficiently long chains, and improved SCG resistance, provided the lamellar thickness remains sufficient to anchor the chains [81, 87, 88, 89]. Moreover, disentanglement of chains from the crystal lamellae is thought to be impeded by high degrees of branching, suggesting that a combination of high M and chain branching ought to result in optimum SCG resistance [90]. It follows that polybutene, which contains one ethyl side-chain for every two main-chain carbon atoms, should also show good SCG resistance, and indeed it is often preferred over PE in applications involving corrosive environments. Similar correlations have also been observed in studies of the impact fracture resistance of PE [91, 92, 93].

The evolution of commercial pipe-grade PEs is easily understood in terms of the above considerations. The 'first generation' (or 'monomodal') HDPEs, introduced in the 1950s, typically contain 3–7 ethyl groups per 1000 carbon atoms, as opposed to 20–30 ethyl or butyl groups in the earlier low density PE (LDPE), produced by high-pressure processes, and which is too compliant to be of interest for high-pressure pipe applications. Improved SCG performance was subsequently obtained using medium density branched polyethylene copolymers (MDPE). SCG in MDPE occurs by essentially the same mechanism of stepwise crack advance as in HDPE, but is associated with increased crack-tip plasticity, often in the form of an epsilon-shaped damage zone, consisting of the central craze accompanied by angled cavitated shear zones of comparable length [61]. However, improvements in SCG resistance in these 'second generation' grades are obtained at the expense of reduced stiffness and poor bulk creep properties. In current 'third generation' (or 'bimodal') short-chain branched HDPE copolymers, therefore, the branched units are concentrated in relatively long chains. The reinforcing effect of the long branched chains and their influence on the crystalline morphology lead to excellent SCG resistance. The overall degree of crystallinity and hence the bulk creep properties remain comparable with those of the monomodal grades [90, 94].

Crosslinking represents the limit of high M and long-chain branching and also disrupts crystallinity. It follows that stage-II fracture is not seen at all in peroxy-crosslinked HDPE [66]. γ irradiation has also been found to result in improved SCG resistance in HDPE, and particularly in MDPE, where very high radiation doses lead to a transition from crazing to shear deformation at the notch tip under appropriate conditions (although in HDPE comparable doses lead to embrittlement) [95, 96].

3.3.5
Cyclic Loading

Qualitatively similar SCG behaviour is seen in testing accelerated by use of cyclic load conditions, namely stick-slip crack propagation at relatively high loads and quasi-continuous crack growth with a mirror-like fracture surface for low level loading [97]. Moreover, although the absolute failure times are significantly reduced at any given temperature, rankings based on fatigue crack propagation have been shown to correlate with those obtained from standard creep tests [61, 62, 98, 99, 100]. This has been explored by systematically increasing the ratio of minimum to maximum stress in the loading cycle, R, at different test temperatures. With increasing R, the fracture mode has been observed to change from stepwise crack propagation to continuous

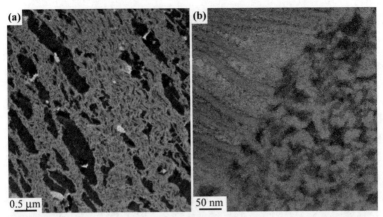

Fig. 13a, b Fibrillar deformation in a third generation HDPE subjected to fatigue testing in air. **a** Overview of the fibrillar zone and **b** detail of the craze–bulk interface

crack growth, although the power law dependence in Eq. 4 persists (if K is taken to be the maximum stress intensity factor, K_{max}) [62]. The SCG rate was also found to increase as the minimum stress intensity factor became more compressive in cyclic fatigue at constant K_{max}. This implies a mechanism in which damage sustained by the craze during the compressive part of the cycle contributes to fibril breakdown in tension [101].

As might be expected, observations of the damage zone associated with fatigue cracks show similar features to those in constant load tests, with progressively finer fibrillar structures being associated with decreasing load levels. However, in fatigue tests, the fibrils show substantial retraction, indicating relatively little stress relaxation prior to fibril breakdown, as shown in Fig. 13. This contrasts with the creep fracture surfaces, in which the fibrils are fully stress relaxed and remain fully extended after failure. Figure 14 shows damage corresponding to the mirror-like region of a fracture surface obtained under low-level loading in an SCG-resistant third generation HDPE, showing features reminiscent of those in Fig. 11, again suggesting interlamellar failure to dominate. It may therefore be inferred that this failure mode is not specific to ESC, with obvious implications for the universality of accelerated testing. Certainly, the effects of M, comonomer content and branch distribution on the kinetics of fatigue crack propagation are broadly consistent with results from creep tests, the SCG rate again decreasing with increasing branch concentration and M. However, in materials possessing high branch concentrations, significant compression has been found to reduce the fatigue crack resistance to the extent that rankings in terms of SCG resistance are reversed with respect to those established from creep testing or at low compression [102].

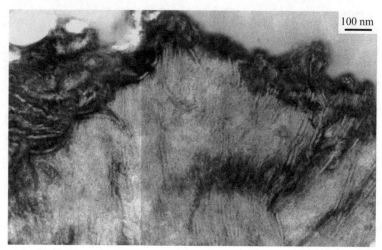

Fig. 14 Detail from the mirror-like region of a third generation HDPE subjected to fatigue testing in air

3.4
Ductile–Brittle Transitions in iPP

In most thermoplastics, transitions from ductile to brittle behaviour may be induced by increasing the test speed. For the reasons already invoked in the introduction, this is of particular concern in iPP, whose impact proper-

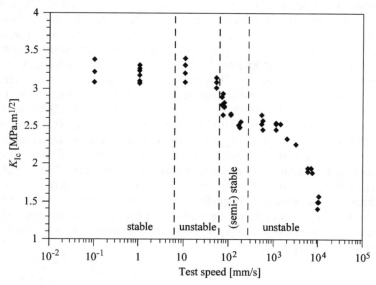

Fig. 15 Variation of K_c with test speed in CT specimens of iPP with M_w of 455 kg mol^{-1} and a polydispersity of 5.4, tested in air, along with an indication of the mode of crack propagation [19]

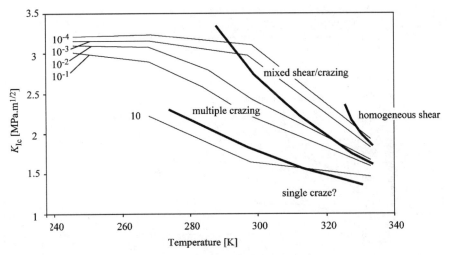

Fig. 16 'Deformation map' for iPP with M_w of 200 kg mol^{-1}. Different regimes of microdeformation behaviour are indicated on a plot of K_c vs. T for various strain rates [19, 26]

ties, especially at low temperatures, are often a limiting factor in applications, whence current interest in the impact modification of iPP (to be discussed in Sect. 4.1). Figure 15 shows K_{Ic} in iPP with M_w=455 g mol^{-1} and a polydispersity of 5.4, as a function of test speed in tensile tests on CT specimens, showing a steep decrease between 1 and 10 mm s^{-1}, associated with the suppression of shear lip formation, and the onset of crazing and unstable crack propagation [19]. (In this case, the main ductile brittle transition is associated with an intermediate regime of stable crack propagation, thought to be due to transient adiabatic heating effects.) The global response is shown in Fig. 16, where the microdeformation behaviour has been mapped in temperature–K_{Ic} space for different strain rates. Consistent with the discussion of Sect. 3.1, crazing and hence brittle behaviour are favoured not only by high speeds, but also by low temperatures at fixed M [26]. A ductile–brittle transition may also be induced by decreasing M, depending on the test conditions. For M_w below about 380 kg mol^{-1}, crazing tends to replace ductile necking in simple tensile tests at room temperature, for example [26], and the regimes of stable crack propagation are progressively displaced to lower speeds and temperatures in fracture tests [103]. This is illustrated further by the results of studies of the kinetics of thermo-oxidative molecular weight degradation in the presence of different stabilisers, which also indicate a strong correlation between the evolution of M and that of the fracture toughness and crack propagation behaviour, as shown in Fig. 17. This is again associated with a transition from homogeneous shear to crazing [104].

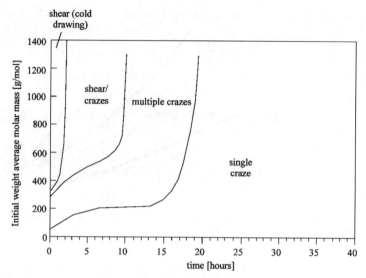

Fig. 17 "Deformation map" for iPP. Different regimes of microdeformation behaviour are plotted as a function of thermal degradation times at 130 °C and the initial M_w [26, 104]

3.4.1
Morphology and Microdeformation

A craze microstructure typical of α iPP is shown in Fig. 18, indicating fibril diameters comparable with those in crazes in glassy polymers and a constant fibril draw ratio, consistent with a surface drawing mechanism for craze widening [28, 105]. Figure 19 shows extensive interlamellar cavitation in the early stages of deformation in a thin film (about 100 nm), thought to be a precursor to the formation of crazes in bulk samples (as described previously for PE), although the lack of plastic constraint in thin films favours subsequent diffuse lamellar shear (Fig. 19b), rather than a well-defined craze morphology.

The relative propensity of iPP for crazing and brittle behaviour may be linked to the unusual microstructure of its relatively stable monoclinic α modification, which is the dominant phase under standard synthesis and processing conditions. The α modification is typically characterised by auto-epitaxial growth of lenticular lamellae, elongated in the [100] direction, giving rise to a 'cross-hatched' morphology in which distinct populations of lamellae are arranged with their long axes roughly perpendicular locally [106, 107]. This cross-hatched morphology has been argued to hinder certain slip mechanisms [108, 109], as in recent work on the compression of bulk α iPP [110]; crystallographic slip on (010)[001] and (100)[001] are activated in the early stages of deformation, although (010)[001] has a lower critical resolved shear stress. σ_y is therefore expected to be higher than if this latter slip system were activated immediately. On the other hand, In the

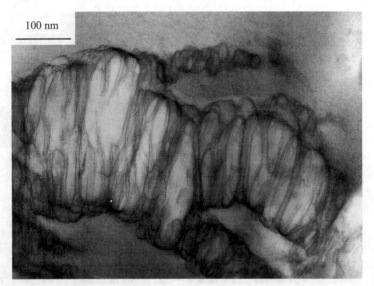

Fig. 18 Fibrillar deformation in bulk α iPP deformed at room temperature, embedded in PMMA and post-stained with RuO$_4$ vapour

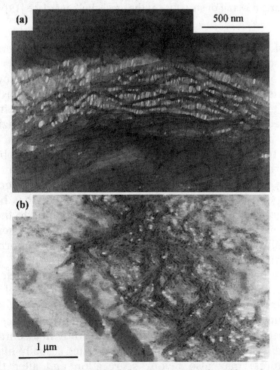

Fig. 19a, b Early stages of irreversible deformation in a thin film of α iPP deformed at room temperature and post-stained with RuO$_4$ vapour, showing **a** interlamellar cavitation and **b** lamellar shear at a later stage in the deformation process [26]

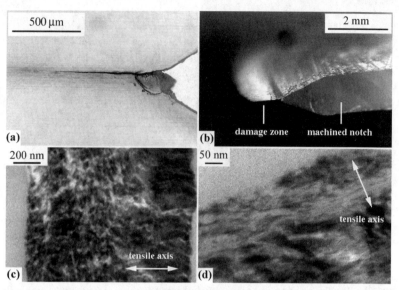

Fig. 20 a Side-view of the crack-tip damage zone in a CT specimen of iPP with M_w of 455 kg mol^{-1} deformed at about 3 m s^{-1}. **b** Oblique view of the damage zone showing the curved deformation front. **c** TEM micrograph of the collapsed fibrillar structure of the crack-tip craze. **d** Detail of structure at the craze–bulk interface [19]

latter stages of yielding, where the initial morphology is less critical, it is the (010)[001] system that dominates, implying a large yield drop, consistent with results from tensile tests. Hence there is a strong tendency for slip processes to localise, which should favour localised crazing in tension.

The trend towards localised deformation increases with increasing test speed, as reflected both by the transition from shear to crazing described above, and by a decrease in the craze density ahead of the crack-tip, as indicated in Fig. 16. Thus, at impact speeds, a single craze is observed at the crack-tip, as shown in Fig. 20, propagating normal to the principal stress axis, implying crazing in this regime to be insensitive to the local morphology, as also reported for crazing at $T<T_g$ [20, 111]. This has been linked to the proximity of the range of temperatures corresponding to the β transition in dynamic measurements on iPP. Under such conditions, there may be insufficient mobility for homogeneous deformation in the amorphous regions to keep pace with the highest deformation rates [112].

The fracture behaviour of iPP is nevertheless generally held to be highly dependent on many other factors. Large spherulites are widely held to promote brittleness, for example, owing to the concentration of structural defects and impurities at their boundaries, and hence the presence of easy crack propagation paths [113, 114]. This is difficult to study in a controlled fashion, since the conditions for obtaining large spherulites (usually high crystallisation temperatures) may also favour segregation and alter the degree of crosshatching and crystallinity; higher crystallinity at fixed spherulite size having been found to reduce fracture toughness [115]. Moreover, high crystallisation temperatures are argued to result in crystallisation-induced entanglement loss, which will again contribute to brittleness [111, 116]. They are also associated with the moulding of nucleated systems, so that any gains owing to the relatively fine spherulitic morphology that results may be offset by the effects of increased crystallinity or entanglement loss [115, 117].

3.4.2
Matrix Polymorphism

In contrast to the α modification, the hexagonal β modification is characterised by relatively broad lamellae that form coplanar stacks, whose plane tends to twist about the growth direction [107, 118]. Although β spherulites appear only sporadically in mouldings of commercial iPP, their proportion can be increased by using certain mineral or organic nucleating agents [119, 120]. It is widely held, mainly on the basis of impact data, that the β phase is more fracture resistant than the α phase [113, 114, 121, 122]. As shown in Fig. 21, in CT tests there is a significant increase in fracture toughness in the presence of the β phase in certain ranges of T and test speed, and an increase in the ductile–brittle transition speed by a least three decades at $T>T_g$ in the

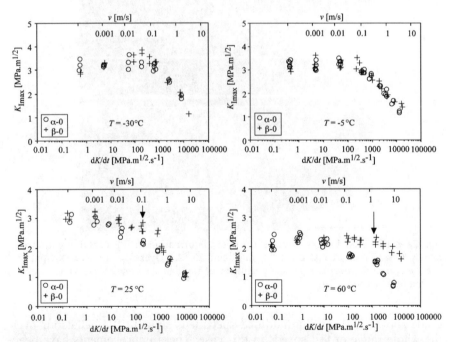

Fig. 21 The critical stress intensity for mode I crack initiation at different temperatures as a function of test speed in α iPP with M_w of 248 kg mol^{-1} and a polydispersity of 5.2 and a similar material containing approximately 80 wt% β phase. The *arrows* mark ductile–brittle transitions in the β modified specimens [24]

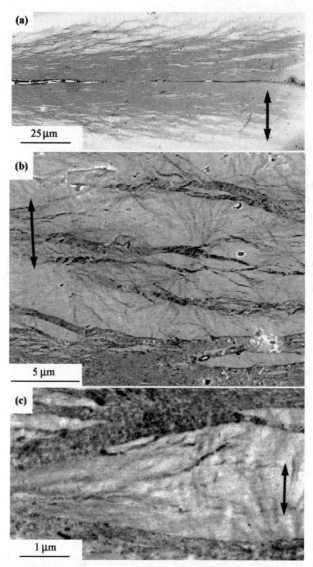

Fig. 22a–c BSE-SEM images of crack-tip microdeformation in iPP containing approximately 80 wt% β phase tested at 0.001 m s^{-1} and 25 °C. **a** Overview of the crack-tip damage zone, **b** and **c** details from the periphery of the damage zone. The *arrows* indicate the loading direction [24]

grade shown [24]. The unmodified specimens show brittle behaviour over the whole range of test speeds in this case. These improvements have been correlated with results from volume extensometry and factography, which show cavitation to be more extensive in β nucleated specimens than in specimens containing predominantly α spherulites [24, 123]. Figure 22 shows

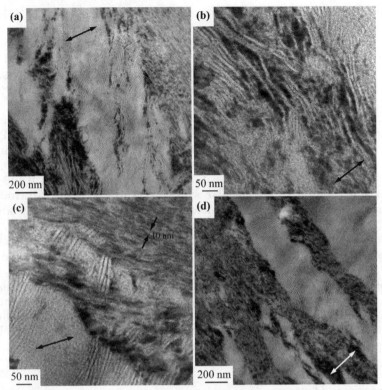

Fig. 23a–d TEM images of microdeformation in iPP containing approximately 80 wt% β phase tested at 0.001 m s^{-1} and 25 °C. **a** Overview of diffuse interlamellar separation and cavitation and part of a fibrillar deformation zone or craze (*top right-hand side* of the image). **b** Detail of the early stages of craze formation in a region in which the direction of the applied load was roughly perpendicular to the lamellar trajectories. **c** Detail of the edge of a fibrillar deformation zone (*top right-hand side* of the image) showing partly relaxed fibrillar structure. **d** Overview of fibrillar deformation zones in a region in which the direction of the applied load was roughly parallel to the lamellar trajectories, again showing partly relaxed fibrillar structure. The *arrows* indicate the direction of the applied load [24]

FEG-SEM of crack-tip deformation in a β nucleated CT specimen and Fig. 23 shows the corresponding TEM images. These confirm deformation to be less localised than in α spherulites at comparable test speeds, with diffuse regions of cavitation and lamellar shear coexisting with relatively well-defined crazes. The craze structures themselves are nevertheless similar to those in α–iPP.

A mechanically induced β to α phase transformation has often been referred to in discussions of the relative ductility of β nucleated materials [121, 124, 125, 126, 127]. That the phase transformation takes place is well established, and since the density of the β phase (0.921 g cm^{-3}) is lower than that of the α phase (0.936 g cm^{-3}), it has been argued that a β to α phase transition should promote cavitation during bulk deformation [124]. However, a stress-induced volume decrease in the crystalline regions of the specimen is unlikely to be at the origin of the interlamellar cavitation in Fig. 23, given that the lamellae undergo very little change in morphology at this stage in the deformation process. Electron diffraction confirms the absence of any significant corresponding phase change, although a clear β to α transition is seen in more highly drawn material [111].

Morphological explanations for the improved ductility of the β nucleated materials have focussed on the lamellar texture. That crazes in α iPP are more localised and better defined than in β iPP may reflect both the influence of the cross-hatched structure on lamellar slip described in the previous section, and the strong correlation between deformation in β iPP and the local orientation of the lamellae with respect to the tensile axis. Indeed, the trend towards more localised deformation in α spherulites may simply reflect the relatively homogeneous lamellar textures of these latter [24].

It nevertheless remains difficult to discriminate morphological effects at the lamellar level from other factors, such as crystallinity and spherulite size [128], on the basis of the available evidence [21, 23, 24, 129, 130, 131, 132] (cf. Sect. 3.4.1). This also makes it difficult to discount alternative explanations for the improved ductility of the β phase above T_g, based on its intrinsically higher molecular mobility, for example [133, 134, 135, 136]. One is therefore forced to conclude that any or all of the factors referred to here may play a significant role in the observed behaviour.

4
Heterogeneous Systems

In this final section, emphasis will be placed on the relationship between the behaviour of the homopolymer and that of heterogeneous systems containing interfaces. Thus, in Sect. 4.2, rather than dwell on the (albeit very important) technological aspects of welding, the discussion centres on the extent to which studies of interfaces might help in understanding the fundamentals of fracture in semicrystalline polymers, as they have in the case of interfaces between amorphous polymers [137].

4.1
Rubber-Toughened iPP

Rubber-toughened grades of iPP are available with a vast range of formulations and microstructures [138]. The most common modifiers are EPR, often introduced by varying the monomer composition during synthesis (reactor blends) and ethylene-propylene-diene monomer elastomer (EPDM), although many other possibilities exist, including linear LDPE (LLDPE) [139]. A correspondingly vast literature has grown up around rubber-toughened iPP over the past 20 years or so [19, 23, 130, 132, 140, 141, 142, 143, 144, 145, 146, 147, 148, 149, 150, 151, 152], but the basic toughening mechanisms remain only partially understood, not least because of the diversity of parameters that contribute to the mechanical performance (modifier T_g and content, particle size and particle size distribution, modifier morphology, interfacial strength, matrix plasticisation, crystallinity and phase behaviour, and the matrix molecular architecture [23, 130, 132, 150, 151, 152, 153]). Although beyond the scope of the present discussion, rigid organic [154] and inorganic particle toughening and hybrid materials have also come under considerable scrutiny [155], since such toughening strategies do not lead to the loss in stiffness generally associated with rubber toughening. Loss in stiffness can be moderated by use of core-shell particle morphologies, such as shown in Fig. 3, for EPR-modified iPP, where the particle cores consist of ethylene-rich semicrystalline inclusions. The ethylene content of such systems is nevertheless limited by the need to maintain compatibility with the iPP matrix. This may be overcome by blending with additional PE homopolymer, with the proviso that the final toughness may be sensitive to the resulting particle morphology [156].

4.1.1
Toughening Mechanisms

It is generally agreed that one important effect of rubber toughening is to delocalise crack-tip deformation. The stiffness mismatch between the matrix and modifier leads to local stress concentrations at the particle equators, promoting craze nucleation, as shown in Fig. 24, or shear banding. Moreover, the associated hydrostatic tension may lead to cavitation of the particles, the accompanying plastic constraint release favouring ductile drawing

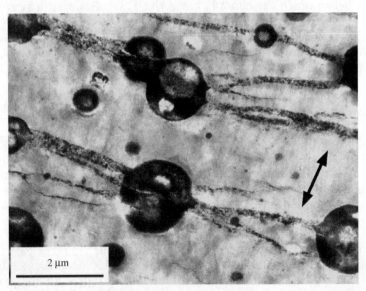

Fig. 24 TEM micrograph showing crazing and rubber particle break-down close to the fracture surface in an iPP/EPR CT specimen deformed at about 7 m s^{-1} [19, 26]

of the inter-particle ligaments. It therefore follows that the presence of large numbers of finely dispersed particles should lead to increased crazing and/ or ductile necking locally and an increase in the extent of the crack-tip damage zone, resulting in increased crack-tip dissipation and hence a higher fracture resistance and improved ductility. The role of matrix drawing is thought to be particularly important, even where crazing is widespread [157]; craze bifurcation and deflection towards cavitated particles lead to the establishment of a continuous three-dimensional network of cavitated material in the vicinity of the crack-tip, significantly reducing the plastic constraint on the intervening matrix ligaments [158]. At high discrete particle contents, achievable in iPP modified with EPDM vulcanates, for example, the constraint release due to particle cavitation may be sufficient for crazing to be entirely replaced by ductile drawing of the inter-particle ligaments over a wide range of conditions, leading to 'super-tough' behaviour. Particle cavitation tends to occur along rows of particles aligned roughly perpendicular to the tensile axis [21], sometimes referred to as 'croids' [159]. The subsequent necking of the intervening matrix ligaments leads to structures analogous to crazes, albeit at very different length scales (see also Sect. 4.2.2).

There has been considerable effort to establish general principles for the morphological control of ductile behaviour with the aim of achieving it at reduced rubber content (to limit stiffness loss). The idea of a critical ligament thickness below which crazing is suppressed [149] has attracted substantial interest, for example, although it is clear that if this were the only relevant factor, indefinitely reducing the particle size at constant rubber content would suffice. In practice, toughening is optimised for particle diameters of a few hundred nm [160] and there is generally a minimum particle size below which cavitation is ineffective [161]. However, it is not clear that the particle size can be optimised for extended ranges of test conditions, since different mechanisms may be involved in toughening, so that tailored bimodal or multimodal particle distributions, say, might provide a more flexible solution than a monomodal distribution [23, 130, 132, 151].

4.1.2
Influence of Matrix Properties

Matrix properties and their dependence on test conditions are also paramount for the performance of toughened iPP [152]. The effect of M closely follows that of unmodified iPP, with low M favouring crazing and brittle behaviour and high M favouring shear deformation [19, 152]. Crazing is again favoured over homogeneous deformation in the modified materials at high speeds and low temperatures, although this does not necessarily lead to brittle behaviour [142, 143, 146, 147]. Indeed, for the modified iPP for which data from CT tests are given in Fig. 25, the toughening effect is particularly manifest in regimes of test speed in which the unmodified matrix is brittle, that is, in regimes in which crazing dominates. Indeed in the example shown, the toughness continues to increase with increasing test speed beyond regimes in which the unmodified matrix is very fragile. At the highest test speeds the behaviour becomes difficult to interpret, owing to significant

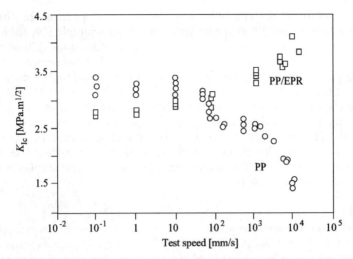

Fig. 25a, b The room temperature critical stress intensity for mode I crack initiation as a function of test speed in **a** iPP with M_w of 455 kg mol^{-1} and a polydispersity of 5.4 and **b** iPP with M_w of 437 kg mol^{-1} and a polydispersity of 5, containing 15 wt% EPR [19, 26]

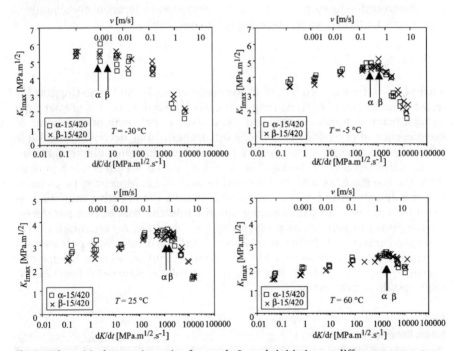

Fig. 26 The critical stress intensity for mode I crack initiation at different temperatures as a function of test speed in α-iPP with M_w of 232 kg mol^{-1} and a polydispersity of 5.3 containing 15 wt% EPR and the same material containing approximately 65 wt% β phase in the matrix. The *arrows* mark the ductile–brittle transition speeds in each case [24]

adiabatic heating at the crack-tip. A 90 K rise in temperature has been reported for toughened iPP at speeds of 10 m s^{-1}, for example [19, 131]. Moreover, β nucleation has little effect on the brittle–ductile transition of rubber-modified iPP at any temperature (Fig. 26). Given that the improvements in toughness of the homopolymer on β nucleation are essentially morphological in origin (one of the possibilities put forward in Sect. 3.4.2), this underlines the important role of the rubber particles in initiating matrix deformation during crack initiation and propagation [24]. It should be borne in mind that LEFM analyses of this type are subject to caution because of the relatively high plasticity of toughened iPP. However, consistent trends have also been observed in EWF measurements on the same and similar materials [26, 112, 162, 163, 164, 165].

4.2
Fracture of Polyolefin Interfaces

Interfaces are not only present in blends, but are also a key feature of structural joints, where it is often necessary to produce interfacial strengths comparable with the cohesive strength of the bulk materials, and of laminates, where a more modest degree of adhesion may suffice [166]. The compatibility between different components can be expressed in terms of the equilibrium interfacial thickness, w, given approximately by:

$$w = \frac{l_o}{3\sqrt{\chi}} \tag{7}$$

where l_o is statistical step length of the chain and χ is the Flory–Huggins interaction parameter [167]. As long as the interfacial thickness is greater than d_e, significant adhesion is expected. With strong anchoring of the chains in the presence of crystalline lamellae, joint strengths may become comparable with cohesive strengths. However, in pairs of incompatible polymers, the interfacial thickness given by Eq. 7 is very low, so that it is necessary to reinforce the interface, as will be discussed in Sect. 4.2.2. Moreover, by controlling the density of interfacial connectors, that is, the effective value of Σ at the interface, one can gain valuable insight into the role of this parameter for toughness in general, as presaged by Eq. 1 and Eq. 2. One might expect controlled interfacial diffusion in fusion bonding of identical polymers to provide an alternative means of controlling Σ, but as will be briefly discussed in the next section, this may be difficult to achieve in practice for semicrystalline polyolefins.

4.2.1
Fusion Bonding

In fusion bonded joints between identical polymers, for which w is infinite, recovery of the cohesive strength is essentially a question of establishing physical contact at the interface (including wetting), self-diffusion of chains

across the interface and co-crystallisation [168, 169]. In fusion bonds between plaques of unmodified polyolefins, joint strengths approaching the cohesive strength generally require heat treatment at $T>T_m$, as shown in Fig. 27
for iPP [170, 171]. Fundamental studies of fusion bonding have tended to
concentrate on glassy polymers heat-treated at T just above T_g, where self diffusion is relatively slow, so that it is generally the rate-controlling step in the
welding process [57]. In semicrystalline polymers at $T>T_m$, on the other
hand, establishment of physical contact, and hence variables such as the applied pressure and the initial surface quality, play an important role in bonding kinetics. This is borne out by non-isothermal bonding experiments in
which a molten plaque with an initial temperature of T_1 is pressed against a
solid plaque with an initial temperature of T_2, with boundary conditions such
that the temperature of the interface can be assumed to be roughly constant
at $T_i=(T_1+T_2)/2$. Under these conditions, strengths comparable with the cohesive strength require $T_i>T_m$, but, as shown in Fig. 27, the kinetics of bond
formation are much more rapid for a given T_i than for non-isothermal bonding [170, 172]. Although this is clearly of practical interest, it also demonstrates that fusion bonding of semicrystalline polymers is far less suited to
the study of the role of molecular connectivity across the interface under controlled conditions than fusion bonding of amorphous polymers [57].

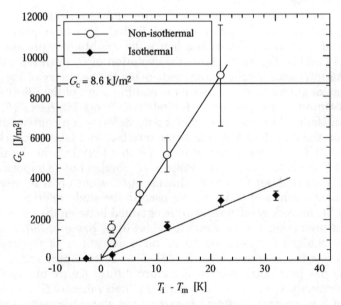

Fig. 27 G_c as a function of T_i-T_m for isothermal iPP bonds (*filled symbols*, t_{hold}=600 s,
p=2 MPa) and non-isothermal bonds (*open symbols*, t_{hold}=40 to 42 s, p=2 MPa) [170]

4.2.2
Reaction Bonding

A strong interface between different, incompatible polymers generally requires either the introduction of a preformed compatibiliser (usually a block copolymer) or formation of covalent bonds across the interface in situ by chemical reaction [55]. Polyolefin graft maleic anhydride (MAH) polymers are commonly used for this purpose, since the MAH reacts readily with a range of functional groups, including the $-NH_2$ terminal groups of polyamides. This route is widely used to form block or graft copolymers at interfaces between iPP and polyamide 6 (PA6) and can be extended to other systems, such as PE/iPP, by combining MAH grafting with $-NH_2$-terminated coupling agents [173]. As long as the molecular weights of the respective constituents are high enough, the resulting covalent bond is well anchored to either side of the interface and is able to play a similar role to entangled strands during cohesive failure [174, 175]. Σ is considerably less than the crossing density of entangled strands in the bulk, owing to the limiting effect of steric hindrance on the total number of copolymers that can accumulate at the interface at equilibrium. For iPP-MAH with M_w=60,000 g mol^{-1}, for example, the effective crossing density, Σ, saturates at about 0.1 nm^{-2}, whereas the number of entangled strands crossing unit surface in bulk iPP, $v_e d_e/2{\approx}1$ nm^{-2} [48, 174, 175].

Since Σ varies with heat treatment time and temperature below this threshold, such bonds offer interesting possibilities for studying the link between Σ and fracture toughness and hence the validity of expressions such as Eqs. (1) and (2). Figure 28 shows the evolution of G_c in (iPP/5 wt% iPP-MAH)–PA6 interfaces obtained from tests using the geometry of Fig. 1 after heat treatment at temperatures above the melting point of the iPP/iPP-MAH [48], confirming G_c to scale as Σ^2 for relatively weak interfaces (G_c up to 100 J m^{-2}). Under these conditions, crack-tip deformation consists of a single craze on the iPP/iPP-g-MA side of the interface and failure occurs at the interface so that cohesive failure of pure iPP should reflect the limiting behaviour as Σ tends to 1 nm^{-2}. The value of G_c obtained by extrapolating the curve in Fig. 28 is about 10^4 J m^{-2}, which is of the same order of magnitude as experimental values of G_c in the iPP used in the study referred to (about $6{\times}10^3$ J m^{-2}). Indeed, good agreement is obtained between the experimental results and predictions of the absolute values of G_c based on Eq. 2, given a reasonable choice of parameters based on observations of the craze microstructure at the interface (Fig. 29) [48]. That G_c no longer scales with Σ^2 at high G_c has been attributed to the more diffuse nature of the damage zone. However, the question of why relatively high values of G_c are obtained for a given Σ when heat treatment is carried out above the melting point of the PA6 and for relatively high molecular weight grafted chains remains open, although tentative explanations in terms of co-crystallisation effects at the interface have been put forward [176].

Microstructural changes may also influence the nature of the damage zone, given that a more diffuse damage zone would generally be expected to

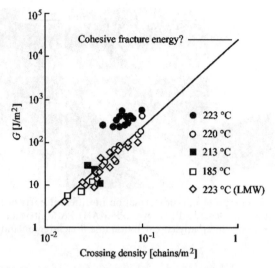

Fig. 28 The evolution of G_c in (iPP/5 wt% iPP-MAH)-PA 6 interfaces on heat treatment at different temperatures above the melting point of the iPP/iPP-MAH [48]

lead to a higher G_c than for a single craze, say. For example, the trajectories of crazes observed on the iPP side of relatively weak reaction-bonded α iPP/ PA6 interfaces subject to mode I opening follow the plane of the interfaces. Isolated β iPP spherulites, on the other hand, may provoke significant devia-

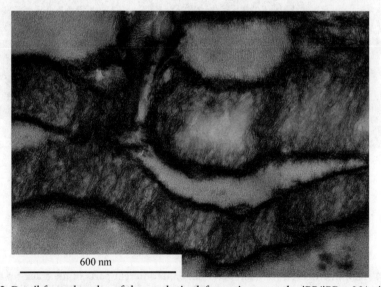

Fig. 29 Detail from the edge of the crack-tip deformation zone the iPP/iPP-g-MA side of (iPP/5 wt% iPP-MAH)-PA 6 interface heat treated at 221 °C for 800 min (G_c=300 J m^{-2}) [48]

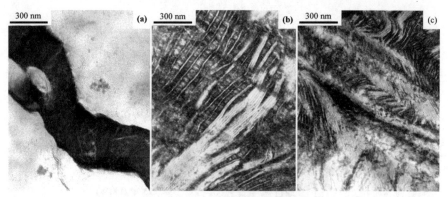

Fig. 30a–c TEM micrograph of microdeformation in different parts of the crack-tip deformation zone at a β nucleated (iPP/5 wt% iPP-MAH)-PA 6 interface. **a** Crazing in an α-rich region, **b** extensive interlamellar cavitation in a β-rich region and **c** lamellar shear in a β-rich region [177]

tion of crazes and/or crack-tips from the interface and nucleation of secondary crazes beyond the interface (in the equatorial or polar regions of the β spherulites) [177]. Certainly, β nucleation is observed to provoke delocalisation of the microdeformation (Fig. 30), although the increases in fracture

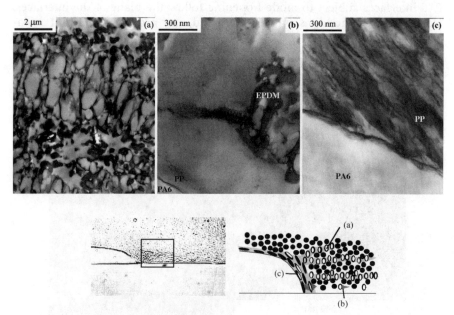

Fig. 31 TEM micrographs of microdeformation in different parts of the crack-tip deformation zone at an interface between iPP containing 52 vol% EPDM modified with 15.3 wt% PP-g-MA and PA6, embedded in PMMA and stained with RuO$_4$. Also shown are an optical image of the entire deformation zone and a schematic indicating the region from which image was taken [49]

toughness are not very marked, in line with the earlier discussion of β modification. Moreover the scaling of G_c with Σ^2 is maintained not only in β modified interfaces, but also in interfaces between iPP/EPDM/iPP-g-MAH and PA6 [49, 176]. As shown in Fig. 31, the crack-tip deformation zone in this latter case consists of rows of cavitated rubber particles, i.e. croids (see Sect. 4.1.1), separated by homogeneously necked ligaments of the iPP matrix extending perpendicular to the principal stress axis, with little evidence for crazing. The layer of iPP immediately adjacent to the PA6 is relatively poor in modifier particles and isolated crazes are observed close to the interface. Nevertheless, immediately adjacent to the crack-tip, the whole of the deformation zone forms a continuous highly drawn structure [49].

5
Conclusions

Globally, the fracture behaviour of semicrystalline polyolefins, as exemplified by PE and iPP, reflects well the general picture established for both glassy and semicrystalline polymers, namely a tendency for crazing and relatively brittle behaviour at (i) high test speeds and low temperatures and (ii) low test speeds and temperatures approaching the softening point (T_m or T_g). This behaviour is also influenced by molecular architecture, with high M and branching thought to limit chain disentanglement, for example, leading to improved fracture properties in all regimes. The range of M and chain branching appropriate for practical applications is nevertheless constrained by their influence on processing and the degree of crystallinity. For example, there are major limitations to the applicability of ultrahigh molecular weight PE (UHMWPE). Practical considerations also provide the focal point for research into polyolefin fracture properties. There is consequently particular interest in the low rate, high-temperature embrittlement associated with SCG in PE, whereas in iPP, whose T_g is just below ambient temperature, impact properties are of most concern. It follows that there has been far more interest in rubber toughening in iPP than in PE.

In the homopolymers, the range of molecular architectures (i.e. the degree of branching) has traditionally been relatively broad in PE, leading to a much wider range of properties than in iPP (leaving aside EP copolymers). However, this situation is evolving rapidly with the development of metallocene catalysis and new polymerisation processes, so that as well as much higher degrees of tacticity, branched or even 'bimodal' iPP are likely to become affordable alternatives to current materials, in which molar mass is the most accessible parameter. The new resins should, in turn, offer improved matrix properties in impact-modified resins, based on similar strategies to those adopted for the improvement of fracture resistance in PE, and also through the influence of tacticity on the degree of crosshatching in the α phase (which is essentially a defect structure).

At the fundamental level, current understanding of crazing and fracture in semicrystalline polymers remains less advanced than in glassy polymers. Even in these latter, phenomena such as disentanglement are generally subject to unverified assumptions concerning their kinetics, or even their existence. In the case of the polyolefins, the reason for this lies not only in the additional level of complexity introduced by the presence of crystalline lamellae, but also the ill-defined molecular architecture of commercial materials. As discussed in Sect. 4.2.2, the study of reaction-bonded interfaces provides an alternative to that of cohesive failure and is expected to provide considerable new insight into these questions.

Acknowledgements The author acknowledges the technical support of the Centre Interdépartemental de Microscopie Electronique (CIME) of the EPFL and H.-H. Kausch, J.-A.E. Månson, C. Grein, P. Béguelin, R. Gensler, C. Creton, F. Kalb, L. Léger, A. Ghanem, A. Goldberg, P.-E. Bourban, G. Smith and many others for the opportunity to be actively involved in various aspects of this work.

References

1. Williams JG (1984) Fracture mechanics of polymers. Ellis Horwood, Chichister
2. Cotterell B, Reddel JK (1977) Int J Fracture 13:267
3. Williams JG, Marshall GP (1975) Proc Roy Soc A 342:55
4. Schapery RA (1984) Int J Fracture 25:195
5. Béguelin P, Barbezat M, Kausch H-H (1991) J Phys III France 1:1867
6. Béguelin P (1996) PhD Thesis, EPFL
7. Béguelin P, Kausch H-H (1995)Techniques for high speed testing. In: Williams JG, Pavan A (eds) Impact and dynamic fracture of polymers and composites. Mechanical Engineering Publications, London
8. Bassett DC (1981) Principles of polymer morphology. Cambridge University Press, Cambridge
9. Butler JH, Joy DC, Bradley GF, Krause SJ (1995) Polymer 36:1781
10. Binning G, Quate CF, Gerber C (1986) Phys Rev Lett 56:930
11. Coulon G, Castelein G, G'Sell C (1998) Polymer 40:95
12. Butler MF, Donald AM, Bras W, Mant GR, Derbyshire GE, Ryan AJ (1995) Macromolecules 28:6383
13. Butler MF, Donald AM, Ryan AJ (1997) Polymer 38:5521
14. Butler MF, Donald AM (1998) Macromolecules 31:6234
15. Bartczak Z, Galeski A (1999) Polymer 40:3677
16. Engstrom P, Fiedler S, Riekel C (1995) Rev Sci Instrum 66:1348
17. Sawyer LC, Grubb DT (1987) Polymer microscopy. Chapman and Hall, London
18. Plummer CJG, Scaramuzzino P, Kausch H-H (2000) Polym Eng Sci 40:1306
19. Gensler R, Plummer CJG, Grein C, Kausch H-H (2000) Polymer 41:3809
20. Friedrich K (1983) Adv Polym Sci 52–53:225
21. van der Wal A, Verheul AJJ, Gaymans RJ (1999) Polymer 40:6067
22. Hamouda HBH, Simaos-Betbeder M, Grillon F, Blouet P, Billon N, Piques R (2001) Polymer 42:5425
23. Grein C, Béguelin P, Plummer CJG, Kausch H-H, Tézé L, Germain Y (2000) Influence of the morphology on the impact fracture behaviour of iPP/EPR blends. In: Williams JG, Pavan A (eds) Fracture of polymers, composites and adhesives, ESIS Publication 27. Elsevier, Oxford
24. Grein C, Plummer CJG, Kausch H-H, Germain Y, Béguelin P (2002) Polymer 43:3279
25. Adams WW, Yang D, Thomas EL (1986) J Mater Sci 21:2239
26. Gensler R (1998) PhD Thesis, EPFL
27. Lin L, Argon AS (1994) J Mater Sci 29:294
28. Narisawa I, Ishakawa M (1990) Adv Polym Sci 91-92
29. Peterlin A (1977) Poly Eng Sci 17:183
30. Huang YL, Brown N (1990) J Polym Sci–Polym Phys Edn 28:2007

31. Huang Y, Brown N (1991) J Polym Sci–Polym Phys Edn 29:129
32. Plummer CJG, Cudre Mauroux N, Kausch H-H (1994) Polym Eng Sci 34:318
33. Kramer EJ (1983) Adv Polym Sci 52–53:1
34. Plummer CJG, Kausch H-H (1996) J Macromol Sci–Phys B35:637
35. Kramer EJ, Berger LL (1990) Adv Polym Sci 91–92:1
36. Donald AM J (1985) Mater Sci 20:2634
37. Plummer CJG, Donald AM (1990) Macromolecules 23:3929
38. McLeish TCB, Plummer CJG, Donald AM (1989) Polymer 30:1651
39. Dugdale DS (1960) J Mech Phys Solid 8:100
40. Stojimirovic A, Chudnovsky A (1992) Int J Fract 57:281
41. Stojimirovic A, Kadota K, Chudnovsky A (1992) J Appl Polym Sci 46:1051
42. Williams JG, Hadavinia H (2002) J Mech Phys Solid 50:809
43. Chudnovsky A, Sehanobish K, Wu S (2000) J Pressure Vessel Technol 122:152
44. Chudnovsky A, Shulkin Y (1999) Int J Fracture 97:83
45. Pandya KC, Ivankovic A, Williams JG (2000) Polym Eng Sci 40:1765
46. Pandya KC, Williams JG (2000) Plast Rubber Comp 29:439
47. Pandya KC, Ivankovic A, Williams JG (2000) Plast Rubber Comp 29:447
48. Plummer CJG, Creton C, Kalb F, Léger L (1998) Macromolecules 31:6164
49. Kalb F, Léger L, Plummer CJG, Creton C, Marcus P, Magalhaes A (2001) Macromolecules 34:2702
50. Brown HR (1991) Macromolecules 24:2752
51. Sha Y, Hui CY, Ruina A, Kramer EJ (1997) Acta Materiala 45:3555
52. Sha Y, Hui CY, Kramer EJ (1999) J Mater Sci 34:3695
53. Sha Y, Hui CY, Ruina A, Kramer EJ (1995) Macromolecules 28:2450
54. Miller P, Buckley DJ, Kramer EJ (1991) J Mater Sci 26:4445
55. Creton C, Kramer EJ, Brown HR, Hui CJ (2002) Adv Poly Sci 156:53
56. Jones RE, Bradley WL (1987) Forensic Eng 1:47
57. Kausch H-H (1987) Polymer fracture, 2nd edn. Springer, Heidelberg Berlin New York
58. Fleissner M (1998) Polym Eng Sci 38:330
59. Richard K, Diedrich G, Gaube E (1959) Kunststoffe 49:616
60. Levers PS (1998) Plast Rubber Comp Proc Appl 27:410
61. Parsons M, Stepanov EV, Hiltner A, Baer E (2000) J Mater Sci 36:5747
62. Parsons M, Stepanov EV, Hiltner A, Baer E (2000) J Mater Sci 35:2659
63. Ward AL, Lu X, Huang Y, Brown N (1991) Polymer 32:2127
64. Lagaron JM, Pastor JM, Kip BJ (1999) Polymer 40:1629
65. Nemeth P, Marosfalvi J (2001) Polym Degrad Stabil 73:245
66. Kramer E, Koppelmann J (1987) Polym Eng Sci 27:945
67. Gedde UW, Viebke J, Leijström H, Ifwarson M (1994) Polym Eng Sci 34:1773
68. Viebke J, Gedde UW (1998) Polym Eng Sci 38:1244
69. Pinter G, Lang RW (2001) Plast Rubber Comp 30:94
70. Beech SH, Mallinson JN (1998) Plast Rubber Comp Proc 27:410
71. G'Sell C, Hiver JM, Dahoun A, Souahi A (1992) J Mater Sci 27:1
72. Favier V, Giroud T, Strijko E, Hiver JM, G'Sell C, Hellinckx S, Goldberg A (2002) Polymer 43:1375
73. Plummer CJG, Goldberg A, Ghanem A (2001) Polymer 42:9551
74. Lu XC, Brown N (1997) Polymer 38:5749
75. Chan MKV, Williams JG (1983) Polymer 24:234
76. Brown N, Lu X (1995) Polymer 36:543
77. Lagaron JM, Capaccio C, Rose LJ, Kip BJ (2000) J Appl Polym Sci 77:283
78. G'Sell C, Favier V, Giroud T, Hiver JM, Goldberg A, Hellinckx S (2000) Proc 11th international conference on deformation, yield and fracture of polymers,10–18th April. Cambridge, UK, p 73
79. O'Connell PA, Bonner MJ, Duckett RA, Ward IM (1995) Polymer 36:2355
80. Duan D-M, Williams JG (1998) J Mater Sci 33:625
81. Huang Y, Kinloch AJ (1992) Polymer 33:1331
82. Lagaron JM, Dixon NM, Gerard DL, Need W, Kip BJ (1998) Macromolecules 31:5845

83. Brown N, Lu XC, Huang YL, Qian RZ (1991) Makromol Chem Makromol Symp 41:55
84. Brown N, Lu X, Huang Y, Harrison IP, Ishikawa N (1992) Plast Rubber Comp Proc Appl 17:255
85. Egan BJ, Delatycki O (1995) J Mater Sci 30:3307
86. Lu XC, Ishikawa N, Brown N (1996) J Polym Sci Part B–Polym Phys 34:1809
87. Lu X, Wang Q, Brown N (1988) J Mater Sci 23:643
88. Bubeck RA, Baker HM (1982) Polymer 23:1680
89. Channell AD, Clutton EQ (1992) Polymer 33:4108
90. Böhm LL, Enderle HF, Fleissner M (1992) Adv Mater 4:231
91. Fleissner M (1982) Angew Makromol Chem 105:167
92. Dayal U, Mathur AB, Shashikant (1996) J Appl Polym Sci 59:1223
93. Stephenne V, Daoust D, Debras G, Dupire M, Legras R, Michel J (2000) J Appl Polym Sci 82:916
94. Hubert L, David L, Seguela R, Vigier G, Degoulet C, Germain Y (2001) Polymer 42:8425
95. Lu XC, Brown N, Shaker M (1998) J Polym Sci Part B–Polym Phys 36:2349
96. Lu XC, Brown N, Shaker M (1995) J Polym Sci Part B–Polym Phys 33:153
97. G'Sell C, Dahoun A (1994) Mater Sci Eng A175:183
98. Shah A, Stepanov EV, Klein M, Hiltner A, Baer E (1998) J Mater Sci 33:3313
99. Shah A, Stepanov EV, Capaccio G, Hiltner A, Baer E (1998) J Polym Sci Part B–Polym Phys 36:2355
100. Parsons M, Stepanov EV, Hiltner A, Baer E (1999) J Mater Sci 34:3315
101. Harcup JP, Duckett RA, Ward IM (2000) Polym Eng Sci 40:635
102. Harcup JP, Duckett RA, Ward IM (2000) Polym Eng Sci 40:627
103. van der Wal A, Mulder JJ, Gaymans RJ (1998) Polymer 39:5467
104. Gensler R, Plummer CJG, Kausch H-H, Kramer E, Pauquet J-R, Zweifel H (2001) Polym Degrad Stabil 67:195
105. Lutz C (1991) PhD Thesis, Stuttgart
106. Olley RH, Bassett DC (1989) Polymer 30:399
107. Norton DR, Keller A (1985) Polymer 26:704
108. Aboulfaraj M, G'Sell C, Ulrich B, Dahoun A (1995) Polymer 36:731
109. Coulon G, Castelein G, G'Sell C (1999) Polymer 40:95
110. Pluta M, Bartczak Z, Galeski A (2000) Polymer 41:2271
111. Plummer CJG, Kausch H-H (1996) Macromol Chem Phys 197:2047
112. Grein C (2001) PhD Thesis, EPFL
113. Karger-Kocsis J, Varga J, Ehrenstein GW (1997) J Appl Polym Sci 64:2057
114. Tjong SC, Shen JS, Li RKY (1996) Polym Eng Sci 36:100
115. Ouederni M, Phillips PJ (1995) J Polym Sci Part B–Polym Phys 33:1313
116. Plummer CJG, Menu P, Cudré Mauroux N, Kausch H-H (1995) J Appl Polym Sci 55:489
117. van der Wal A, Mulder JJ, Gaymans RJ (1998) Polymer 39:5477
118. Keith HD, Padden FJJ, Walter NM, Wickhoff HW (1959) J Appl Phys 30:1485
119. Garbaraczyk J, Paukzta D (1981) Polymer 22:562
120. Huang MR, Li XG, Fang BR (1995) J Appl Polym Sci 56:1323
121. Karger-Kocsis J (1996) Polymer Bull 36:117
122. Gahleitner MW (1996) J Appl Polym Sci 61:649
123. Li JX, Cheung WL, Chan CM (1999) Polymer 40:3641
124. Chu F, Yamaoka T, Ide H, Kimura Y (1994) Polymer 35:3442
125. Riekel C, Karger-Kocsis J (1999) Polymer 40:541
126. Li JX, Cheung WL (1998) Polymer 39:6935
127. Karger-Kocsis J, Varga J (1996) J Appl Polym Sci 62:291
128. Labour T (1999) PhD Thesis, Ecole Centrale de Lyon
129. Dompas D, Groeninckx G, Isogawa M, Hasegawa T, Kadokura M (1995) Polymer Commun 36:437
130. van der Wal A, Nijhof R, Gaymans RJ (1999) Polymer 40:6031
131. van der Wal A, Gaymans RJ (1999) Polymer 40:6045
132. van der Wal A, Verheul AJJ, Gaymans RJ (1999) Polymer 40:6057

133. Ramsteiner F (1983) Kunststoffe 72:148
134. Karger-Kocsis J, Kulesnev VN (1982) Polymer 23:699
135. Vincent PI (1974) Polymer 15:111
136. Jacoby P, Bersted BH, Kissel WJ, Smith CE (1986) J Polym Sci 24:461
137. Creton C, Kramer EJ, Brown HR, Hui CY (1992) Macromolecules 25:3075–3088
138. Karger-Kocsis J (ed) (1999) Polypropylene—an A to Z reference. Kluwer, Dodrecht
139. Kukaleva N, Jollands M, Cser F, Kosior E (2000) J Appl Polym Sci 76:1011
140. Bucknall CB (1977) Toughened plastics. Appl Sci Pub
141. Jang BZ, Uhlmann DR, Van der Sande JB (1985) Polym Eng Sci 25:98
142. Jang BZ, Uhlmann DR, Van der Sande JB (1985) J Appl Polym Sci 30:2485
143. Jang BZ, Uhlmann DR, Van der Sande JB (1984) J Appl Polym Sci 29:3409
144. Jang BZ, Uhlmann DR, Van der Sande JB (1984) J Appl Polym Sci 29:4377
145. Karger-Kocsis J, Csikai I (1997) Polym Eng Sci 241
146. Chou CJ, Vijayan K, Kirby D, Hiltner A, Baer E (1988) J Mater Sci 23:2521
147. Chou CJ, Vijayan K, Kirby D, Hiltner A, Baer E (1988) J Mater Sci 23:2533
148. Lazzeri A, Bucknall CB (1995) Polymer 36:2895
149. Wu S (1990) Polym Eng Sci 30:753
150. Dukanszky B, Tudos F, Kallo A, Bodor G (1989) Polymer 30:1399
151. Ramsteiner F (1991) Acta Polymerica 42:584
152. van der Wal A, Nijhof R, Gaymans RJ (1998) Polymer 39:6781
153. Kalfoglou NK (1985) Angew Makromol Chem 129:103
154. Wei GX, Sue HJ, Chu J, Huang CY, Kong KC (2000) J Mater Sci 35:555
155. Liang JZ (2002) J Appl Polym Sci 83:1547
156. Fortelny I, Navratilova E, Kovar J (1991) Angew Makromol Chem 118:195
157. Magalhães AML, Borggreve RJM (1995) Macromolecules 28:5841
158. Plummer CJG, Béguelin P, Kausch H-H (1999) Colloids Surf A 153:551
159. Sue H-J (1992) J Mater Sci 27:3098
160. Dwyer SM, Boutini OM, Shu C (1996) Compounded polypropylene products. In: Moore EP (ed) Polypropylene handbook. Hanser, München, p 211
161. Dompas D, Groeninckx G (1994) Polymer 35:4743
162. Karger-Kocsis J (1996) Polym Eng Sci 36:203
163. Zheng W, Leng Y, Zhu X (1996) Plast Rubber Comp Proc Appl 25:490
164. Li WD, Li RKY (1997) Polym Testing 16:563
165. Ferrer-Balas D, Maspoch ML, Martinez AB, Ching E, Li RKY, Mai YW (2001) Polymer 42:2665
166. Cole PJ, Macosko CW (2000) J Plast Film Sheeting 16:213
167. Helfand E, Tagami Y (1971) J Polym Sci Part B–Polym Phys 9:741
168. Kinloch AJ (1987) Adhesion and adhesives. Chapman and Hall, New York
169. Xue YQ, Tervoort TA, Rastogi S, Laemstra PJ (2000) Macromolecules 33:7084
170. Smith GD, Plummer CJG, Bourban P-E, Månson J-AE (2001) Polymer 42:6247
171. Smith GD, Toll S, Månson J-AE (1994) Interface healing in PP in Proc Flow processes in composite materials. 7–9 July, Galway, Ireland
172. Zanetto J-E, Plummer CJG, Bourban P-E, Månson J-EA (2001) Polym Eng Sci 41:890
173. Gerard JF, Kotek J, Colbeaux A, Fenouillot F, Wautier H (2003) Abst AM Chem Soc (in press)
174. Boucher E, Folkers JP, Hervet H, Leger L, Creton C (1996) Macromolecules 29:774
175. Boucher E, Folkers JP, Creton C, Hervet H, Léger L (1997) Macromolecules 30:2102
176. Kalb F (1998) PhD Thesis, Collège de France
177. Plummer CJG, Kausch H-H, Creton C, Kalb F, Léger L (1998) Proc IUPAC world polymer congress—Macro 98. 13-17 July, Brisbane, Australia

Received: July 2003

Adv Polym Sci (2004) 169:121–150
DOI: 10.1007/b13521

Perspectives in Stabilisation of Polyolefins

S. Al-Malaika

Polymer Processing and Performance Research Unit,
School of Engineering and Applied Science, Aston University, Birmingham, B4 7ET, UK
E-mail: S.Al-Malaika@Aston.ac.uk

Abstract Polyolefins, which normally undergo high temperature manufacturing and fabrication operations, are susceptible to oxidation during each stage of their lifecycle. Stabilisers and antioxidants are used to inhibit the oxidative damage that is ultimately responsible for loss of physical properties, embrittlement and premature failure. Environmental awareness and health and safety considerations have spurred intense searches for new approaches to procure improved, safe, and more efficient antioxidants and stabiliser systems for polymers. Current activities have concentrated on two approaches: the first advocates the use of biological (naturally occurring) antioxidants, and the second relies on the use of reactive antioxidants that are chemically attached onto the polymer backbone for greater permanence and safety. Stabilisation of polyolefins through the use of vitamin E and some reactive, non-migratory antioxidants is the subject matter of this chapter.

Keywords Stabilisation · Polyolefins · Vitamin E · Reactive antioxidants

1
Introduction

Commercially produced polyethylene and polypropylene have inherently limited stability to high temperatures and UV light and require stabilisation to survive the melt processing, fabrication and end-use conditions.

Rapid expansion in the polyolefin markets, including new markets gained by the more recent commercialisation of metallocene-catalysed polyolefins, has been made possible by the parallel development of a mechanistic understanding of the chemistry of stabilisation and the development of important classes of antioxidants and 'new' efficient stabilisers for inhibiting polymer oxidation. The advent of highly efficient stabiliser systems has opened newer markets with more demanding applications of polyolefins involving harsher processing/fabrication conditions and extreme service conditions.

The terms stabilisers and antioxidants are used in the polyolefin industry to describe chemical agents that inhibit the degradative effects of oxygen, which are normally combined with other factors such as temperatures and light (e.g. melt or processing antioxidants, heat and light stabilisers or photo-antioxidants). The arbitrary use of the terms primary and secondary antioxidants is unhelpful as they do not relate to their mechanistic role. Although antioxidants and stabilisers are normally used at low concentrations (generally <1%) they are key ingredients in the compounding of polyethylene and polypropylene, which make up 60% of the total antioxidant consumption.

Fundamental understanding of the chemistry of polyolefin oxidation and the mechanisms of their stabilisation has been pivotal in achieving the current body of knowledge and the development of the vast structural diversity of stabiliser molecules that are available today. This chapter aims to project, with mechanistic perspectives, a technological overview of recent progress made in the stabilisation of polyolefins that are suitable for human-contact applications.

2
Oxidation Mechanisms for Polyolefins: An Outline

Oxidative degradation of polyethylene (PE) and polypropylene (PP) can occur at all stages of their lifecycle (polymerisation, storage, processing, fabrication and in-service). The auto-oxidation process of polyolefins is best described by the classical free-radical-initiated chain reaction outlined in Scheme 1 [1]. Impurities initially present in the polymers during polymerisation or melt processing, exert profound effects on the behaviour of the final polymer article in service.

The *initiation* reactions are still controversial and many factors can contribute to the formation of the first macro-alkyl radicals, e.g. heat, mechanical stress, light and transition metal impurities. *Propagation* reactions involve the very fast reaction of oxygen (a biradical) with polymer alkyl radi-

$$P\text{-}H + X^\bullet \longrightarrow P^\bullet + HX \qquad (1) \quad \left.\vphantom{\begin{array}{c}a\end{array}}\right\} \text{ Initiation}$$

$$P^\bullet + O_2 \longrightarrow POO^\bullet \qquad (2)$$
$$POO^\bullet + P\text{-}H \longrightarrow POOH + P^\bullet \qquad (3) \quad \left.\vphantom{\begin{array}{c}a\\b\end{array}}\right\} \text{ Propagation}$$

$$POOH \xrightarrow{\Delta,\ hv,\ M^+} PO^\bullet + {}^\bullet OH \qquad (4) \quad \left.\vphantom{\begin{array}{c}a\end{array}}\right\} \text{ Chain branching}$$

$$2POO. \longrightarrow \begin{array}{l}\text{Molecular Products} \\ \text{(ketones \& alcohols)}\end{array} \quad (5) \quad \left.\vphantom{\begin{array}{c}a\\b\\c\end{array}}\right\}$$
$$POO^\bullet + P^\bullet \longrightarrow POOP \qquad (6) \quad \text{ Termination}$$
$$2P^\bullet \longrightarrow P\text{-}P \qquad (7)$$

Scheme 1 Simplified auto-oxidation mechanism of polyolefins (PH is polymer substrate)

cals leading to the formation of macro-alkylperoxyl radicals (Scheme 1, reaction 2). This is followed by a slower (rate determining) reaction involving the abstraction of a hydrogen from another polymer molecule giving rise to the formation of macro-hydroperoxides (reaction 3), the first molecular product in the chain reaction; the rate of H-abstraction is a function of both the C-H bond dissociation energy and the stability of the final macro-alkyl radical [1]. Hydroperoxides formed in this reaction are associated with further initiation reactions (e.g. Scheme 1 reaction 4) resulting in detrimental changes to molar mass and properties of the polymer, ultimately leading to catastrophic failure, see Fig. 1. The nature of the *termination* reaction is highly dependent on the molecular structure of the polymer and the prevailing conditions. For example, under normal oxygen pressure (oxygen saturation), conditions dominating in the course of end-product use, the $POO^\bullet$ radicals are the predominant reactive species, i.e. $[POO^\bullet] > [P^\bullet]$, and termination reactions 5 and 6 (Scheme 1) predominate. However, in the presence of limited amounts of oxygen (oxygen deficiency), conditions occurring mainly during polymer processing as well as in thick samples where oxidation rate is controlled by oxygen diffusion, alkyl radicals predominate, i.e. $[P^\bullet] > [POO^\bullet]$, and bimolecular termination reactions involving $R^\bullet$ assume greater significance. This leads to polymer cross-linking and increased molar mass, and /or disproportionation without change to molar mass.

Although PE and PP oxidise by similar propagation reactions, there are some marked differences in their oxidative degradation processes. For example, propagation reaction (Scheme 1, reaction 2) is approximately 20 times faster at the tertiary carbon atom of PP than at a secondary carbon atom of PE; similarly, the propagation step (Scheme 1, reaction 3) is about six times faster at a tertiary carbon atom than at a secondary one. On the other hand,

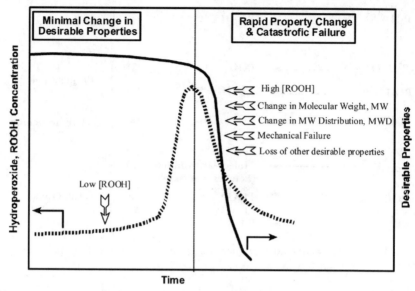

Fig. 1 Generalised scheme relating hydroperoxide concentration and loss of polymer's useful properties

the 'normal' termination reaction 5 is approximately three orders of magnitude faster in the case of secondary peroxyl radicals in PE (producing ketones and alcohols) than in the case of tertiary peroxyl radicals in PP. Hence, the kinetic chain length of auto-oxidation of PE is much shorter than that of PP (about 10 in PE and about 100 in PP) [2, 3, 4]. Scheme 2 gives an overall view of the cyclical oxidation process in polyolefins and the detrimental reactions and products that may be generated by the various propagating species.

Thermo-oxidative degradation of polyolefins is most pronounced during the high-temperature, high-shear conversion processes (e.g. extrusion, injection moulding, blow moulding, internal mixing) used to produce the final fabricated article, including further re-processing and recycling operations. The high temperatures required for these processing operations are detrimental to the stability of the macromolecular structure. The high mechanical shearing forces in an extruder, for example, lead to C–C chain scission and the formation of macro-alkyl radicals. Both alkyl and macro-alkoxyl radicals are important intermediates that are ultimately responsible for the formation of chemical impurities such as macro-hydroperoxides (a potential source of oxidative instability in the fabricated product), carbonyl groups, double bonds, and changes in the molecular weight of the macromolecule. In the case of PE, crosslinking reactions predominate under typical processing conditions (e.g. extrusion temperatures typically up to about 260 °C), see Scheme 3. The presence of low concentrations of various unsaturated groups, e.g. terminal (vinyl), in-chain (vinylidene), and pendant (trans-vinylene), as impurities play an important role in the overall oxidation and crosslinking of the polymer. Different polymerisation conditions for poly-

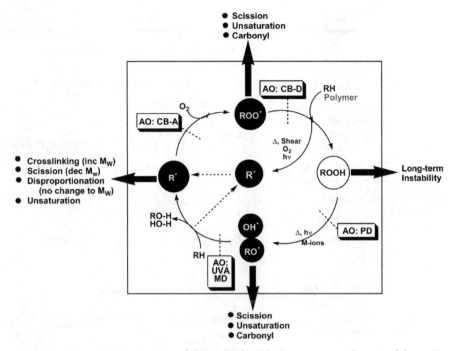

Scheme 2 Schematic presentation of the cyclical oxidation process and some of the main reactions/products formed from the propagating radicals. The antioxidant mechanisms interrupting the oxidative cycles are also shown. *AO* antioxidant, *CB-A* chain breaking acceptor, *CB-D* chain breaking donor, *PD* peroxide decomposer, *UVA* UV-absorber, *MD* metal deactivator

ethylenes result in different amounts and types of unsaturation. For example, the relatively high concentration of vinyl groups (with respect to other types of unsaturation) present in Zeigler-catalysed linear low density polyethylenes (LLDPE), see Table 1, would contribute, at least in part, to its crosslinking through addition of macro-alkyl radicals to the vinylic double bonds (see Scheme 3) [5, 6, 7, 8]. Degradation by chain scission in PE is less favourable and would become more important under more severe processing conditions, e.g. at higher extrusion temperatures, and proceeds to a large extent by β-scission of alkoxyl radicals (Scheme 3).

Polypropylene (PP), on the other hand, undergoes predominantly chain scission under all processing conditions [7, 8, 9, 10, 11] with associated reduction in the molar mass and melt viscosity (see Scheme 3). The propagation reaction (Scheme 1, reaction 3) in PP is particularly facilitated by intramolecular hydrogen abstraction leading to the formation of adjacent hydroperoxides along the polymer chain that are less stable than isolated hydroperoxides and lead to an increased rate of initiation.

Photo-oxidation of PE and PP is mainly due to slight absorption of light above 285 nm by the commercial grades due to the presence of light-absorbing impurities, particularly oxygen-containing species, and trace levels of

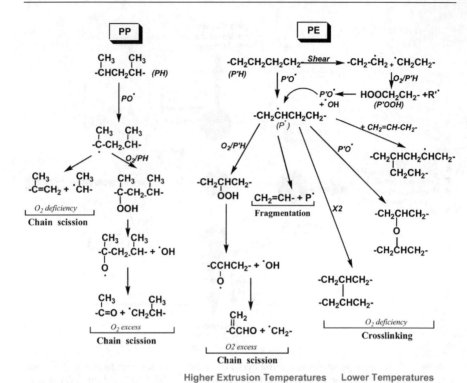

Scheme 3 Reactions of macro alkyl radicals during high temperature processing of poly-olefins [33]

metals and adventitious species arising mainly from the production process-es, i.e. polymer manufacture, melt processing and fabrication. Hydroperox-ides formed during melt processing (and to a lesser extent during manufac-turing and storage) are the most important photo-initiators during the early stages of polymer photo-oxidation [12]. There is a fundamental difference in the behaviour of hydroperoxides in PE and PP: they do not accumulate dur-ing photo-oxidation in PE, whereas the concentration of PP hydroperoxides increases steadily during photo-oxidation . This is in contrast with that ob-served in thermal oxidation where accumulation of ROOH groups occurs in both polymers). Carbonyl-containing products become more important dur-ing the later stages of photo-oxidation: for example ketonic photolysis un-dergo Norrish types I and II photo-cleavage processes resulting in backbone scission to give free radicals, which can initiate photo-oxidation (Norrish-I) or undergo rearrangement to give molecular products (inefficient photo-ini-tiators) with backbone scission (Scheme 4) [12, 13, 14, 15]. Products of PE photo-oxidations include acids, ketones, γ-lactones, esters and vinyl-alkenes, with an overall higher concentrations of vinyl groups and acids formed in PE when compared to PP.

Table 1 Characteristics of a metallocene (m) and a Ziegler (Z) octane-based LLDPE polymers

Polymer	Density	MI	MFR[#]	1-octane (from ^{13}C NMR)	Unsaturation			Total double bond
					Vinyl	t-Vinylene	Vinylidene	
	g cm^{-3}	g/10 min		W%	mol L^{-1}	mol L^{-1}	mol L^{-1}	
Z-LLDPE	0.92	1.0	36	17.1	0.25	0.06	0.08	0.39
m-LLDPE PL1840	0.91	1.0	38	13.4	0.03	0.27	0.22	0.52

[#] MFR defined as the ratio of the high load melt index to the low load melt index

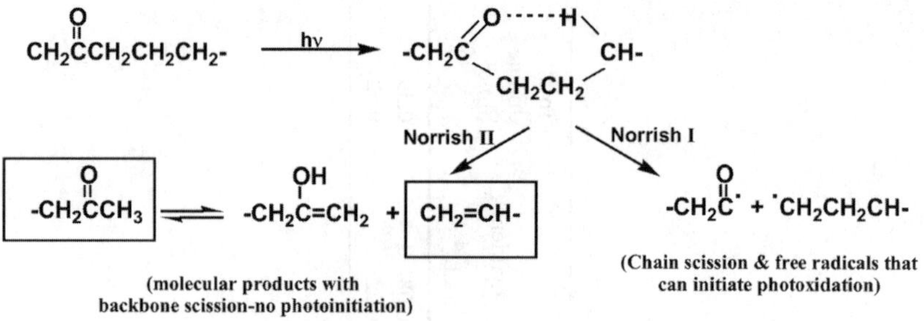

Scheme 4 Photolysis of polyethylene (the Norrish reactions)

3
Stabilisation of Polyolefins Relevant to Human-Contact Applications

The structural differences in polyolefins give rise to differences in stability towards oxygen. For example, PP is much more oxidisable than PE and cannot be processed without adequate stabilisation. Antioxidants are normally incorporated into PP and PE during high temperature processing operations to impart either melt stability (melt stabilisers) or/and to provide in-service protection against heat (long term thermal stabilisers) and UV light (UV stabilisers).

The effectiveness of stabilisers and antioxidants in real polymer applications depends on several factors: chemical, physical, toxicological, and economic. Under aggressive service conditions their performance can be dominated by physical characteristics (e.g. physical loss from polymers through volatilisation during processing and fabrication) or by exudation and solvent extraction during end-use. Physical loss is more serious in semi-crystalline polymers and in polymer artefacts with high surface-to-volume ratio, e.g. coatings, fibres, films [16, 17].

Chemical factors include structure, hydrolytic-, thermal- and photochemical stabilities. The intrinsic chemical activity of stabilisers and antioxidants is a function of their molecular structure which can be determined from studies in model substrates. Some stabilisers, notably aromatic amines and some hindered phenols, impart discoloration or yellowing to the polymer substrate during processing and fabrication or during long-term use. This is highly undesirable in plastics and is largely due to the formation of quinonoid-type transformation products of the parent antioxidant molecule [18, 19]. Aromatic amines cause extensive discoloration, hence they are generally restricted to the stabilisation of rubbers. Some stabilisers are also sensitive towards hydrolysis, especially phosphites, and to a lesser extent, phosphonites. This results not only in the loss of their antioxidant activity but also to the formation of acidic species that can initiate corrosion in processing machines [20], hence special attention is needed for storing and handling hydrolytically sensitive antioxidants and stabilisers.

Under normal conditions of melt processing, most antioxidants undergo oxidative transformations during the course of their action [21, 22, 23, 24, 25]. The overall stabilisation imparted to polymers by antioxidants, therefore, is determined not only by chemical and physical factors controlling the performance of the parent antioxidants, but also by additional contributions from their oxidative transformation products. This can either be beneficial (when transformation products are themselves antioxidants and are non-sensitising and non-discolouring) or detrimental (when products exert pro-oxidant effects or/and are discolouring) to the overall polymer stability.

In an effort to control the use of antioxidants in polymers that are intended for human-contact applications, e.g. food packaging, medical, many health authorities have sought to enforce strict regulations, with costly mandatory toxicity tests imposed by many governments for their clearance. The use of naturally occurring (biological) antioxidants or grafted (non-migratory) antioxidants is an attractive strategy for the stabilisation of polyolefins for use in such applications.

3.1
The Biological Antioxidant Vitamin E

The use of antioxidants in human-contact applications (e.g. food-contact, medical and pharmaceutical) present a challenge in terms of their safety and level of migration into the contact media, e.g. food, body fluids. As an antioxidant, vitamin E is a suitable candidate to explore for such areas of application.

Vitamin E is a fat-soluble sterically hindered phenol antioxidant which play a vital role in biological systems. The naturally occurring vitamin E is a mixture of tocopherols (I) and tocotrienoles (II) with the most bioactive form being the α-tocopherol (Ia). Tocopherols have three chiral carbons and the natural materials have 2R, 4'R, 8'R configuration with the RRR α-tocopherol being the most bioactive form of the vitamin [26]. Synthetic vitamin E, on the other hand, is a mixture of equal amounts of all eight possible optical isomers of α-tocopherol and is referred to as dl-α-tocopherol. The biological antioxidant function of vitamin E in-vivo centres around the protection of unsaturated lipids of the cells from the damaging effect of peroxidation [27, 28].

Tocol Structure (I)
α-tocopherol: R5=R7=R8=CH3 (Ia)
β-tocopherol: R5=R8=CH3; R8=H (Ib)
γ-tocopherol: R7=R8=CH3; R5=H (Ic)
δ-tocopherol: R8=CH3; R5=R7=H (Id)

Trienol Structure (II)
α-tocotrienol: R5=R7=R8=CH3 (IIa)
β-tocotrienol: R5=R8=CH3; R8=H (IIb)
γ-tocotrienol: R7=R8=CH3; R5=H (IIc)
δ-tocotrienol: R8=CH3; R5=R7=H (IId)

Synthetic hindered phenols (Ph-OH) are one of the oldest and most important class of antioxidants for polymers. They act as efficient chain breaking antioxidants by trapping alkyl peroxyl radicals and interrupting the chain propagating step, reactions 1 and 2.[18, 23, 29]:

$$RO \cdot O + Ph - OH \xrightarrow{k_{inh}} Ph - \cdot O + ROOH \tag{1}$$

$$RO \cdot O + Ph - \cdot O \rightarrow Non - radical\ oxidation\ products \tag{2}$$

Similarly, the antioxidant activity of vitamin E is centreed on its chain-breaking donor activity; in-vitro rate studies on α-tocopherol have shown that it is one of the most efficient alkylperoxyl radical traps, far better than commercial hindered phenols such as BHT, 2,6-di-*tert*.butyl-4-methylphenol. Its efficiency was attributed [30, 31] to the highly stabilised structure of tocopheroxyl radical (which is formed during the rate-limiting step, reaction 3) because of favourable overlap between the *p*-orbitals on the two oxygen atoms.

Vitamin E (α-tocopherol)   ROO$^\bullet$ CB-D   tocopheroxyl radical   + ROOH   (3)

In order to assess the suitability of vitamin E in polymer applications, we have addressed the following questions:

1. Can vitamin E withstand the high melt processing temperatures of polymers, e.g. polyolefins, and, if it does, how efficient would it be compared to the commercial norms such as Irganox 1010 (III) and Irganox 1076 (IV)?
2. What is the effect of vitamin E on polymer discoloration and, if it is a problem, what can be done to reduce it?
3. How does it function as a melt stabiliser?

Irganox 1010, III   Irganox 1076, IV

3.1.1
Effect of Polyolefin Processing Temperatures on the Melt Stabilising Activity of Vitamin E

To address the question of whether the biological antioxidant would withstand the high melt processing temperatures, dl-α-tocopherol (synthetic vitamin E) was extruded at different concentrations in PE and PP under different processing temperatures ranging from 210 °C to 285 °C. Similar experi-

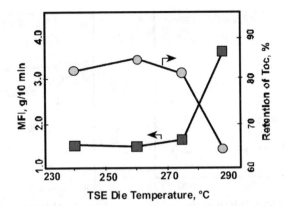

Fig. 2 Effect of extrusion temperature on melt stability of PP processed with 300 ppm tocopherol and on its level of retention in the polymer (Reproduced with kind permission of Polym Degrad Stab, 1999, 64:145)

ments were carried out with the synthetic hindered phenols Irganox 1010 and 1076 for comparison. The extrudates were cooled and pelletised and subsequently subjected to four further multi-pass extrusion processes under the same conditions. Melt flow (MF) measurements were conducted under standard conditions (using 2.16°kg, 0.118°cm die at 190 °C for PE or 230 °C in the case of PP). High pressure liquid chromatography (HPLC) was employed to determine the concentrations of additives remaining in the polymers after processing [32, 33, 34, 35].

Figure 2 shows that the stabilising performance of α-tocopherol, determined from the extent of change in melt flow, is not greatly affected by typical processing temperatures of PE and PP; a drop in stability is observed at much higher processing temperatures (around 290 °C) which is accompanied by a greater 'loss' of the antioxidant [33].

α-Tocopherol is a very effective melt processing antioxidant for both PE and PP, especially at low concentrations, with superior performance at all concentrations compared to the conventional hindered phenols, e.g. Irganox 1010 (Fig. 3) [32, 33]. In PP, for example, tocopherol may be used cost-effectively at only one-quarter of the concentration typically required for polymer stabilisation by traditional hindered phenols, such as Irganox 1010. In all the samples examined, the overall tocopherol retention level has been shown to be generally better than 80% of the original amount added; the remaining amount of ≤20% can be almost completely accounted for by various transformation products formed: dimeric, trimeric, quinonoid and aldehydic structures [34] (see Sect. 3.1.3). Examination of the melt stability of PP containing each of the identified oxidation products of tocopherol (which were synthesised [36] to match those isolated and characterised from processed tocopherol-containing polymers) have shown that these oxidation products are themselves very good processing antioxidants for PP [35]. The melt stabilising activity of tocopherol in polyolefins is, there-

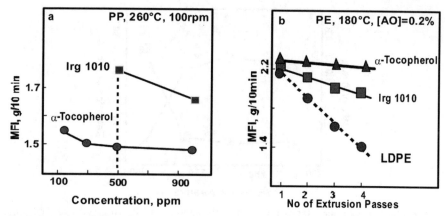

Fig. 3a,b Melt stability of **a** PP and **b** PE extruded with antioxidants (Reproduced with kind permission of Polym Degrad Stab, 1999, 64:145)

fore, attributed not only to the high rate of deactivation of radicals responsible for PP chain scission and PE crosslinking, but also to the excellent antioxidant effect of its oxidation products formed during its antioxidant function.

3.1.2
Effect of Vitamin E on Colour Stability of PP and Approaches to Minimise Polymer Discoloration

In general, sterically hindered phenols incorporated during processing of polyolefins contribute to discoloration (yellowing). This yellowing has been attributed to a number of factors including the formation of coloured oxidation products, e.g. quinonoid structures, and to interactions between the phenols and transition metal ion catalyst residues from the polymerisation stage [37, 38]. The extent of discoloration depends on the chemical structure of the parent antioxidant, its oxidation products, and the type and amount of catalyst residues in the polymer. It was shown [33] that at low concentration, both tocopherol and Irganox 1010 cause comparable levels of discoloration during melt processing of PP, but at increasingly higher concentrations, the extent of discoloration affected by tocopherol is noticeably higher (Fig. 4). The formation of certain tocopherol products during its antioxidant action are responsible for the observed discoloration (see Sect. 3.1.2).

Phosphites are known to act by a preventive mechanism, i.e. preventing the formation of initiating radicals from hydroperoxides by reducing the latter to alcohols, see reaction 4 [20]. In addition to their peroxidolytic activity (PD), sterically hindered aromatic phosphites, e.g. Ultranox U626, act also by chain breaking (CB) mechanism. These phosphites react with the propagating alkylperoxyl (ROO·), reactions 5, and alkoxyl (RO·) radicals, reactions

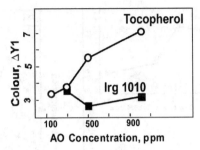

Fig. 4 Effect of hindered phenol concentration on extent of discoloration of extruded PP (Reproduced with kind permission of Polym Degrad Stab, 1999, 64:145)

6, leading to termination of the auto-oxidation chain process, reaction 7 [39]. Thus the peroxidolytic antioxidant function of phosphites (and the additional CB-activity in the case of aromatic phosphites) complements synergistically the chain breaking donor (CB-D) activity of hindered phenols to give effective melt stabilisation of PP.

$$(OR)_3P + ROOH \overset{PD}{\rightarrow} (OR)_3P=O + ROH \tag{3}$$

$$(OPh)_3P + ROO^{\cdot} \overset{CB}{\rightarrow} [ROO \cdot P(OPh)_3] \rightarrow (OPh)_3P=O + RO^{\cdot} \tag{4}$$

$$(OPh)_3P + RO^{\cdot} \overset{CB}{\rightarrow} [RO \cdot P(OPh)_3] \rightarrow (RO)(OPh)_2P + PhO^{\cdot} \tag{5}$$

$$PhO^{\cdot} + ROO^{\cdot} \rightarrow \text{Inactive Products(Chain Termination)} \tag{6}$$

Due to their peroxidolytic activity, phosphites are normally used in combination with hindered phenols in order to deactivate hydroperoxides formed as a consequence of the antioxidant function of hindered phenols, see reaction 1. The enhanced melt stability of tocopherol-containing PP extruded in the presence of a phosphite is clearly illustrated in Fig. 5a with an optimum melt stabilising performance at a 1:2 w/w ratio in favour of the phosphite.

In addition to their use in combination with other antioxidants as melt stabilisers, phosphites are also used to reduce discoloration of polymer melts [37, 39, 40, 41, 42]. Using the phosphite Ultranox U-626 with tocopherol at a 2:1 w/w ratio, respectively, has resulted in a considerable colour reduction in the polymer (Fig. 5a-inset) [33]. This has also led to a 30% reduction in the total oxidation products formed in the polymer with only three products formed: trimers, tocoquinone and aldehydes. This significant reduction in product formation results in an excellent retention of tocopherol in the polymer of over 95%. The preservation of the initial tocopherol con-

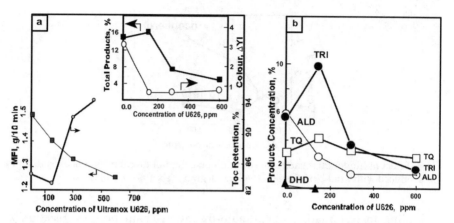

Fig. 5a,b Effect of concentration of U-626 in combinations containing 300 ppm tocopherol on **a** melt stability of PP and tocopherol retention, and **b** on colour stability of PP and total product formation (Reproduced with kind permission of Polym Degrad Stab, 1999, 64:145 and 65:143)

centration in the processed polymer is not, however, matched when lower phosphite concentration in the combination is used, e.g. at a 0.5:1 w/w ratio (i.e. 150 ppm phosphite). In spite of this apparent lack of effectiveness in enhancing the extent of tocopherol retention in the polymer, the use of such low phosphite concentration with tocopherol has resulted in a considerable reduction in colour of the polymer (Fig. 5a-inset). Quantitative analysis of the products has revealed that the use of 150 ppm U626 (in combination with 300 ppm tocopherol) leads to a major change in the distribution of the tocopherol products in favour of the less coloured products (trimers); with a large reduction in the concentration of the most discolouring product (aldehydes); see Fig. 5b and Sect. 3.1.3.

The higher levels of tocopherol retention in polymers processed with vitamin E in the presence of small concentrations of phosphites was shown [33, 43], at least in part, to be due to a regeneration of the tocopherol by one of the C–C coupling products (a hindered phenol dimer) of the phosphite (Scheme 5). It has also been suggested [4041, 42] that the role of phosphite stabilisers in reducing or preventing discoloration of polymers containing hindered phenols is due to their reactions with coloured quinonoid-type products of the parent hindered phenols, hence interrupting the conjugated pi-electron system and resulting in less coloured products. Similar reactions have also been proposed [33] to take place in the presence of the biological hindered phenol, α-tocopherol (Scheme 5).

Another approach to achieving an effective colour suppression in polymers containing tocopherol is through the use of small concentrations of non-phosphorous-containing compounds, such as polyhydric alcohols [33]. The effect of polyhydric alcohols on colour reduction shows that increasing the concentration of alcohols (e.g. trimethylolpropane, TMP) in the antioxidant combination leads to greater reduction in colour, concomitant with

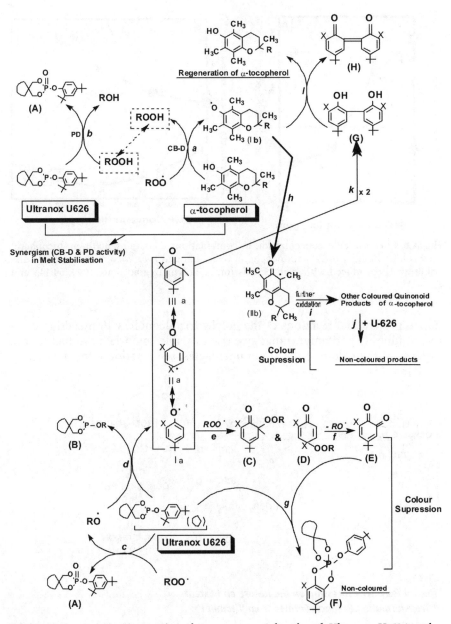

Scheme 5 Co-operative interactions between α-tocopherol and Ultranox U-626 under melt processing conditions [33]

higher levels of retention of the tocopherol in the polymer and with no adverse effect on melt stability, (see Fig. 6a) as well as a favourable product distribution resulting in lower concentrations of the highly discolouring products of tocopherol (Fig. 6b). A further cause of colour suppression may be

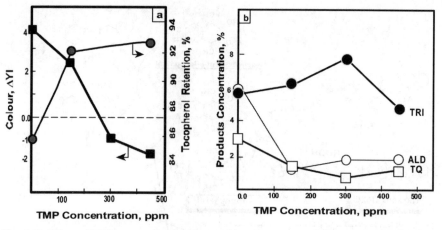

Fig. 6a,b Effect of TPM concentration in combinations containing 300 ppm tocopherol on **a** colour stability of PP and tocopherol retention, and **b** on tocopherol products distribution (Reproduced with kind permission of Polym Degrad Stab, 1999, 64:145 and 65:143)

due to preferential reactions of the polyhydric alcohols with metal ion catalyst residues (e.g. titanium) that give rise to stable colourless products at the expense of the coloured metal ion phenolates, see reaction 8 for the case of TMP [33, 43].

$$ (8) $$

3.1.3
Effect of Polymer Processing on the Nature and Extent of Transformation Products Formed from Vitamin E

The synthetic form of vitamin E exists as a mixture of eight possible optical isomers. The use of the synthetic form of the vitamin (dl-α-tocopherol) in the polymer would be expected to yield oxidative transformation products with a complex mixture of stereoisomeric structures. This was investigated by examining the oxidation of dl-α-tocopherol in solution by a mild oxidant, using HPLC and spectroscopic analysis to isolate and characterise the iso-

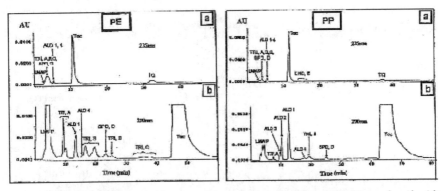

Fig. 7 HPLC chromatograms of some of Toc products in PE and PP (Reproduced with kind permission of Polymer, 2001, 42:2915)

meric structures of the main coupling oxidative products of tocopherol [36]. Results from these studies were subsequently used to confirm the identity of the different compounds corresponding to a large number of HPLC peaks isolated from processed PE and PP stabilised with the synthetic form of the vitamin [35].

Figure 7 gives an example of HPLC chromatograms of extracts of PE and PP samples stabilized with dl-α-tocopherol showing various peaks associated with some of the stereoisomeric forms of the main antioxidant transformation products [36]. However, in spite of the large number of HPLC peaks detected in tocopherol-stabilised PP and PE, it was shown that these corresponded to only small number of transformation products, each having many stereoisomeric forms. Five main products were formed in the polymers corresponding to three oxidative coupling products (trimer TRI, spirodimer SPD, dihydroxydimer DHD) in addition to tocoquinone TQ, and aldehydes ALD (with the main aldehyde being the 5-formyl-γ-tocopherol), see structures below. Thus, for the purpose of understanding the effect of processing on the nature of tocopherol-transformation products formed in polyolefins stabilized with vitamin E, the different isomers of each major product were collected together (from preparative HPLC) and analysed as a single entity. All the five products that were formed from tocopherol during melt processing of PP and PE, are more coloured than tocopherol itself, with the aldehydes being the most coloured and the trimers the least coloured [34].

Dihydroxy dimer (**DHD**)

Trimer (**TRI**)

Tocoquinone (**TQ**)

Spirodimer (**SPD**)

Aldehyde (**ALD**)

The overall effect of the initial tocopherol concentration on the product distribution in PE was somewhat similar to that observed in PP, in that increasing the tocopherol concentration resulted in the lowering of the concentrations of the trimers, aldehydes, and in only little change in the concentration of tocoquinone. The main difference found in the case of PP was the absence of the spirodimer, under all extrusion conditions examined (see e.g. Fig. 9) [34]. This difference is attributed to the greater oxidisability of PP at the higher extrusion temperatures used for this polymer (compared to PE). This is supported by the observation that the SPD, when used as an authentic (synthesised) sample in extruded PP, was much less thermally stable under PP processing conditions than the other tocopherol products such as the TRI and DHD [33].

Based on the evidence obtained from the amount and nature of transformation products formed, a mechanism of melt stabilising action of tocopherol in PP and PE has been proposed, see Scheme 6 [34]. It is well known that, like other hindered phenols, α-tocopherol is rapidly oxidised by alkylperoxyl radicals to the corresponding tocopheroxyl radical (α-Toc, Scheme 6a). Further oxidation of the tocopheroxyl radical in the polymers leads to the formation of coupled and quinonoid-type products, e.g. SPD, TRI, DHD (see Figs. 8 and 9). Dimerisation of the intermediate o-quinone methide (QM) leads to the formation of the quinonoid-type dimeric coupled product, SPD (Scheme 6 reaction d).

The trimers are one of the major oxidative coupling products formed in extruded tocopherol-containing PP and PE. Trimers may either form by direct oxidation of the SPD by alkylperoxyl radicals (Scheme 6, reaction e) or through a reaction of SPD with QM in a Diels-Alder type process (reaction f). Oxidation studies of tocopherol in model compounds provided a clear evidence [44] in support of the latter reaction. It appears reasonable to suggest, therefore, that in the case of PP the major route for the formation of trimers (having much higher stability than SPD at PP processing tempera-

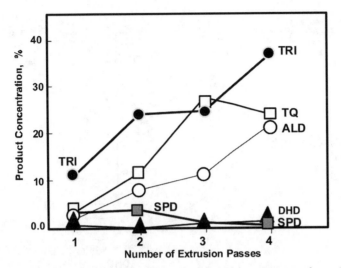

Fig. 8 Effect of multi-extrusion of PE on tocopherol transformation products; initial Toc concentration is 0.2%. (Reproduced with kind permission of Polym Degrad Stab, 1999, 65:143)

tures) in tocopherol-stabilised polymer samples is through the further reaction of the product SPD with QM.

The tocopheroxyl radical may also isomerise to the corresponding tocopheryl-benzyl radical (Scheme 6, reaction g) which can undergo a further radical–radical coupling reaction to give rise to the dihydroxydimer, DHD (reaction h). The DHD can also be formed through the reduction of TRI and/or SPD (reactions i and j) under the reducing conditions prevalent during polymer melt processing where $[R^·]>>[ROO^·]$ due to the low oxygen pressure in the closed environment of an extruder. However, the low levels of the DHD formation (compared to other products) in tocopherol-sta-

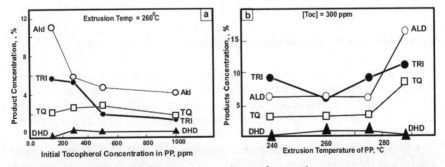

Fig. 9 Effect of initial tocopherol concentration and extrusion temperature on nature and distribution of Toc products formed in PP. (Reproduced with kind permission of Polym Degrad Stab, 1999, 65:143)

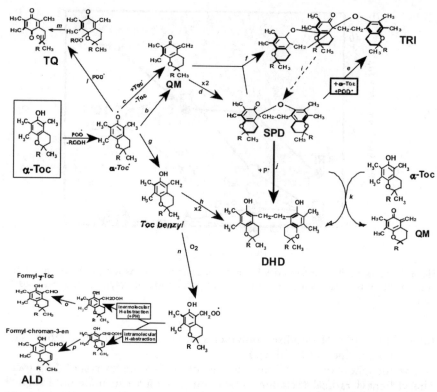

Scheme 6 Mechanism of melt stabilising action of α-tocopherol in polyolefins [34]. *TRI* trimer, *SPD* spirodimer, *DHD* dihydroxydimer, *QM* quinine methide, *ALD* aldehyde, *TQ* tocoquinone

bilised polyolefin samples (see Figs. 8 and 9) lends support to the suggestion that the DHD, which is highly stable at the temperatures used, does not play a major role in the stabilisation chemistry of α-tocopherol in PP under high temperature melt extrusion conditions.

3.2
Reactive Antioxidants as Non-Migratory Stabilisers

A strategy that is based on the use of reactive antioxidants can also be explored to achieve stabilisation of polymers suitable for human-contact applications. Reactive antioxidants which become an integral part of the macromolecular chain can result in non-migratory stabiliser systems that would be unaffected by extractive hostile contact media.

In general, reactive antioxidants are compounds that contain one or more antioxidant functions (the antioxidant, AO, component) and one or more chemical functions capable of reacting either with monomers (same or different) or with polymers (the reactive, Rv, component). The AO function is based on any antioxidant moiety, see examples A–D below, whereas the reactive moiety can either be a polymerisable (e.g. Rv functions 1–4) or non-polymerisable (e.g. Rv functions 5–7) groups and may or may not contain a spacer (an inert flexible and short chemical link connecting the antioxidant moiety to the reactive function).

$$CH_2=\overset{R}{\underset{|}{C}}-\text{(AO)} \quad ; \quad CH_2=\overset{R}{\underset{|}{C}}-\overset{O}{\overset{||}{C}}O\text{-(CH}_2)_n\text{-(AO)} \quad ; \quad CH_2=\overset{R}{\underset{|}{C}}-\overset{O}{\overset{||}{C}}NH\text{-(CH}_2)_n\text{-(AO)} \quad ; \quad CH_2=CR-\overset{O}{\overset{||}{C}}O\text{-(AO)}-\overset{O}{\overset{||}{C}}-CR=CH_2$$

Rv function   Rv function   spacer   Rv function   spacer   Rv function   Rv function

(1)   (2)   (3)   (4)

where R = H ; R = CH$_3$; n = 0 - 4

$$\text{(AO)}-O(CH_2)_n\overset{O}{\overset{||}{C}}O\text{-(CH}_2)_n\text{-SH} \quad ; \quad \text{(AO)}-O\overset{O}{\overset{||}{C}}C\text{-CH=CH-}\overset{O}{\overset{||}{C}}O\text{-(AO)} \quad ;$$

spacer   Rv function   Rv function

(5)   (6)

Rv function

(7)

Examples of AO functions: A ; B ; C ; D

Reactive antioxidants may either be copolymerised with monomers during polymer synthesis or grafted on preformed polymers; they are therefore linked to the polymer. Although the copolymerisation route has been successfully exploited [45, 46, 47, 48, 49], it has not received greater attention because of cost incurred in the synthesis and production of tailor-made 'speciality' materials for low-volume specific applications. On the other hand, grafting of antioxidants on preformed polymer melts can offer a more flexible and versatile approach where standard compounding and processing machines are used to conduct chemical grafting reactions. Both routes, however, offer tremendous advantages in terms of the physical persistence of antioxidants in the polymer. In both cases, the process of chemical attachment (target reaction) of antioxidants onto the polymer backbone proceeds in competition with other undesirable processes. The main pre-requisite here, therefore, is to achieve the target reaction without detriment to the overall polymer properties and the fabrication process.

Both thermal- and photo-antioxidant functions have been grafted onto polyolefins during melt processing in the presence of a free radical initiators [50, 51, 52, 53, 54, 55, 56, 57, 58]. The practical success of in-situ melt grafting of antioxidants on polyolefins, however, depends on the correct choice of chemical systems and processing variables that would reduce the interference of side reactions without altering significantly the polymer characteristics, e.g. molar mass, morphology and physical properties.

The three different types of reactive antioxidant systems typically used for grafting reactions on polyolefins are briefly as described below.

3.2.1
Monofunctional Polymerisable Antioxidants

The use of polymerisable monofunctional antioxidants with one reactive group per antioxidant molecule are considered here. Production of these antioxidants is generally straightforward. It can offer, therefore, a broad, versatile and economic route for the production of a range of polymer-grafted antioxidants and antioxidant concentrates. Different monofunctional antioxidants have been reactively grafted on polyolefins, e.g. PP, LDPE, HDPE, poly(4-methyl-1-ene), in the presence of free radical initiators using single- or twin-screw extruders or internal mixers.

It has been demonstrated, however, that the efficiency of chemical attachment of such monofunctional polymerisable antioxidants on polyolefins (reaction 9) is always low [50, 56]. This is due mainly to the highly competitive homopolymerisation reaction of the reactive antioxidant (reaction 10). For example, studies on the effect of processing variables on the extent of melt grafting on PP of different mono-acryloyl-containing hindered phenol (DBBA) and hindered amine (AOTP) antioxidant functions have shown that grafting efficiency is less than 50% [55, 56]. The remaining ungrafted antioxidants were recovered, almost completely, as homopolymers of the parent antioxidants which were incompatible with the host polymer and were readily removed by extractive solvents. Furthermore, the performance of these homopolymers when incorporated in the polymer matrix as conventional antioxidants is very poor. The problems of homopolymer formation and low efficiency of grafting of mono-functional polymerisable antioxidants in polyolefins were subsequently addressed by alternative approaches (see below).

3.2.2
Monofunctional Non-Polymerisable Antioxidants

Non-polymerisable monofunctional antioxidants were subsequently used to avoid the problem of homopolymerisation of the antioxidant. For example, melt grafting of the two maleated antioxidants, BPM and APM, on PP was shown to lead to high grafting efficiencies (up to 75% in the former and >90% in the latter) which were attributed to the non-polymerisable nature of the maleate (maleimide) functions [57, 59, 60]. The performance of these antioxidants, especially under extractive organic solvent conditions, was also shown to far exceed that of conventional antioxidants with similar antioxidant functions. Table 2, for example, shows the advantages of the grafted

Table 2 Thermo-oxidative stability (at 150 °C) of PP without and with 30% glass fibre (GF) in the absence and presence of antioxidants. Soxhlet extraction with chloroform and acetone [61]. UV-stability of PP films processed with BMP is also shown [57]

Antioxidant	Induction period at 150 °C (h)		Days to craze formation	
	PP films, no GF		PP plaques (1 mm) with 30% GF	
	Unextracted	Extracted	Unextracted	Extracted
g-APM	2250	2400	55	60
Irg 1010	1200	5	42	5
Irg 1076	350	5	22	3
Control (No AO)	1	1	2	2
	UV embrittlement time (h)			
g-BMP	2850	1400	-	-

HAS, BPM, and aromatic amine APM in PP samples under extractive and thermo-oxidative and photo-oxidative conditions [57, 59, 60, 61].

3.2.3
Bifunctional Polymerisable Antioxidants

The use of reactive antioxidants containing two polymerisable polymer-reactive functions in the same antioxidant molecule is outlined here. Careful choice of the processing parameters, the type, and the amount of free radical initiator can lead to very high levels of antioxidant grafting [53, 57]. For example, melt grafting of concentrates (e.g. 5–20 wt%) of the di-acrylate hin-

Table 3 Grafting efficiency of 10% concentrates (masterbatches) of bi- and mono-functional HAS antioxidants in PP. Reactions were conducted during reactive processing in the presence of 0.005 molar ratio of dicumyl peroxide (initiatior)

Antioxidant Structure	Code	% Grafting
$CH_2=CHCOO$—N(Me)(Me)(Me)(Me)—$CO-CH=CH_2$	AATP	100
$CH_2=CHCOO$—N(Me)(Me)(Me)(Me)—$CO-(CH_2)_{12}CH_3$	MyATP	32
$CH_3(CH_2)_{12}-COO$—N(Me)(Me)(Me)(Me)—$CO-CH=CH_2$	AMyTP	14
$CH_2=CHCOO$—NH(Me)(Me)(Me)(Me)	AOTP	48

dered piperidine, AATP, on PP in the presence of a peroxide initiator has led to almost 100% grafting. This exceptional grafting efficiency of AATP is in marked contrast with the much lower grafting levels achieved with the mono-functional HAS analogues, e.g. MyATP and AMyTP (see Table 3) [55]. Examination of the mechanisms involved in the grafting process of such bifunctional antioxidants has shown that the grafting reaction occurs through the intermediacy of a crosslinked structure involving the polymer and the reactive antioxidant, leading finally to an antioxidant-grafted polymer product which remains comparable in its general characteristics (e.g. solubility, crystallinity, molar mass) to a conventionally stabilised sample [55, 57].

3.2.4
Monofunctional Polymerisable Antioxidants in Presence of a Comonomer

The use of a reactive di- or poly-functional comonomer (non-antioxidant) which can co-graft with a monofunctional polymerisable antioxidant on polymers can improve the grafting efficiency from as low as 10–40% to an excess of 80–90%. This strategy, however, presents immense challenges due to the presence of more than one polymerisable group in the comonomer which could lead to additional undesirable (competing) side reactions com-

Table 4 Comparison of the antioxidant performance (accelerated UV ageing) of synergistic mixture (melt grafted in presence of Tris) with a conventional antioxidant mixture based on the same antioxidant functions (at 1:1 w/w ratio)

Antioxidant	UV embrittlement time (h)	
(0.4% in PP films)	Unextracted	Extracted
None	75	70
DBBA #	205	80
HAEB #	330	70
PP-g(DBBA-HAEB)$_{Tris}$	1160	1130
Irg 1010 (AO2)+UV531	1750	70

plicated by the possibility of comonomer-induced crosslinking reactions of the polymer. The success of this 'one-pot' synthetic approach lies in the ability to achieve a delicate balance between the composition of the chemical system (antioxidant, comonomer, free radical initiator) and reaction conditions (e.g. temperature, residence time) with the aim of promoting the target grafting reaction at the expense of all competing side reactions [52, 55, 56].

In practice, the success of this method has been clearly illustrated [52, 55, 56]. The novelty of this approach lies in the fact that co-grafting of polymerisable polyfunctional agents (traditionally used as crosslinkers, e.g. the trimethylol propane triacrylate, Tris) with mono-vinyl antioxidants (and other additives) in extruders or mixers leads to the production of highly grafted antioxidants in a non-crosslinked polymer. This co-grafting method can be applied to a wide range of antioxidant functions (e.g. HAS, UVA, hindered phenols, aromatic amines) to achieve outstanding levels of antioxidant grafting. Table 4 shows an example which illustrates the superior performance, especially under extractive conditions, of a highly bound synergistic antioxidant system (hindered phenols plus UV absorber) produced by this method in PP compared to a conventional (unbound) commercial antioxidant system.

The chemistry of grafting of monofunctional antioxidant in the presence of multifunctional comonomer (e.g. Tris) was examined both in the polymer and in model hydrocarbons at high temperatures and was found to be similar to the chemistry of grafting bifunctional antioxidants such as AATP [55, 56, 57]. A transient branched (or crosslinked) antioxidant-polymer material was shown to be formed initially and to contain graft antioxidant-Tris-copolymer tied-up in a crosslinked polyolefin (PO), see Scheme 7. Further melt mixing at high temperatures appears to initiate preferential chain scissions of the graft copolymer leading to restructuring of the crosslink and to further grafting reactions. This re-structuring process continues with further processing to yield, ultimately, a xylene-soluble polymer with highly grafted (in most cases in the order of 80–95%) antioxidant moieties. Such antioxidant systems have been shown to give superior performance under extractive conditions compared to ungrafted analogues [52].

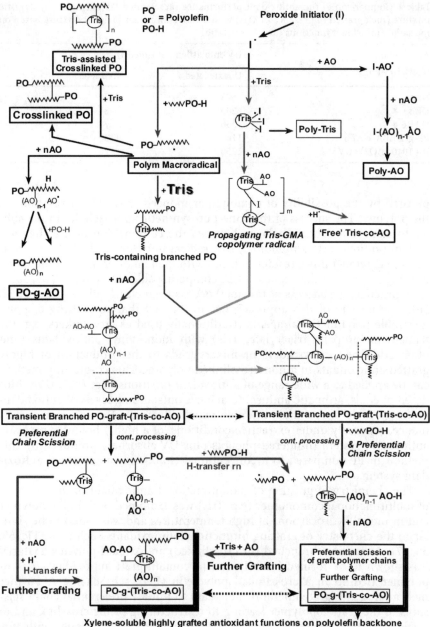

Scheme 7

4
Conclusion and Future Prospects

Stabilisers and antioxidants are essential ingredients for ensuring long-term durability of polyolefins. Stabilisers are chosen for target applications on the basis of chemical, physical, toxicological and economic factors. The final selection of an antioxidant package must take into consideration the performance requirements of the end-use polymer article including toxicity, compatibility, appearance and colour. Issues of efficiency and safety (including migration) of stabilisers has been at the forefront of recent progress made in the areas of biological, reactive, and macromolecular antioxidants.

Vitamin E was shown to have high melt stabilising efficiency in polyolefins, which can be achieved in the presence of a very small concentration of the antioxidant. Furthermore, in vivo studies have shown that the major products of tocopherol detected in animal tissues include tocoquinone, the dihydroxydimer, the spirodimer and, to a lesser extent, the trimer [62]. We have shown that similar products are also formed in PE and PP when processed with α-tocopherol [34]. However, in spite of the similarity in the nature of transformation products formed, there is one important difference between the overall role of α-tocopherol in vivo and in vitro and this is based on the effectiveness of the products in the two systems. Whereas none of the oxidation products seem to have any biological activity in vivo, our in vitro studies clearly show [33] that TRI, SPD, DHD and TQ have excellent melt stabilising activities in both PP and PE. The excellent melt stabilising activity of tocopherol in polyolefins is, therefore, due to the great capability of tocopherol and that of its major transformation products to efficiently deactivate all damaging free radicals, mainly alkylperoxyl and alkyl radicals, as outlined in Scheme 6. It was also illustrated that the problem of polymer discoloration caused by the use of higher concentrations of tocopherol can be eliminated by using certain co-additives that can interfere with the coloured tocopherol-derived quinonoid products and with polymer impurities responsible for their discoloration.

The outstanding melt stabilising efficiency of tocopherol in polyolefins, coupled with the safety aspect associated with the use of such a biological antioxidant, should improve its future position in the antioxidant market and encourage its wide-spread use in many applications, especially those related to medical and other human-contact applications.

Compared to conventional antioxidants, reactive antioxidants that are capable of becoming covalently bound to the polymer backbone have been shown to be much less readily (or not) lost from polymers during fabrication and in-service. There is now a large volume of evidence that demonstrates the performance (in terms of polymer protection) of 'immobilised' antioxidants in practice, especially when polymer products are subjected to harsh environments, e.g. exposure to high temperatures, UV light and leaching solvents. This clearly illustrates the fact that high mobility of low molar mass antioxidants is not a NECESSARY pre-requisite for achieving stabilisation, as has been previously thought, and attachment of antioxidants on polymers can be industrially exploited to advantage.

Reactive antioxidants grafted on polymer melts behave in a similar way to low molar mass conventional antioxidants but offer many additional advantages. The polymer-linked antioxidants (especially those with high grafting efficiency) do not suffer from the problem of compatibility and are non-volatile, non-migratory and hence cannot be physically lost from the polymer even under highly aggressive and extractive environments. Such antioxidant systems would be much more risk-free and environmentally friendly. The ability to produce highly grafted antioxidant concentrates (masterbatches) which can be used in conventional (the same or different) polymers as 'normal' additives extends the use of reactive antioxidants in a range of applications.

The future directions on work in the field of polymer stabilisation will be dictated by the demands for high chemical efficiency, physical retention and safety of antioxidants in polymers, especially when polymers are targeted for human-contact applications. It can be expected that the enforcement of more stringent regulations and legislations for certain applications of stabilised polymers (such as in food, toys, medicine and other health-related areas) would ensure a higher level of industrial interest and development in the use of both biological antioxidants and reactive antioxidants for producing safe and 'permanently' stabilised polymer compositions. Current emphasis on sustainable development and green chemistry should encourage greater utilisation of renewable resources and environmentally benign synthetic routes to produce new stabilisers or to replace existing ones, as well as greater drive to exploring other naturally occurring antioxidants.

References

1. Al-Malaika S (1993) In: Scott G (ed) Atmospheric oxidation and antioxidants-1. Elsevier, New York, p 45
2. Decker C, Mayo FR (1973) J Polym Sci, Polym Chem Ed 11:2847
3. Decker C, Mayo FR, Richardson H (1973) J Polym Sci, Polym Chem Ed 11:2879
4. Howard JA (1972) Adv Free Radical Chem 4:49
5. Johnston RT, Morrison EJ (1996) In: Clough RL, Billingham NC, Gillens KT (eds) Polymer durability: degradation, stabilisation and lifetime prediction, Advances in Chem Series-249. American Chemical Society, Washington, p 651
6. Al-Malaika S, Peng X, Issenhuth S (2002) Unpublished Internal Report, PPP Research Unit, Aston University
7. Al-Malaika S, Scott G (1983) In: Allen NS (ed) Degradation and stabilisation of polyolefins. Elsevier Applied Science, London, p 247
8. Stivala SS, Kimura J, Gabbay SM (1983) In: Allen NS (ed) Degradation and stabilisation of polyolefins. Elsevier Applied Science, London, p 63
9. Hinsken H, Moss S, Pauquet J-R, Zweifel H (1991) Polym Degrad Stab 34:279
10. Moss S, Zweifel H (1989) Polym Degrad Stab 25:217
11. Drake WO, Pauquet JR, Todesco RV, Zweifel H (1990) Angew Makromol Chem 176/177:215
12. Scott G (1993) In: Scott G (ed) Atmospheric oxidation and antioxidants-2. Elsevier, New York, p 385

13. Carlsson DJ, Graton A, Wiles DM (1979) In: Scott G (ed) Developments in polymer stabilisation-1. Elsevier Applied Science, London, p 219
14. Gugumus F (1990) In: Pospisil J, Klemchuk P (eds) Oxidation inhibition of organic materials-2. CRC, Boca Raton, p 29
15. Al-Malaika S, Scott G (1983) In: Allen NS (ed) Degradation and stabilisation of polyolefins. Elsevier Applied Science, London, p 283
16. Billingham NC (1990) In: Pospisil J, Klemchuk P (eds) Oxidation inhibition of organic materials-2. CRC, Boca Raton, p 249
17. Billingham NC, Calvert PD (1980) In: Grassie N (ed) Developments in polymer degradation-3. Elsevier Applied Science, London, p 139
18. Pospisil J (1979) In: Scott G (ed) Developments in Polymer Stabilisation-1. Elsevier Applied Science, London, p 1
19. Pospisil J (1984) In: Scott G (ed) Developments in polymer stabilisation-7. Elsevier Applied Science, London, p 1
20. Schwetlick K, Habicher WD (1996) In: Clough RL, Billingham NC, Gillens KT (eds) Polymer durability: degradation, stabilisation and lifetime prediction, Advances in Chem Series-249. American Chemical Society, Washington, p 349
21. Pospisil J (1990) In: Pospisil J, Klemchuk P (eds) Oxidation inhibition of organic materials-1. CRC, Boca Raton, p 33
22. Duynstee EFJ (1994) Proceeding of 6th international conference on advances in stabilisation and controlled degradation of polymers, Lucern
23. Zweifel H (1996) In: Clough RL, Billingham NC, Gillens KT (eds) Polymer Durability: degradation, stabilisation and lifetime prediction, Advances in Chem Series-249. American Chemical Society, Washington, p 375
24. Al-Malaika S (1993) In: Scott G (ed) Atmospheric oxidation and antioxidants-1. Elsevier, New York, p 161
25. Al-Malaika S, Coker M, Scott G, Smith P (1992) J App Polym Sci 44:1297
26. Diplock AT (1985) In: Diplock AT (ed) Fat soluble vitamins. Technomic, Lancaster, Pennsylvania, p 154
27. Burton GW, Joyce A, Ingold KU (1982) Lancet 2:327
28. Burton GW, Ingold KU (1983) Arch Biochem Biophys 221:281
29. Scott G (1993) In: Scott G (ed) Atmospheric oxidation and antioxidants-1. Elsevier, New York, p 121
30. Burton GW, Le Page Y, Gabe EJ, Ingold KU (1980) J Am Chem Soc 102:7791
31. Burton GW, Ingold KU (1986) Acc Chem Res 19:194
32. Al-Malaika S, Ashley H, Issenhuth S (1994) J Polym Sci, Part A: Polym Chem 32:3099
33. Al-Malaika S, Goodwin C, Issenhuth S, Burdick D (1999) Polym Degrad Stab 64:145
34. Al-Malaika S, Issenhuth S (1999) Polym Degrad Stab 65:143
35. Al-Malaika S, Issenhuth S, Burdick D (2001) Polym Degrad Stab 73:491
36. Al-Malaika S, Issenhuth S (2001) Polymer 42:2915
37. Pospisil J (1980) Adv Polym Sci 36:69
38. Kresta J E and Majer J (1969), J Appl Polym Sci, 13:1859
39. Schwetlick K (1990) In: Scott G (ed) Mechanisms of polymer degradation and stabilisation. Elsevier Applied Science, London, p 23
40. Schiers J, Pospisil J, O'Connor MJ, Bigger SW (1996) In: Clough RL, Billingham NC, Gillens KT (eds) Polymer durability: degradation, stabilisation and lifetime prediction, Advances in Chem Series-249. American Chemical Society, Washington, p 359
41. Klender GJ, Glass RD, Kolodchin W, Schell RA (1985) Antec conference proceedings. Society of Plastics Engineers, p 989
42. Klender GJ, Glass RD, Juneau MK, Kolodchin W, Schell RA (1987) Retec polyolefins V. Society of Plastics Engineers, p 225
43. Al-Malaika S, Khayat M, Pullright I, Hurley P (2001) Unpublished Internal Report, PPP Research Unit, Aston University
44. Nillson JLG (1969) Acta Pharm Suecica 6:1

45. Fu S, Gupta A, Albertsson AC, Vogl O (1985) In: Allen NS, Rabek JF (eds) New trends in the photochemistry of polymers. Elsevier Applied Science, London, p 247
46. Vogl O, Albertsson AC, Janovic Z (1985) In: Klemchuk P (ed) Polymer stabilisation and degradation, American Chemical Society Symposium Series-280. American Chemical Society, Washington, p 197
47. Bartus J, Goman P, Sustic A, Vogl O (1993) Polym prepr. American Chemical Society, Div Polym Chem 34:158
48. Sustic A, Albertsson AC, Vogl O (1987) Polym Mat Sci Eng 57:231
49. Li SJ, Bassett W Jr, Gupta A, Vogl O (1983) J Macromol Sci-Chem A20:309
50. Munteanu D (1987) In: Scott G (ed) Developments in polymer stabilisation-8. Elsevier Applied Science, London, p 179
51. Al-Malaika S (1997) In: Al-Malaika S (ed) Reactive modifiers in polymers. Blackie Academic, London, p 266
52. Scott G, Al-Malaika S (1995) US Pat 5,382,633
53. Scott G, Al-Malaika S, Ibrahim A (1990) US Pat 4,959,410
54. Scott G (1987) In: Scott G (ed) Developments in polymer stabilisation-8. Elsevier Applied Science, London, p 209
55. Al-Malaika S, Scott G, Wirjosentono B (1993) Polym Degrad Stab 40:233
56. Al-Malaika S, Suharty N (1995) Polym Degrad Stab 49:77
57. Al-Malaika S, Ibrahim AQ, Rao J, Scott G (1992) J App Polym Sci 44:1287
58. Al-Malaika S, Ibrahim AQ, Scott G (1988) Polym Degrad Stab 22:233
59. Al-Malaika S (1988) In: Benham JL, Kinstle JF (eds) Chemical reactions on polymers, American Chemical Society Symposium Series-364. American Chemical Society, Washington, p 409
60. Al-Malaika S, Sheena HH, Wirjosentono B (unpublished work)
61. Al-Malaika S, Quinn N (unpublished work)
62. Draper HH (1993) In: Scott G (ed) Atmospheric oxidation and antioxidants-3. Elsevier, New York, p271

Received: July 2003

Adv Polym Sci (2004) 169:151–176
DOI: 10.1007/b13522

Chemiluminescence as a Tool for Polyolefin Oxidation Studies

Karin Jacobson[2] · Petter Eriksson[1] · Torbjörn Reitberger[3] · Bengt Stenberg[1]

[1] Department of Fibre and Polymer Technology, Royal Institute of Technology, Stockholm, Sweden
E-mail: pettere@polymer.kth.se
E-mail: stenberg@polymer.kth.se
[2] Swedish Corrosion Institute, Stockholm, Sweden
E-mail: karin.jacobson@corr-institute.se
[3] Department of Chemistry, Nuclear Chemistry, Royal Institute of Technology, Stockholm, Sweden
E-mail: torbreit@nuchem.kth.se

Abstract The oxidation of polymers such as polypropylene and polyethylene is accompanied by weak chemiluminescence. The development of sensitive photon counting systems has made it comparatively easy to measure faint light emissions and polymer chemilumines- cence has become an important method to follow the initial stages in the oxidative degrada- tion of polymers. Alternatively, chemiluminescence is used to determine the amount of hy- droperoxides accumulated in a pre-oxidised polymer. Chemiluminescence has also been ap- plied to study how irradiation or mechanical stress affects the rate of polymer oxidation. In recent years, imaging chemiluminescence has been established as a most valuable technique offering both spatial and temporal resolution of oxidation in polymers. This technique has disclosed that oxidation in polyolefins is non-uniformly distributed and proceeds by spreading.

This review is the result of several investigations performed by the authors and other re- search groups where chemiluminescence has been used to study oxidative degradation of polyolefins, either as the main technique or as a complement to other techniques.

Keywords Chemiluminescence · Imaging chemiluminescence · Polyethylene · Polypropylene · Oxidation · PP · PE

1
Introduction

1.1
Luminescence Phenomena

Excess energy of an excited substance is usually lost as heat, through vibrations and by collisions with surrounding molecules, but occasionally it is emitted as radiation. When this radiation is emitted in the visible light region, it is referred to as luminescence. The mode of excitation can vary and luminescence phenomena are often named after the source of excitation, e.g. electroluminescence from electric energy excitation and sonoluminescence from excitation by sound waves. Fluorescence and phosphorescence, which are both due to excitation by light, are widely used as analytical tools in polymer science [1]. Chemiluminescence was first observed in biological specie such as the fire fly, luminescent bacteria and marine organisms. This type of luminescence is now termed bioluminescence and is a phenomenon that has been known and studied for a long time and has been thoroughly reviewed by Brolin and Wettermark [2]. The luminescence phenomena of main interest in this chapter are chemiluminescence, originating from chemical reactions, thermoluminescence, originating from thermal excitation and triboluminescence originating from mechanical excitation.

1.2
Chemiluminescence

The history of chemiluminescence from polymers is relatively short. In 1961 Ashby [3] reported that light was emitted from polymers heated in air. Because of the relationship between light and oxidation the term 'oxyluminescence' was used. Ashby saw the possibility of using chemiluminescence as a tool for testing the stability of polymers and studied the role of oxygen concentration and the effect of antioxidants. A few years later, Schard and

Russell [4] continued these studies and made proposals concerning the origin of the emitted light.

The development of very sensitive photon counting systems and a better understanding of the phenomena have now made it possible to use the chemiluminescence (CL) technique as a valuable method of establishing degrees of degradation and of studying degradation mechanisms. In their review Matisová-Rychlá and Rychly [5] summarise the variety of possible chemical and physical pathways leading to the appearance of light from thermal oxidation of polymers.

Experiments with filters have shown that CL emission from polymers is in the blue-violet region (~400–500 nm). The low intensity of emission indicates a 'forbidden' transition from a triplet to a singlet state, i.e. phosphorescence. The emission spectra of CL often agree with that of carbonyl cromophores [6, 7, 8]. Deactivation of an excited carbonyl is generally believed to be the source of chemiluminescence from polymers (see Scheme 1).

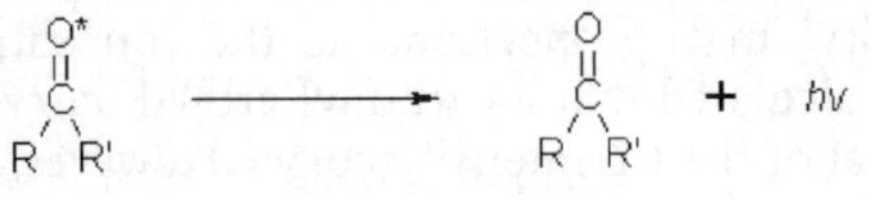

Scheme 1

Several mechanisms have been suggested to produce the energy required to populate an excited carbonyl, which is at least 290–340 kJ mol^{-1} [8]. Direct homolysis of hydroperoxides [9, 10], disproportion of alkoxy radicals [11] and β-scission of alkoxy radicals [12] are all exothermic enough. However, the most widely accepted mechanism has been the highly exothermic (460 kJ mol^{-1}) bimolecular termination of primary or secondary alkyl peroxyl radicals, i.e. the Russell mechanism (Scheme 2). It proceeds via an intermediate tetroxide to give an excited carbonyl, an alcohol, and oxygen [13, 14].

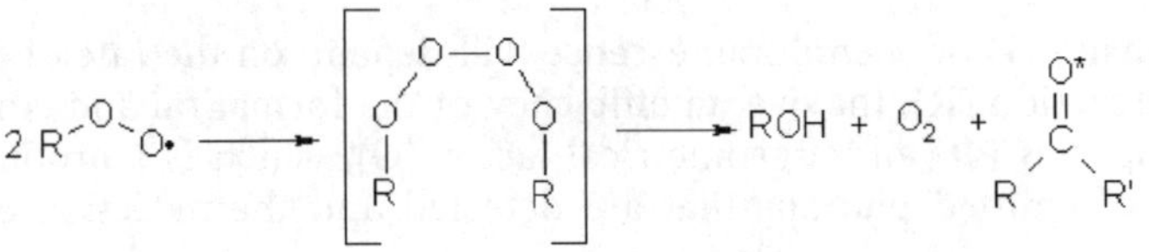

Scheme 2

Polypropylene, in which tertiary radicals predominate, nevertheless gives CL. This has been an argument against the validity of the Russell mechanism, which requires at least one of the peroxy radicals to be primary or secondary. However, Mayo and co-workers [15, 16] showed that termination reactions are accompanied by production of alkoxy radicals that will cleave to produce, ultimately, primary and secondary alkyl peroxy radicals. These in their turn could react with tertiary peroxy radicals via the Russell mechanism and thus give CL.

George [8] suggests that there may be several light emission processes occurring in the polymer and Lacey and Dudler [17] suggest that the identity of the CL emitting species changes with oxidation time.

Results by Achimsky et al. [18] showed a linear relationship between the maximum chemiluminescence intensity and the oxygen concentration. The conclusion was drawn that the light emission comes from decomposition of hydroperoxides rather than from the Russell mechanism shown in Scheme 2. Matisová-Rychlá and Rychlý have recently published several papers in favour for decomposition of hydroperoxides as being responsible for the light emission during oxidation of PP [19, 20, 21].

Blakey et al. have recently critically examined the mechanism responsible for the light emission in PP [22]. They showed that kinetically, termination of two peroxy radicals (Russell mechanism) and bimolecular, as well as mono molecular, decomposition of hydroperoxides would yield a chemiluminescence intensity curve proportional to the concentration of hydroperoxides. They also calculated that a carbonyl growth curve would be proportional to the integral of the CL intensity curve. However, when they simultaneously monitored the chemiluminescence and the FTIR emission spectra for oxidising unstabilised PP they instead found that the CL–time curve was proportional to a carbonyl growth–time curve. It was therefore suggested that neither the classical Russell mechanism nor the decomposition of hydroperoxides were responsible for the light emission in oxidising PP. The light emission was instead suggested to occur via a chemically induced electron exchange luminescence (CIEEL) mechanism. In a later paper further results were presented in favor of the CIEEL mechanism as being responsible for the light emission during oxidation of PP [23]. The mechanism responsible for the light emission is most likely to remain a matter of discussion.

1.2.1
Chemiluminescence in Oxidative Atmosphere

The intensity (I) of chemiluminescence will depend on the rate of the luminescent reaction (R), the overall efficiency of the formation and emission of excited species (Φ) and a geometrical factor (G), which is a product of the fraction of emitted photons that are detected and the detection efficiency [24]:

$$I = G\Phi R$$

Φ is thus the probability that the termination reaction leads to the emission of a quantum from the excited carbonyl oxidation product. Φ is typically 10^{-9} for deactivation of an excited carbonyl but is not necessarily constant throughout the measurement due to, for example, yellowing.

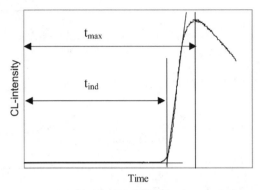

Fig. 1 Schematic CL curve of a stabilised sample of polypropylene oxidised in an oxidative atmosphere. The induction time, t_{ind} and time to maximum, t_{max} are indicated in the figure

The intensity of chemiluminescence generated upon degradation of a polymer has been found to be enhanced by various degradative stimuli, e.g. UV light, heat and stress. The spectral distribution in all cases was found to be similar. This suggests that the overall mechanism of the reaction is similar and that only the type and relative proportions of the initiation and propagation steps are different due to the various stimuli that bring about the degradation [8, 25]. A schematic CL curve of a stabilised polypropylene sample is shown in Fig. 1. The curve shows an induction time, t_{ind} during which the stabiliser is consumed. An auto-accelerating phase that reaches a maximum at t_{max}, after which the CL signal decreases. The mechanical properties of a stabilised polymer are generally believed to be lost at the end of the induction period. An unstabilised polypropylene sample would generate a similar CL curve without an induction time.

1.2.2
Chemiluminescence in Inert Atmosphere

In an inert atmosphere, no further oxidation of the polymer is possible but it was early observed that polymers heated in an inert atmosphere still emit light [6, 24, 26, 27, 28]. Upon heating a polymer specimen in an inert atmosphere, the CL intensity will increase up to a maximum and then decrease down to the level of the background noise. If the specimen is cooled down, still in an inert atmosphere, and then re-heated, only a very low signal can be detected (see Fig. 2).

Since the CL emission decreases upon treatment of polypropylene with peroxide-destroying agents, such as sulphur dioxide, it was concluded that the emission originated from hydroperoxides present in the material [28]. The area under the CL emission peak is denoted total luminescence intensity (TLI) and has been found to be proportional to the hydroperoxide concentration in the early stages of the oxidation of polypropylene [28].

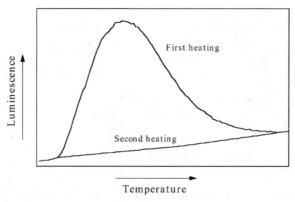

Fig. 2 A schematic CL curve resulting from the first and second heating cycles in an inert atmosphere

Other explanations of the origin of luminescence from polypropylene heated in an inert atmosphere have also been suggested [6, 26].

1.3
Chemiluminescence from Polyolefins

Polypropylene is much more sensitive to oxidative degradation than polyethylene. This is due to the tertiary hydrogens present in polypropylene. Polypropylene would not even be commercially available if it had not been for antioxidants, since it cannot be processed without stabilisers. Polyethylene, on the other hand, is sometimes sold more or less unstabilised. Compared to PP, PE contains a much lower concentration of hydroperoxides after oxidation under identical conditions. Instead, a larger fraction of the absorbed oxygen is incorporated as carbonyl groups in PE than in PP [29]. Moreover, in contrast to the hydroperoxides in PP which tend to occur in sequences the PE hydroperoxides are generally isolated from each other. These two differences in oxidation features are due to a lower reactivity of peroxyl radicals towards secondary hydrogens, and to a faster termination of secondary than tertiary peroxyl radicals [30]. In addition to being less oxidatively stable, polypropylene also gives off more chemiluminescence than polyethylene. The reason for this is not completely established. The oxidative degradation and CL emission from polyethylene is also much more complex than that of polypropylene [31]. Polypropylene is, partly due to this, by far the most studied polymer in relation to oxidation and CL but some studies where CL has been used to study polyethylene degradation are presented later in this chapter. In addition to these studies there are a number of other authors that have used CL to study polyethylene oxidation [32, 33, 34, 35, 36, 37, 38]. One of the main differences found between the CL curves from PP and PE is the double sigmoidal shape of the isothermal CL curve of PE. This is further discussed in Sect. 2.4.3.

1.4
Thermoluminescence

Thermoluminescence (TL) (also referred to as radiothermoluminescence or charge recombination luminescence) is the thermally stimulated emission of light that follows previous absorption of energy from radiation. It is distinguished from photoluminescence phenomena, such as fluorescence and phosphorescence, by its ability to emit light upon heating long after irradiation [1, 39, 40, 41, 42, 43]. Kron et al. [44] have shown that for gamma and beta irradiation of HDPE and LLDPE this time can be as long as several years. The probable source of thermoluminescence in organic solids is the recombination of trapped charges. Upon irradiation of polymers not all electrons ejected from their parent molecules recombine. Instead some of them are trapped either physically or chemically. A physically trapped electron is localised in cavities or voids created by imperfections in the structure of the polymer, such as crystal imperfections in semi-crystalline polymers. In a chemical trap the electron is bound to some particular molecule and resides within the characteristic molecular orbitals of that molecule [41]. The irradiation is usually carried out at a low temperature, and upon subsequent heating the electrons are released from their traps due to molecular motions. α, β and γ transitions would be expected to break up electron traps and thus give rise to luminescence. A number of authors have used TL to study the physical properties of polymers [39, 41, 45, 46, 47].

1.5
Triboluminescence

The first report of triboluminescence in the literature dates back to Francis Bacon's *The Advancement of Learning* from 1605 [48]. Bacon found that lumps of sugar emitted light when scraped. Triboluminescence takes its name from Greek *tribein* 'to rub' and is defined as 'the emission of light caused by the application of mechanical energy to a solid', i.e. triboluminescence is produced by direct mechanical excitation, e.g. crushing of crystals or tearing open gummed envelopes. Since the definition of the phenomenon is very broad, mechanical methods and mechanisms of excitation can be very different. Of the three luminescence phenomena described here triboluminescence is the least studied and the least understood. This is especially true for polymers. In his review on triboluminescence, Zink [49] states that the borderline between triboluminescence, thermoluminescence and chemiluminescence is often fuzzy. For example, the mechanical energy applied to a crystal in a triboluminescence experiment could be converted to heat, which in its turn could cause either thermoluminescence or, through a chemical reaction, chemiluminescence. A number of authors have discussed the occurrence of triboluminescence during CL measurements of stressed specimens [50, 51, 52].

2
Use of Chemiluminescence in Oxidation Studies

Chemiluminescence technique has been used extensively at the department of Fibre and Polymer Technology at The Royal Institute of Technology (KTH) in Stockholm. Since the beginning of the 1990s when the first CL equipment was introduced at the department more than 31 papers [44, 53, 54, 55, 56, 57, 58, 59, 60, 61, 62, 63, 64, 65, 67, 68, 69, 70, 71, 72, 73, 74, 75, 76, 77, 78, 79, 80, 81, 82, 83]and eight doctoral theses [84, 85, 86, 87, 88, 89, 9091] have been published where the chemiluminescence technique has been used, either as the main analytical technique or as a complement to other techniques. In this section, results from some of the investigations performed on polyolefins will be summarised.

2.1
Chemiluminescence Used to Investigate Peroxides

The formation and decomposition of peroxides are key reactions in the oxidation process. The peroxides are primary oxidation products which decompose to form new radicals. It is desirable to know the identities and reactivities of the different types of peroxides formed in order to understand and predict the oxidation process. However, it is not trivial to distinguish between different types of peroxides or to estimate their respective concentrations. Techniques for measuring hydroperoxides have been reviewed by Carlsson et al. [92] and more recently by Scheirs et al. [30].

As mentioned in the introduction, hydroperoxides can be measured by recording the area under the CL curve in an inert atmosphere i.e. the total luminous intensity (TLI). Kron et al. found that when measuring CL in inert atmosphere together with peroxide concentration, as measured by iodometry, for oxidised polypropylene, proportional relationships were obtained when the TLI was plotted versus peroxide concentration (see Fig. 3) [60]. In addition, changes in melting temperature and polydispersity index with aging time have also been found to correlate with changes in the TLI [59].

In the literature it has been stated that exposure to dimethylsulfide (DMS) can be used as a tool to distinguish between different types of polypropylene (PP) peroxides [93]. The reasons for the difference in PP peroxide reactivity towards DMS was further investigated by us. This was accomplished using iodometry, infrared spectroscopy, measurements of total emitted chemiluminescence, chemiluminescence imaging and measurements of glass transition and melting temperatures [59]. On the basis of these results, it was proposed that DMS reacts preferentially with peroxides that are adjacent to each other, rather than with isolated peroxides. The higher reactivity of adjacent PP peroxides towards DMS is believed to be due to a catalytic action of protic species. A correlation was also found between the concentration of DMS-resistant peroxides and both the glass transition temperature (T_g) and the melting temperature (T_m). This indicates that the reactivity towards DMS

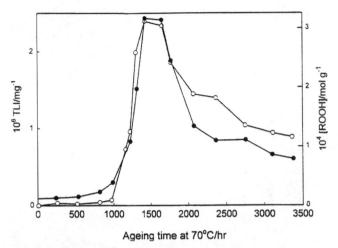

Fig. 3 Concentration of peroxides (*open circles*) and total luminous intensity, TLI, (*filled circles*) of PP powder aged at 70 °C for different periods of time

between different peroxides is affected by the segmental mobility. As a result of the oxidation, the segmental mobility was found to decrease, whereas the concentration of DMS-resistant peroxides increased. We propose that the isolated peroxides can be hindered from bimolecularly decomposing and from being catalysed in their reaction with DMS by the restricted mobility in the material.

2.2
Chemiluminescence to Investigate Biodegradable Polymers

In some applications degradation can be desirable, such as in biomedical, hygienic and packaging products. In order to enhance degradation of poly-olefins, a pro-oxidant system can be added which decreases the overall oxidative stability. The chemiluminescence technique has proven to be a very suitable method to follow degradation in such systems. In 1992 an investigation was made where chemiluminescence was used to detect hydroperoxides during oxidation of low density polyethylene (LDPE) and LDPE filled with starch or different loadings of a pro-oxidant [53]. The pro-oxidant consisted of corn-starch, linear low density polyethylene (LLDPE), styrene-butadiene copolymer (SBS) and manganese stearate. The unsaturations in the SBS were believed to generate hydroperoxides. The decomposition of these hydroperoxides was then catalysed by the manganese salt, which resulted in radicals that eventually induced oxidation in the LDPE matrix and thereby enhanced degradation. The influence of different bacteria or fungi in the surrounding media on the degradation of these systems was also investigated using chemiluminescence [58]. Recently Khabbaz et al. evaluated different test methods for analysing LDPE containing two different pro-oxidant systems. The

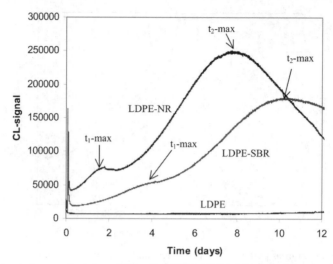

Fig. 4 CL intensity as a function of time for LDPE, LDPE-SBR and LDPE-NR. Isothermal measurements made at 100 °C in air. Figure from [81]. Copyright (Wiley 2001). Reprinted by permission of Wiley, New York

two pro-oxidant systems consisted of manganese stearate mixed with styrene-butadiene copolymer (SBS) or natural rubber (NR). Chemiluminescence and thermogravimetry were found to be the most effective techniques for detecting differences in the pro-oxidative effect between the systems when mixed into LDPE. Figure 4 shows CL measurements of pure LDPE and LDPE with pro-oxidant based on SBS (LDPE-SBS) or NR (LDPE-NR).

The measurements were performed in air at 100 °C. As can be seen, the oxidation of LDPE-SBR and LDPE-NR takes place in two stages, the first stage most likely corresponds to the oxidation of the pro-oxidant system and the second to the oxidation of the LDPE matrix. The radicals generated during degradation of the pro-oxidant system considerably enhance degradation in the LDPE matrix since the pure LDPE only shows a very small increase in CL signal during the experiment.

2.3
Luminescence from γ and β Irradiated HDPE and LLDPE

Ionising irradiation is used in graft polymerisation, for making heat-shrinkable products and to crosslink polymers in the cable and wire industry. It is also widely used for sterilisation of medical polymeric commodities [94]. The two types of ionising radiation used in industrial irradiation of polymers are fast electrons from accelerators and gamma-rays, the latter are most often from a cobalt-60 radioisotope source. Ionising radiation has sufficient energy to eject valence electrons from molecules in the absorbing material. The subsequent recombination of electrons and ions in polymers leads to the formation of excited molecules, and eventually to free radicals. The free radicals are initiation sites for the subsequent reactions in the polymer. Depending on the chemical structure and morphology of the irradiated polymer, the free radicals formed can cause reactions leading to crosslinking or backbone scission [43]. If oxygen is present during the irradiation, it reacts readily with formed carbon-centred radicals to give peroxyl radicals, thereby suppressing the extent of crosslinking reactions.

A number of research groups have shown that the effect of ionising radiation critically depends on the dose rate when irradiations are carried out in the presence of oxygen [95, 96, 97]. A high dose rate produces a large number of radicals per unit time and quickly causes a depletion of oxygen in the material. Due to the relatively low rate of oxygen diffusion, the radicals formed later react under essentially inert conditions to form crosslinks or by disproportionation. For low dose rate radiation, the oxygen diffusion is sufficiently rapid to maintain a certain concentration of oxygen in the sample. In this case, the newly formed radicals will react with oxygen to a larger extent than under high dose rate irradiation. Therefore, for a given dose, a low dose rate causes more oxidative degradation of a polymer than does a high dose rate. The oxidative degradation is a cyclic chain of radical reactions and propagates in the presence of oxygen. It leads to undesirable physical effects like embrittlement and discoloration of the polymeric material. Oxidation of PE during and after ionising irradiation has been studied by several research groups [95, 96, 97, 98, 99].

The luminescence of polyethylene (HDPE and LLDPE) induced by ionising irradiation has been investigated by Kron et al. [44]. The effects of filling materials, irradiation doses and dose rates on the luminescence were investigated. For freshly irradiated samples, both unfilled and filled, the luminescence was concluded to predominantly consist of thermoluminescence (TL). Upon subsequent aging, the contribution of TL was found to decrease due to its decay with time after irradiation. The light emission was found to increase in the later stages of the ageing, for irradiated and also for un-irradiated samples. This light was interpreted as chemiluminescence (CL), emitted as a result of polymer oxidation. The maximum of the CL was higher for irradiated than un-irradiated PE, indicating that the ionising radiation gave rise to an increased amount of oxidation products. Thus, CL has been shown to be a feasible technique in studies of radiation-induced oxidation.

A diagnostic three-step luminescence measurement was designed to compare oxidation and crosslinking in a series of irradiated samples [86]. The validity of this measurement was tested by applying it to two series of samples. Established techniques, such as tensile testing, gel permeation chromatography, IR spectroscopy and thermal analysis, were used to corroborate the conclusions drawn from the three-step luminescence measurement.

2.4
Chemiluminescence and Mechanical Stress

There has been a prevailing theory that oxidative degradation is accelerated by mechanical stress [100]. This theory is based on fracture kinetic work by Tobolsky and Eyring [101], Bueche [102, 103, 104], and Zhurkov and co-workers [105, 106, 107]. Their work resulted in an Arrhenius-type expression [108] sometimes referred to as the Zhurkov equation. This expression caused Zhurkov to claim that the first stage in the microprocess of polymer fracture is the deformation of interatomic bonds reducing the energy needed for atomic bond scission to $U=U_0-\gamma\sigma$, where U_0 is the activation energy for scission of an interatomic bond, γ is a structure sensitive parameter and σ is the stress.

Since chemiluminescence is a very sensitive method of studying oxidative degradation, it has been used to measure the effect of stress on oxidation of polymers, i.e. stress-induced chemiluminescence (SCL). SCL is by definition a type of triboluminescence, and it is likely that SCL and other forms of tri-boluminescence can occur at the same time. SCL is, however, the only type of tribo-induced luminescence that is oxygen dependent and can therefore be sorted out by measurements in inert and oxidative atmospheres.

One of the earliest reports of SCL considered elastomers and was published by Butyagin and co-workers in 1970 [50]. They measured the total luminescence from films which had been deformed by air pressure. In 1977 Mendenhall [109] recognised that the accelerating effect of stress on polymer-aging reactions could be studied by using CL, and in 1979 Fanter and Levy [51] were among the first to design a special system for studying SCL. They used a miniature tensile stress device, a controlled environment chamber and a luminescence detection system with associated optics, integrated into a single unit. The tensile unit could exert forces up to 800 N. The materials studied were an epoxy system and polyamide 66. Since then Monaco et al. [52] and George et al. [25, 27] have measured stress-induced luminescence from epoxies and polyamide 66, respectively. The equipment used by Monaco et al. was of the same general design as Fanter and Levy's but George et al. used a different approach by attaching a photo multiplier tube to an Instron tensile testing machine. Dickinson et al. [110] have measured the photon emission as well as the electron emission from polycarbonate samples and Hosoda et al. [111] have used an imaging chemiluminescence (ICL) technique to study oxidation of stressed samples. In order to thoroughly investigate further how chemiluminescence could be used for studying stressed polymers, the authors constructed a special SCL instrument [73].

2.4.1
Chemiluminescence during Tensile Testing

The first phenomenon to be investigated was the light which is emitted during tensile testing of injection moulded PP and polyamide 66 specimens

[72]. This luminescence was shown to be from all three origins of light, i.e. chemiluminescence, thermoluminescence and triboluminescence. It could also be a superposition of two or all three phenomena. The chemiluminescence was shown to originate mainly from hydroperoxide decomposition due to stress-induced adiabatic heating, which after necking is localised in the travelling neck shoulders. This CL was thus not an effect of direct stress. However, some CL was also found to be due to radical formation at fracture, i.e. directly caused by stress, as discussed below. The thermoluminescence proved to be a result of radiation from fluorescent light tubes and was negligible if the specimens were kept in the dark prior to a measurement. The triboluminescence was found to be emitted from irradiated specimens in which stress helped to recombine separated charges.

2.4.2
Chemiluminescence from Polymers Oxidising under Stress

From the previously described investigation it was clear that the light emitted after the yield point was not due to the stress in itself. It was now time to answer a much more important question: How do sub-yield stresses influence the oxidative degradation of polymers? In order to attain significant oxidation within a reasonable time, the process was accelerated by an elevated temperature. The materials studied were injection moulded parts of high density polyethylene, polypropylene and polyamide. Polyamide was also studied as film, in which there is no restricted oxygen diffusion. SCL was used to study the oxidation in situ of specimens exposed to loads between 0% and 100% of load at yield. In addition, the TLI as measured by conventional CL, the carbonyl index as measured by FTIR, the elongation at break, the UV absorption and the mass crystallinity as studied by DSC were investigated from creep-aged specimens. Imaging chemiluminescence (ICL) was also used to monitor the spatial distribution of oxidation across a stress profile in pre-aged specimens. It was shown in all these measurements that stress did not change the course of oxidation in any of the materials [74, 78].

2.4.3
Chemiluminescence Used to Study Cutting of UHMWPE

Ultra-high molecular weight polyethylene (UHMWPE) has been used in orthopaedic prosthetic surgery for many years due to its excellent mechanical properties and frictional resistance. A large number of studies on both retrieved prostheses and raw material have, however, been necessary in order to understand and prevent degradation of the prostheses. The shape of the prostheses and the compression moulded blocks from which they are cut is usually not suitable for examination. In a number of studies microtomy has therefore been used in order to produce pieces suitable for further studies [112, 113, 114, 115, 116, 117, 118, 119, 120]. However, very often when microtomy is used, it is without any consideration of the fact that the process

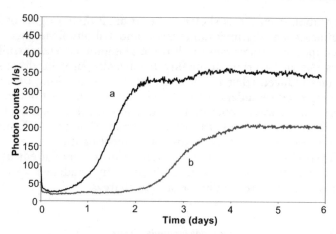

Fig. 5 CL from UHMWPE oxidising in air at 120 °C. Curve *a* represents an untreated sample and *b* a sample that has been treated with SO₂ to decompose hydroperoxides prior to the experiment

can affect the properties of the material to be studied. For example, both orientation and crystallinity have been found to be affected by microtoming [121]. It has also been found that the rate of oxidation, as measured by growth of the carbonyl band in FTIR spectra of oven aged microtomed UHMWPE films, was dependent on the velocity with which they were microtomed [55]. The lower the cutting velocity, the faster was the oxidation. When performing real time chemiluminescence measurements in air at 120 °C an additional peak was observed during the induction period (see curve a in Fig. 5). A couple of different explanations to this double CL peak observed for PE, can be found in the literature. Zlatkevich [41] proposed that it reflected two kinds of peroxides with different stability and Setnescu et al. [122] interpreted it as being a result of oxidation of very reactive sites and oxidative crystalline-phase conversion, respectively. Broska and Rychlý [31] investigated the double stage oxidation of polyethylene using chemiluminescence. Based on isothermal experiments with different oxygen concentrations they presented the theory that the double sigmoidal CL–time curve was caused by oxygen starvation of the sample. The results were supported by FTIR measurements and determination of gel content, which also revealed that cross linking only took place during the second oxidation step.

When examining the first peak (see curve a in Fig. 5) it was found that it was proportional to the hydroperoxide contents as measured by FTIR, after NO derivatisation of un-aged films. When the sample was treated with SO₂, to decompose the hydroperoxides, the peak disappeared as shown (see curve b in Fig. 5). The CL curve of the SO₂-treated sample has the typical sigmoidal appearance, just like an oxidising polypropylene sample. It was thus concluded that the first peak in the CL curve of the oxidising polyethylene in this study is from hydroperoxides already present in the material before the measurement starts. The second peak is the auto-oxidation. It might be noteworthy that the measurements were performed below the melting point and that the UHMWPE is completely free of stabilisers or additives since this is strictly regulated when using the material in orthopaedic implants.

Since the hydroperoxide content was proportional to the cutting rate it was concluded that the hydroperoxides had been formed during microtoming. It was found that the tungsten carbide knife contained about 6% cobalt, which is known to be a very potent hydroperoxide decomposer [123] and thus an initiator for the oxidative degradation. It was found that oxidation in the films was accelerated by just placing the knife on top of the film during oven aging [55].

2.4.4
Chemiluminescence to Study Fracture of some Polymers

CL can also be used to directly detect radical formation. Radicals produced by e.g. mechanical rupture of a chain will react with oxygen and form peroxyl radicals. Since the CL signal is presumed to originate from the bimolecular termination of these peroxyl radicals, the light emitted will be proportional to the number of peroxyl radicals formed [1, 5, 8, 124]. Polymer chains are, however, known to withstand high stresses without rupturing [125]; this is achieved by disentangling and chain slipping. When this type of relaxation is not possible, due to e.g. cross-linking or other restrictions of movement such as in highly oriented fibres, radical formation as a result of chain scission has been detected [126, 127]. A good correlation between electron spin resonance (ESR or EPR) and CL as peroxyl radical detectors has then been found, with CL being the more sensitive of the two techniques [25]. Chemiluminescence has, in addition to its sensitivity, the advantage over ESR that it can be used at a broader temperature range.

An investigation was made where the same tensile test specimens as were used to study the luminescence after yield were treated with SO_2 to destroy the hydroperoxides and thus prevent luminescence from hydroperoxide decomposition. Luminescence was, however, still found at fracture but only in an oxidative atmosphere. In pure nitrogen no luminescence could be detected. From this, the conclusion was drawn that some chains are ruptured at fracture and that radicals are formed [77].

A question related to this is whether radicals are created while cutting polymer samples with a knife or a pair of scissors. There is no unambiguous evidence for this to be found in the literature. A number of research groups have, however, found that their samples oxidise differently depending on whether they were cut with a sharp or a blunt scalpel or if the samples has been stamped or cut [128]. If radicals were created during the sample preparation this could act as an initiation for the subsequent oxidative degradation. The oxidation is often observed to start from the edge of a specimen.

In order to investigate the possible formation of radicals at cutting, the SCL instrument was modified to measure the photon emission during cutting of polymers with different chain mobility. The luminescence emitted while a knife from a scalpel cut through films of different polymers was monitored. When cutting a film of natural rubber, which was cross linked but did not contain any carbon black that would absorb the light, very high photon counts were recorded. This indicated that a large number of radicals were formed due to rupture of chains. UHMWPE is not cross linked like the rubber but has a very high molecular weight which could prevent efficient chain slip and thus result in chain fracture. When cutting unstabilised UHMWPE, about half of the measurements did not show any luminescence above the background noise. However, in the other half of the measurements, sharp peaks of luminescence were recorded. The absolute values of this luminescence were, however, much lower than for the rubber. When cutting films of unstabilised LDPE no luminescence was recorded in any of the experiments. It thus seems as if radicals are only formed occasionally while cutting and that chain slipping prevails [55].

3
Imaging Chemiluminescence (ICL)

3.1
ICL Applied to Polymer Oxidation Studies

A further development of the chemiluminescence technique has been to use a CCD camera or a position-sensitive photomultiplier to monitor the site of the emitted light, to generate an image of the ongoing oxidation. The technique is called imaging chemiluminescence (ICL) and makes it possible to follow the location of oxidation in a polymer sample, to study diffusion-controlled oxidation or to visualise heterogeneity in oxidative stability. In addition to this, the imaging ability makes it possible to run several samples in the same experiment, which increases the productivity. However, care must be taken since oxidative spreading via the gas phase can influence the results.

The use of ICL in the field of polymers is rather new and the first papers describing results obtained by ICL were presented in the early 1990s by Fleming and Craig [129]. They used a position-sensitive photon counting photomultiplier to monitor the oxidation of hydroxyterminated polybutadiene/isophorone diisosyanate resin. They showed that it was possible to produce an image of the ongoing oxidation and that the light intensity was most intense at cracks and edges. This ICL instrument was also used in an article [82] by Mattson et al. where it was observed that oxidation started a 'hotspot' and then spread to the rest of the sample. The material used was a mixture of polystyrene and a triblock SBS rubber. It was also shown that an uneven distribution of antioxidant resulted in a heterogeneous distribution of the emitted chemiluminescence signal in natural rubber.

Celina et al. [130] used a commercially available charge-couple device (CCD) camera instead of a position-sensitive photomultiplier. The reason for choosing a CCD camera is that a CCD camera is simpler and less expensive. Furthermore, a photomultiplier is easily damaged by high levels of light and are therefore less suitable when obtaining conventional images as well as luminescence emission images. In their work, they proved the usefulness of a CCD camera to obtain images of oxidising polypropylene. They also showed that oxidising polypropylene particles are able to spread oxidation to other polypropylene particles in close physical contact. Oxidative spreading was observed in a polypropylene film when a part of the film had been initiated with UV light prior to ICL measurement. Their work confirmed the proposed heterogeneous model for oxidation in PP [131, 132, 133].

The influence of artefacts such as temperature gradients, light guiding effects and oxygen deficiency was reported by Lacey and Dudler in experiments where they used a cryogenically cooled, slow scan CCD camera [134]. They also showed that location of oxidation correlated with the morphology in semi-crystalline PP. In a second paper they evaluated the ICL technique as a new testing procedure to screen antioxidant effectiveness in PP. They compared the ICL technique with oven aging and found that it offered several advantages, such as quicker results and better discrimination between samples. They also used unstabilised PP film to initiate oxidation in a stabilised PP film, which resulted in a propagating oxidation front that could be followed using the ICL technique. It was shown that the rate of spreading depends on the type of antioxidant.

As a result of Mattsons' et al. initial work mentioned above, an ICL apparatus was constructed at the Department of Polymer Technology at KTH in 1996. The instrument used a position-sensitive photomultiplier to create images on a phosphorous screen, which is recorded by a CCD camera. A detailed description can be found in the literature [63]. In a following publication, the ICL instrument was used to study oxidation profiles in aged PA 6,6 and the results were compared to the oxidation profiles obtained from FTIR measurements and were found to correlate [64]. The oxidative behavior of HTPB was further studied and the results were reported in two papers [67, 68]. In a later publication [66] the difference in oxidative stability between EPDM particles was visualised. Spreading of oxidation between EPDM particles was observed and the conclusion was drawn that the gas phase contributed to the spreading. The instrument has also been used to study artificially recycled PP. The relation between fraction of oxidising material and CL intensity was discussed and the chemiluminescence technique was found to be one of the most sensitive techniques to detect degradation in recycled PP [135]. The usefulness and possibilities, but also the limitations, of the ICL technique in the field of polymer oxidation have been discussed in a review article [65].

Kohler et al. discussed the potential of the chemiluminescence technique as an industrial test method. Imaging chemiluminescence was used to assess antioxidant performance. An advantage over oven aging was found to be the possibility for evaluation of the oxidative stability of samples with unusual geometries, such as fibres and powder particles [136]. A correlation was also found between oven aging and chemiluminescence measurements on stabilised PP and it was shown that chemiluminescence measurements done at

150 °C can reproduce oven aging data taken at such low temperature as 120 °C.

Recently ICL was used to follow the diffusion-limited oxidation of hydroxyl terminated polybutadiene at various temperatures and oxygen pressures. The CL intensity profiles were found to correlate well with theoretical oxygen concentration profiles [137].

3.2
Initiation and Spreading of Oxidation as Studied by ICL

In two recent papers [79, 80] results were presented where oxidation was initiated in stabilised polypropylene (PP) films using a piece of copper, unstabilised PP or UV light. After the initiation, the oxidation spread as a propagating front that was followed using the ICL technique. The same spreading behavior in stabilised PP films has earlier been observed by Dudler et al. [138]. The speed of the oxidation front was found to depend on film thickness, type and concentration of antioxidant and also the experimental conditions such as temperature [79]. An advantage of the forced initiation as a technique to study oxidative stability is that it reduces experimental scatter since it minimises the influence of randomly distributed weak points, such as defects in the samples due to sample preparation or local areas of low antioxidant concentration. It also considerably shortens the experimental times due to elimination of the induction period. However, in unstabilised PP oxidative spreading could not be observed. It appears as if the presence of stabilisers is a prerequisite for oxidative spreading to be detectable in the form of a propagating oxidative front.

A PP film stabilised with AO1 and P1 (structure shown in Fig. 6) in the ratio 1:2 that has been initiated using UV light according to a previously described procedure [79] can be seen in Fig. 7. Only the area between the dotted lines has been irradiated, the rest of the film was covered with aluminium foil during initiation.

The spreading of the oxidation was followed in air atmosphere at 150 °C. In Fig. 7, the first picture has been taken with external light that shows the location of the PP-film. The following images show the location of the oxidation front at five different times.

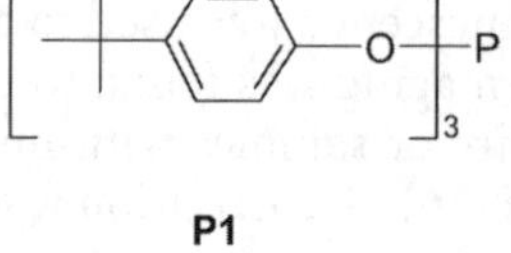

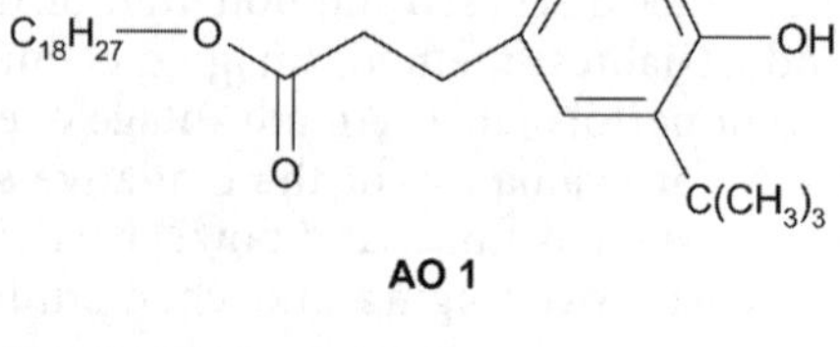

Fig. 6 Structures of P1 and AO1

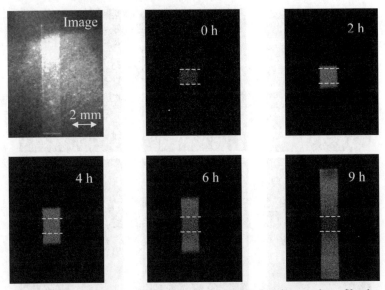

Fig. 7 ICL experiment made at 150 °C in air. A stabilised polypropylene film has been UV-initiated in-between the two *dotted lines*. The spreading of the oxidation to the rest of the film was followed

The location of the oxidation front, taken as the average of the two oxidation fronts caused by initiating the sample in the middle of the PP film, was then plotted versus time. The resulting curve shows that the spreading of oxidation accelerates with time (see Fig. 8). It has been reported that the volatile species produced during oxidation of PP can contribute to the spreading of oxidation in-between different samples of PP [139, 140]. A device was constructed that enabled the air flow to be directed over the oxidising

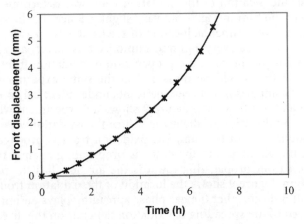

Fig. 8 Location of the oxidation fronts in Fig. 7 plotted against time

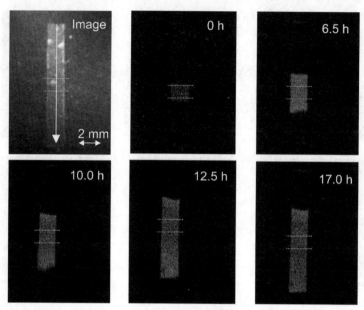

Fig. 9 ICL experiment made at 150 °C in air. A stabilised polypropylene film has been UV-initiated in-between the two *dotted lines*. The spreading of the oxidation to the rest of the film was followed. The *arrow* indicates the direction of the air flow

PP-film surface. The device was constructed so that the PP film could be placed in a tunnel that left 1 mm of free space over the PP film. Air was then flushed into the tunnel and thereby immediately removed any gaseous products formed . Care was taken not to inadvertently alter the sample temperature due to the gas flow. The device could not be placed in the ICL sample chamber for technical reasons. Therefore, the whole device was placed in an oven at 150 °C. The PP film was then taken out and put into the ICL sample chamber and the location of the oxidation front was determined; therefore the location of the film in Fig. 9 can vary slightly. Since the directed air flow was interrupted every time the location of the front was measured, measurements were only made every 2–3 h to minimise this disturbance. By initiating in the middle of the film, the propagation of oxidation fronts with, and against, the air flow could be measured in the same experiment. This enabled a direct comparison between the fronts under identical conditions, i.e. temperature and gas flow rate. Figure 9 shows an experiment where the air is flushed in the direction of the arrow. Since the oxidation was initiated in an area in the middle of the film, one oxidation front is propagating in the direction of the air flow, and the other is propagating against the air flow. The oxidation front propagating towards the air flow has to spread in the bulk of the film. Figure 9 shows the location of the oxidation fronts at different times and indicates that the gas phase spreading plays an important role since the speed of the spreading of oxidation depends on the direction of the

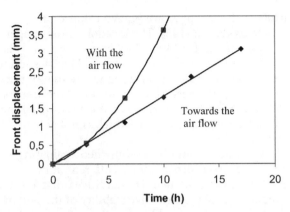

Fig. 10 The location of the oxidation fronts in Fig. 9 plotted against time

air flow. The spreading of oxidation towards the airflow is considerably slower and is not accelerating (see Fig. 10). It has been suggested earlier that the propagation of oxidation is mediated through the gas phase [141] and that PP oxidised in a closed system, where oxidation products are not removed, oxidises faster than in an open system [142]. The results in Figs. 9 and 10 confirm these conclusions.

Initiation of a polypropylene sample was also done by placing a piece of copper in contact with a PP film stabilised with the antioxidant system in

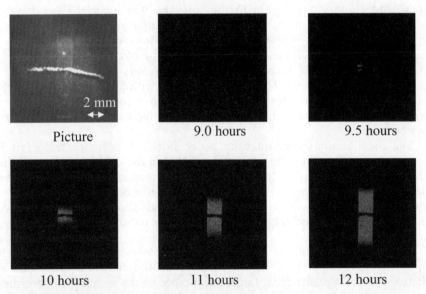

Fig. 11 ICL experiment made at 150 °C in air. A piece of a copper wire is placed in contact with a stabilised PP film. The initiation and spreading of oxidation originating from the copper wire is clearly seen

Fig. 6 [79]. Metals such as copper are known to decompose hydroperoxides and thereby initiate oxidation [143]. The location of the copper wire is clearly seen at the centre of the PP film. The copper wire was placed in direct contact with the PP film surface. The images in Fig. 11 show the initiation and spreading of oxidation from the location where the PP film is in contact with the copper wire to the rest of the film; the experiments were run at 150 °C in air. The initiation by copper was not caused by an increase in heat transfer since initiation with aluminium, which is not able to decompose hydroperoxides, failed.

This kind of experiment could be useful when evaluating stabilisers that are designed to act as metal deactivators. Different additives could also be placed in contact with the polymer. The additives that initiated oxidation could be expected to reduce the oxidative stability of the polymer.

To see if oxidative spreading could take place between different polymers, initiation of stabilised PP using a piece of unstabilised polyamide 6 (PA 6) was performed [79]. However, PA 6 can not initiate oxidation in PP even if oxidising PA 6 is placed in direct contact with PP. To clarify the processes of cross spreading between different polymers more experimental work is necessary.

4
Conclusions

The chemiluminescence technique has been used in the field of polymer research for more than 40 years but has not yet been widely accepted as a standard technique. During the first decades this was due to the fact that the photomultiplier systems of that time did not offer a very high sensitivity and hence no advantage over other techniques. However, the development of highly sensitive photon counting techniques now offer a sensitivity that is superior to many of the techniques used today to follow and detect oxidative degradation in polymers. This, in combination with an instrument capable of running several samples at the same time to increase the productivity, would be an interesting technique to scan oxidative stability in a large number of samples.

The chemiluminescence intensity is dependent on polymer, temperature and geometry. However, the relationship between the oxidation rate and the chemiluminescence intensity is not known and very few conclusions can be drawn from comparing light intensities between samples. In order to use chemiluminescence to determine the degree of oxidation, a correlation curve using a quantitative technique has to be constructed under the same ageing condition.

The formation of light is still a matter of discussion and further research to clarify and understand the mechanism behind the light emissions is therefore a prerequisite for a larger acceptance of the technique. Nevertheless, the results presented in this chapter show the usefulness of the technique to investigate oxidation of polyolefins.

Acknowledgements We thank Professor Ann-Christine Albertsson, Professor Ulf Gedde at our department and Professor Björn Terselius, now at Kristianstad University, for their contribution to this article.

We also thank the former PhD students at the department: Dr. Bengt Mattson, Dr. Camilla Wiberg, Dr. Anna Kron, Dr. Gustav Ahlblad, Dr. Dan Forsström and Dr. Farideh Khabbaz, for their hard work to increase our knowledge of polymer degradation using chemiluminescence.

References

1. George GA (1989) In: Zlatkevich (ed) Luminescence techniques in solid state polymer research. Dekker, New York, chap 1
2. Brolin S, Wettermark G (1992) Bioluminescence analysis. VCH, Weinheim
3. Ashby GE (1961) J Polym Sci L:99
4. Schard MP, Russel CA (1964) J Appl Polym Sci 8:985
5. Matisová-Rychlá L, Rychlý J (1996) In: Clough RL, Billingham NC, Gillen KT (eds) Polymer durability. Adv Chem Ser 1996, chap 12
6. Matisová-Rychlá L, Rychlý J, Vavrekova M (1978) Eur Polym J 14:1033
7. Mendenhall GD, Nathan RA, MA G (1978) In: EG B (ed) Applications of polymer spectroscopy. Academic Press, New York, chap 8
8. George GA (1989) In: Zlatkevich (ed) Luminescence techniques in solid state polymer research. Dekker, New York, chap 3
9. Reich L, Stivala SS (1967) Die Macromol Chemie 103:74
10. Zlatkevich L (1985) Polym Sci Polym Phys Ed 23:1691
11. Quinga EMY, Mendenhall GD (1983) J Am Chem Soc 105:6520
12. Audouin-Jirackova L, Verdu J (1987) J Polym Sci Pol Chem 25:1205
13. Russell GA (1957) J Am Chem Soc 79:3871
14. Vasil'ev RF (1970) Russ Chem Rev 39:529
15. Mayo FR (1978) Macromolecules 11:942
16. Niki E, Decker C, Mayo FR (1973) J Polym Sci Pol Chem 11:2813
17. Lacey D, Dudler V (1996) Polym Degrad Stab 51:109
18. Achimsky L, Audouin L, Verdu J, Rychlá L, Rychlý J (1999) Eur Polym J 35:557
19. Rychlý J, Matisová-Rychlá L, Jurcák D (2000) Polym Degrad Stab 68:239
20. Matisová-Rychlá L, JR (2000) Polym Degrad Stab 67:515
21. Rychlý J, Matisová-Rychlá L, Tiemblo P, Gomez-Elvira J (2001) Polym Degrad Stab 71:253
22. Blakey I, George GA (2001) Macromolecules 34:1873
23. Blakey I, George GA, Billingham NC (2001) Macromolecules 34:9130
24. Billingham NC, Then ETH, Gijsman PJ (1991) Polym Degrad Stab 34:263
25. George GA, Egglestone GT, Riddell SZ (1982) J Appl Polym Sci 27:3990
26. Matisová-Rychlá L, Fodor ZS, Rychlý J, Iring M (1980) Polym Degrad Stab 3:371
27. George GA, Egglestone GT, Riddell SZ (1983) Polym Eng Sci 23:412
28. Billingham NC, Burdon JW, Kakulska IW, O'Keefe ES, ETH T (1988) Proc Int Symp, Lucerne 2:11
29. Iring M, Laszlo-Hedvig S, Barabas K, Kelen T, Tudos F (1978) Eur Polym J 14:439
30. Scheirs J, DJ C, SW B (1995) Polym Plast Technol Eng 34:97
31. Broska R, Rychly J (2000) Polym Degrad Stab 72:271
32. Setnescu R, Jipa S, Setnescu T, Podina C, Osawa Z (1998) Polym Degrad Stab 61:109
33. Kihara H, Yabe T, Hosda S (1992) Polym Bull 29:369
34. Osawa Z, Kuroda H (1982) J Polym Sci Pol Lett 20:577
35. Jipa S, Setnescu R, Setnescu T, Dumitru M, Mihalcea I, Podina C, Osawa Z (1998) J Mater Sci-Pure Appl Chem A35:1103
36. Jipa S, Zaharescu T, Setnescu R, Setnescu T, Brites M, Silva A, Marcelo-Curto M, Gigante B (1999) Polym Int 48:414

37. Tiemblo P, Gómez-Elvira JM, Teyssedre G, Massines F, Laurent C (1999) Polym Degrad Stab 64:59
38. Tiemblo P, Gómez-Elvira JM, Teyssedre G, Massines F, Laurent C (1999) Polym Degrad Stab 64:67
39. Fleming RJ, Hagekyriakou (1984) Radiat Protect Dosim 8:99
40. McKeever SWS (1985) Thermoluminescence of solids. Cambridge University Press, Cambridge, chap 1
41. Zlatkevich L (1989) In: Zlatkevich L (ed) Luminescence techniques in solid state polymer research. Dekker, New York, chap 7
42. Fleming RJ (1990) Radiat Phys Chem 36:59
43. Charlesby A (1991) In: Clegg DW, Collyer AA (eds) Irradiation effects on polymers. Elsevier Applied Science, London, chap 2
44. Kron A, Stenberg B, Reitberger T (1997) Polym Int 42:131
45. Charlesby A, Partridge RH (1963) Proc R Soc London A 271:188
46. Boustead I, Charlesby A (1970) Proc R Soc London A 316:291
47. Markiewicz A, Fleming RJ (1988) J Phys D. Appl Phys 21:349
48. Bacon F (1978) In: Kitchin GW (ed) The Advancement of learning. Rowan and Littlefield, London
49. Zink JI (1978) Accts Chem Res 11:289
50. Butyagin P Yu, Yerofeyev VS, Musalyelyan IN, Patrikeyev GA, Streletskii AN, Shulyak AD (1970) Polym Sci USSR 12:330
51. Fanter DL, Levy RL (1979) ACS Symp Ser 95:211
52. Monaco SB, Richardson JH (1989) In: Zlatkevich L (ed) Luminescence techniques in solid state polymer research. Dekker, New York, chap 6
53. Albertsson A-C, Barenstedt C, Karlsson S (1992) Polym Degrad Stab 37:163
54. Malmström J, Engman L, Bellander M, Jacobson K, Stenberg B, Lönnberg V (1998) J Appl Polym Sci 70:449
55. Jacobson K, Costa L, Bracco P, Augustsson N, Stenberg B (2001) Polym Degrad Stab 73:141
56. Mattson B, Reitberger T, Stenberg B, Terselius B (1991) Polym Test 10:399
57. Mattson B, Gillen KT, Clough RL, Östman E, Stenberg B (1992) Polym Degrad Stab 41:211
58. Albertsson A-C, Barenstedt C, Karlsson K (1994) J Appl Polym Sci 51:1097
59. Kron A, Stenberg B, Reitberger T (1996) Polym Degrad Stab 48:89
60. Kron A, Stenberg B, Reitberger T, Billingham NC (1996) Polym Degrad Stab 53:119
61. Kron A, Stenberg B, Reitberger T (1997) Prog Rubber Plast Technol 13:81
62. Ahlblad G, Jacobson K, Stenberg B (1996) Plast Rubber Compos Proc Appl 25:464
63. Ahlblad G, Stenberg B, Terselius B, Reitberger T (1997) Polym Test 16:59
64. Ahlblad G, Forsström D, Stenberg B, Terselius B, Reitberger T, Svensson L-G (1997) Polym Degrad Stab 55:287
65. Ahlblad G, Reitberger T, Terselius B, Stenberg B (1998) Angew Makromol Chem 262:1
66. Ahlblad G, Reitberger T, Terselius B, Stenberg B (1999) Polym Degr Stab 65:169
67. Ahlblad G, Reitberger T, Terselius B, Stenberg B (1999) Polym Degrad Stab 65:185
68. Ahlblad G, Reitberger T, Terselius B, Stenberg B (1999) Polym Degrad Stab 65:179
69. Forsström D, Kron A, Mattson B, Reitberger T, Stenberg B, Terselius B (1992) Chem Technol 65:736
70. Forsström D, Kron A, Stenberg B, Terselius B, Reitberger T (1994) Polym Degrad Stab 43:277
71. Forsström D, Reitberger T, Terselius B (2000) Polym Degrad Stab 67:255
72. Jacobson K, Stenberg B, Terselius B, Reitberger T (1999) Polym Degrad Stab 64:17
73. Jacobson K, Färnet G, Stenberg B, Terselius B, Reitberger T (1999) Polym Test 18:523
74. Jacobson K, Stenberg B, Terselius B, Reitberger T (1999) Polym Degrad Stab 65:449
75. Jacobson K, Stenberg B, Terselius B, Reitberger T (1999) Polym Degrad Stab 65:107
76. Jacobson K, Stenberg B, Terselius B, Reitberger T (2000) Prog Rubber Plast Techn 16:135

77. Jacobson K, Stenberg B, Terselius B, Reitberger T (2000) Polym Int 49:654
78. Jacobson K, Stenberg B, Terselius B, Reitberger T (2000) Polym Degrad Stab 68:53
79. Eriksson P, Reitberger T, Ahlblad G, Stenberg B (2001) Polym Degrad Stab 73:177
80. Eriksson P, Reitberger T, Stenberg B (2002) Polym Degrad Stab 78:183
81. Khabbaz F, Albertsson A-C (2001) J Appl Polym Sci 79:2309
82. Mattson B, Kron A, Reitberger T, Craig AY, Fleming RH (1992) Polym Test 11:357
83. Mattson B, Stenberg B (1992) Rubber Chem Technol 65:315
84. Mattson B (1993) Thermo-oxidative degradation and stabilization of rubber materials. PhD Thesis, Royal Inst of Technology, Stockholm
85. Barenstedt C (1994) Environmental degradation of starch-modified polyethylene. PhD Thesis, Royal Inst of Technology, Stockholm
86. Kron A (1996) Chemiluminescence applied to oxidation of polyolefins. PhD Thesis, Royal Inst of Technology, Stockholm
87. Ahlblad G (1998) Imaging chemiluminescence applied to oxidation of rubber materials and polymers. PhD Thesis, Royal Inst of Technology, Stockholm
88. Forsström D (1999) Novel techniques for characterisation of the oxidative stability of polyamides. PhD Thesis, Royal Inst of Technology, Stockholm
89. Jacobson K (1999) Oxidation of stressed polymers as studied by chemiluminescence. PhD Thesis, Royal Inst of Technology, Stockholm
90. Khabbaz (2001) Environmentally degradable polyethylene. PhD Thesis, Royal Inst of Technology, Stockholm
91. Eriksson P (2002) New Approaches to investigate and prevent oxidation of polypropylene. PhD Thesis, Royal Inst of Technology, Stockholm
92. Carlsson DJ, Lacoste J (1991) Polym Plast Technol Eng 32:377
93. Zahradnicková A, Sedlar J, Dastych D (1991) Polym Degrad Stab 32:155
94. Singleton RW, RJTC (1986) Prog Rubber Plast Techn 2:10
95. Yoshii F, Sasaki T, Makuuchi K, Tamura N (1985) J Appl Polym Sci 30:3339
96. Wilski H (1987) Radiat Phys Chem 29:1
97. Spadaro G (1993) Eur Polym J 29:1247
98. Kashiwabara H, Hori Y (1981) Radiat Phys Chem 18:1086
99. Tabb DL, Sevcik JJ, Koenig JL (1975) J Polym Sci 13:815
100. White JR, Nya R (1994) Trends Polym Sci 6:197
101. Tobolsky A, Eyring HJ (1943) J Chem Phys 11:125
102. Bueche J (1955) J Appl Phys 26:1133
103. Bueche J (1957) J Appl Phys 28:784
104. Bueche J (1958) J Appl Phys 29:1231
105. Zhurkov SN, Narzullayev BN (1953) Zhur Techn Fiz 23:1677
106. Zhurkov SN, Tomashevskii E (1955) Zhur Techn Fiz 25:66
107. Zhurkov SN, Sanfirova TP (1955) Dokl Akad Nauk SSSR 237:101
108. Zhurkov SN, Zakrevskii VA, Korsukov VE, Kuksenko VS (1972) Soviet Physics-Solid State 13:1680
109. Mendenhall GD (1977) Angew Chem Int Ed Engl 16:225
110. Dickinson JT (1990) In: Summerscales (ed) Non-destructive testing of fibre-reinforced plastics composites, vol 2. Elsevier Applied Science, London
111. Hosoda S, Seki Y, Kihara H (1993) Polym Degrad Stab 34:22
112. Shen FW, Yu YJ, McKellop H (1999) Biomed Mater Res 48:203
113. Costa L, Luda MP, Trossarelli L, Brach Del Prever EM, Crova M, Gallinaro P (1998) Biomaterials 19:659
114. Costa L, Luda MP, Trossarelli L, Brach Del Prever EM, Crova M, Gallinaro P (1998) Biomaterials 19:1371
115. Deng M, Shalaby SW (1997) Biomaterials 18:645
116. Brach Del Prever EM, Crova M, Costa L, Dallera A, Camino G, Gallinaro P (1996) Biomaterials 17:873
117. Chenery DH (1997) Biomaterials 18:415
118. Eyerer P, Kurth M, McKellop HA, Mittlemeier T (1987) J Biomed Mater Res 21:275

119. Yu YJ, Shen FW, McKellop HA, Salovey R (1999) J Polym Sci Pol Chem 37:3309
120. Yeom B, Yu YJ, McKellop HA, Salovey R (1998) J Polym Sci Pol Chem 36:329
121. Costa LKJ, Brunella V, Bracco P (2001) Polym Test 20:649
122. Setnescu R, Jipa S, Osawa Z (1997) Polym Degrad Stab 60:377
123. Osawa Z, Tsurumi K (1989) Polym Degrad Stab 26:151
124. Billingham NC (1990) J Polym Sci Pol Phys 28:257
125. Kausch HH (1978) Polymer fracture. Springer, Berlin Heidelberg New York
126. Wang J, Smith Jr KJ (1999) Polymer 40:7261
127. DeVries KL (1979) J Appl Polym Sci Pol Phys:439
128. Stockholm (September 1999) Oral discussion at the Polymer degradation and stabilisation workshop
129. Fleming RH, Craig AY (1991) Polym Preprints 32
130. Celina M, George G A, Lacey DJ, Billingham NC (1995) Polym Degrad Stab 47:311
131. Celina M, George GA (1993) Polym Degrad Stab 40:323
132. Celina M, George GA (1995) Polym Degrad Stab 50:89
133. George GA, Celina M, Lerf C, Cash G, Weddel D (1997) Macromol Symp 115:69
134. Lacey DJ, Dudler V (1996) Polym Degrad Stab 51:101
135. Ahlblad G, Gijsman P, Terselius B, Jansson A, Möller K (2001) Polym Degrad Stab 73:15
136. Kohler DR, Krohnke C (1998) Polym Degrad Stab 62:385
137. Sinturel C, Billingham NC (2000) Polym Int 49:937
138. Dudler V, Lacey DJ, Kröhnke Ch (1996) Polym Degrad Stab 51:115
139. Sedlar J, Pac J (1974) Polymer 15:613
140. Celina M, George GA, Billingham NC (1993) Polym Degrad Stab 42:335
141. Matisová-Rychlá L, Rychlý J, Verdu J, Audouin L, Csomorova K (1995) Polym Degrad Stab 49:51
142. Gijsman P, Verdun F (2001) Polym Degrad Stab 74:533
143. Zweifel H (1998) Stabilisation of polymeric materials. Springer, Berlin Heidelberg New York

Received: July 2003

Adv Polym Sci (2004) 169:177–199
DOI: 10.1007/b13523

Environmental Degradation of Polyethylene

Minna Hakkarainen · Ann-Christine Albertsson

Department of Fibre and Polymer Technology, The Royal Institute of Technology (KTH),
100 44 Stockholm, Sweden
E-mail: *minna@polymer.kth.se*
E-mail: *aila@polymer.kth.se*

Abstract The environmental degradation of polyethylene proceeds by synergistic action of photo- and thermo-oxidative degradation and biological activity. Since biodegradation of commercial high molecular weight polyethylene proceeds slowly, abiotic oxidation is the initial and rate-determining step. Enhanced environmentally degradable polyethylene is prepared by blending with biodegradable additives or photo-initiators or by copolymerisation. One of the key questions for successful development and use of environmentally degradable polymers is to understand the interaction between degradation products and nature. Polymer fragments and degradation products should be environmentally assimilable and should not accumulate or negatively affect the environment. Determination of abiotic and biotic oxidation products is an important step towards establishing the environmental degradation mechanism and environmental impact of the material. More than 200 different degradation products including alkanes, alkenes, ketones, aldehydes, alcohols, carboxylic acid, keto-acids, dicarboxylic acids, lactones and esters have been identified in thermo- and photo-oxidised polyethylene. In biotic environment these abiotic oxidation products and oxidised low molecular weight polymer can be assimilated by microorganisms. In future we will probably see a development of new polyethylenes with tailor-made structures specially designed for environmental degradation through different pathways. Paralleled with the development of these new materials we need to obtain better understanding of the environmental impact of degradable polymers and the interactions between nature and degradation products in a dynamic system.

Keywords Polyethylene · Environmental degradation · Environmental impact · Degradation products

Abbreviations

BHT	Butylated hydroxy toluene
E/CO	Ethylene-carbon monoxide
FTIR	Fourier transform infrared spectroscopy
HDPE	High density polyethylene
LDPE	Low density polyethylene
LDPE-MB	Polyethylene modified with masterbatch
LDPE-PO	Polyethylene modified with pro-oxidants
LDPE-starch	Polyethylene modified with starch
Mn	Number average molecular weight
Mw	Weight average molecular weight
NMR	Nuclear magnetic resonance
PP	Polypropylene
PS	Polystyrene
PVA	Polyvinyl alcohol
PVC	Polyvinyl chloride
SBS	Styrene-butadiene copolymer
SEM	Scanning electron microscopy
UF	Urea formaldehyde resin
UV	Ultraviolet light
XRF	X-ray fluorescence

1
Introduction

The end-of-life management of post-consumer plastic materials plays an important role in the development of sustainable polymer products. Sustainable waste management can include source reduction, product reuse, product recycling, energy recycling and development of environmentally degradable polymeric materials in order to biologically recycle plastic waste. Environmentally degradable polymeric materials can degrade via physical, chemical, mechanical or biological action or by a combination of these mechanisms. The loss of structural integrity results from the combined action of rain, wind, sunlight, insects, animals, microorganisms and so on. There are many demands that the environmentally degradable materials have to fulfil. We have to be able to control the service-life to ensure safe use of the products with no possibility for premature failure or release of contaminants from the material. After the service-life, the material should degrade to harmless products in a controlled way. The environmental conditions are very important for the degradation to occur. Biodegradation is catalysed by enzymes and its rate depends greatly upon environmental conditions such as temperature, moisture, oxygen, and a population of suitable microorganisms. Photo-oxidation does not take place if the material is not subjected to sunlight and so on. This means that a material may degrade rapidly in one environment but remain unaffected in another environment. The statement that a material is degradable, photo-degradable, thermodegradable or biodegradable should, thus be connected to the specific disposal pathway.

2
Enhanced Environmentally Degradable Polyethylene

The biodegradation and environmental degradation of commercial high molecular weight polyethylene proceeds very slowly. This slow degradation rate can be partly attributed to the presence of antioxidants in the material as it has been shown that polyethylene films, from which a small amount of antioxidant (BHT) had been removed, bio-eroded rapidly, whereas under the same conditions polyethylene films containing BHT were completely inert to microorganisms [1, 2]. Polyethylene can be rendered more susceptible to environmental degradation by blending, copolymerisation or grafting. The main market areas for degradable polyethylenes are as packaging material, composting bags, agricultural or landscaping applications (mulch film), hygiene products (e.g. disposable diapers) and other disposable articles (cups, food trays, shopping bags).

2.1
Enhanced Photo-Degradable Polyethylene

The sensitivity towards photo-degradation is enhanced by introducing UV-absorbing groups along the polyethylene chain or by blending polyethylene with UV-absorbing additives. Carbonyl groups have been copolymerised directly into the polyethylene main chain (ethylene-carbon-monoxide) (E/CO) [3, 4] or into the side chain (vinyl ketone copolymer) (Ecolyte) [5, 6]. Carbonyl groups facilitate oxidation of the copolymer and make it more susceptible to biodeterioration [7]. These systems do not introduce any new chemicals into the environment, they simply accelerate the natural photo-oxidative degradation. A disadvantage is that these materials have no induction period and they are, thus, mainly suitable for short term applications such as disposable cups and food trays, loop carriers of beverage-can six-packs. Degradation rate can, however, be modified by changing the concentration of carbonyl groups.

Metal ions catalyse the dissociation of hydroperoxides and thereby enhance the rate of oxidative chain scission in polyolefins [8]. A disadvantage is that the metal compounds have to be added during compounding, which makes the polymer sensitive to thermal degradation during processing. Scott overcame this difficulty by using sulphur complexed transition metal ions, particularly dithiocarbamates, which do not promote thermal degradation [9, 10, 11, 12]. The thermal antioxidant and UV-stabilising behavior of the dithiocarbamates are due to their ability to destroy hydroperoxides in a non-radical process. After fulfilling their stabilising function during the manufacture and service life, these additives undergo a sharp inversion of activity and become catalysts for photo- and thermo-oxidation. A rapid reduction of the molecular weight results in loss of mechanical and physical properties. The useful lifetime can be adjusted by varying the initial concentration of the antioxidant or by using a combination of photo-antioxidant and photo-initiator [13, 14]. The large geographic and seasonable differences in the amount of UV exposure has to also be taken account when designing for controlled degradation time.

2.2
Blending, Grafting and Copolymerisation with Biodegradable Polymers and Additives

Griffin first introduced the idea to replace part of the PE matrix with a biodegradable additive such as starch [15, 16]. In addition to offering at least partly biodegradable material, blending with starch replaces part of the polyethylene matrix with a polymer from renewable resources. A microbial consumption of starch leads to porosity and void formation leading to increased surface area and oxygen permeability. The amount of starch is typically 6–15%. The materials are marketed under the trademark Ecostar. In 1988, Griffin patented an improved process in which starch was used together with an unsaturated polymer and a pro-oxidant, an organic salt of a transition metal [17]. The overall disintegration of starch-filled materials demands the use of transition metal compounds to accelerate the abiotic oxidation step. Starch can also be added to PE in a gelatinised form [18, 19]. These products generally contain over 40% starch are termed starch-based materials.

Graft copolymerisation of polyethylene onto starch provides another method for preparing starch-polyethylene composites [20]. An important advantage of graft copolymerisation is the fact that starch and polyethylene are held together by chemical bonds. Polyethylene with chemically bonded starch showed excellent biodegradability when starch content was more than 10 wt% [21]. Maleic anhydride, methacrylic anhydride or maleimide were used as coupling agents to bond starch and polyethylene. Polyethylene has also been blended with other natural polymers, e.g. lignin [22, 23], and with biodegradable aliphatic polyesters [24, 25, 26]. Bailey prepared biodegradable polyethylene by introducing ester groups into the polyethylene chain [27, 28]. Austin followed Bailey's idea and introduced ester groups into the polyethylene chain by copolymerisation of ethylene, 2-methylene-1,3-dioxepane and optionally carbon monoxide in the presence of a free radical catalyst [29, 30]. The existing commercial enhanced environmentally degradable polyethylenes are, however, all based on the ideas presented by Guillet, Scott and Griffin in the 1970s [3, 5, 9, 15].

3
Environmental Degradation of Polyethylene

The environmental degradation of polyethylene proceeds by synergistic action of photo- and thermo-oxidative degradation and biological activity. Since biodegradation of high molecular weight polyethylene generally proceeds slowly, the abiotic oxidation of the polyethylene matrix is the initial and also rate-determining step, crucial to the ultimate destruction of the polymer. Not only does it destroy the physical integrity of the polymer, causing it to become brittle and to fragment, but the oxygenated species, especially carboxylic acids, are hydrophilic and more susceptible to biodegradation than the high molecular weight polymer.

3.1
Abiotic Degradation of Polyethylene

The oxidation of hydrocarbon polymers begins during processing and the formation of hydroperoxides during fabrication affects the rate of thermo- and photo-oxidation during subsequent use. The radical chain mechanisms involved in thermo- and photo-oxidation are essentially similar. Generally the degradation mechanism and rate are largely dependent on the availability of oxygen [31]. In the presence of oxygen the chain scission reactions dominate over the molecular enlargement reactions. The alkyl radicals react rapidly with oxygen and form peroxyl radicals, which can hydrogen abstract inter- or intramolecularly to form polymeric hydroperoxides. Intramolecular abstraction results in the formation of volatiles but no detectable decrease in molecular weight. Intermolecular abstraction on the other hand leads to rapid diminishing of molecular weight. Thermo-oxidation and oxidation products of polyethylene at 150–250 °C, i.e. at processing conditions, have been a subject of several studies [32, 33, 34, 35]. The high temperature degradation process and mechanisms differ from the mechanisms taking place during long-term ageing at moderate temperatures. Higher temperature means faster reactions and larger amount of free radicals. The availability of oxygen will become limited because diffusion rate and solubility of oxygen is too low [36]. At lower oxygen concentration, the probability that two neighbouring alkyl radicals will survive long enough to react with each other instead of reacting with oxygen is higher and the molecular enlargement reactions are more dominant leading to broadening of the molecular weight distribution. At lower reaction temperatures, which are actual during environmental degradation, the reaction times are much longer, the number of radicals is smaller and oxygen has more time to diffuse to reaction sites.

Large broadening of the molecular weight distribution was seen for starch/polyethylene blends during thermo-oxidation at 190 °C and 230 °C [37], while the molecular weight distribution narrowed during ageing at 80 °C [38]. The effect of limited oxygen concentration was also seen during long-term ageing in water as broadening of molecular weight distributions was observed [38]. In accordance, a larger increase in the carbonyl index was seen during ageing at 80 °C in air, compared to ageing in water at the same temperature, because at low oxygen concentrations crosslinking competes more effectively with the hydroperoxide formation. Hydroperoxides are the key intermediates in polyethylene oxidation. Hydroperoxides decompose when heated or irradiated. The decomposition may occur by homolysis of the peroxide bond or bimolecularly. The bimolecular reactions can involve reactions of a hydroperoxide with another hydroperoxide or reactions with alkyl, alkoxy or peroxy radicals. Hydroperoxides may also react with alcohols, ketones or carboxylic acids. The homolysis of the peroxide bond becomes less important during thermo-oxidation below 150 °C, while the bimolecular reactions have been shown to be valid during the oxidation of paraffins under 150 °C [39, 40] and are probably important even for oxidation of polyethylene at low temperatures. At low temperatures, polyethylene hydroperoxides remain isolated along the polymer chain making reactions between two neighbouring hydroperoxides in the same chain improbable [41, 42]. As a result of hydroperoxide decomposition, oxygen-containing groups are formed into the polyethylene chain. The role of hydroperoxides and the most important reactions of PE oxidation have been summarised and discussed by several authors [43, 44, 45, 46, 47, 48, 49, 50, 51].

The presence of pro-oxidants is critical for the initiation of degradation at moderate temperatures in dry air [52, 53, 54, 55]. LDPE modified with starch and pro-oxidants (LDPE-MB) degraded rapidly even at moderate temperatures, while regular LDPE and LDPE modified with starch (LDPE-starch) were rather stable towards low temperature thermo-oxidation [38]. The degradation of LDPE-MB proceeded faster in air than in water. In accordance, LDPE films containing Co acetylacetone, Co stearate and Mn stearate as pro-oxidants degraded rapidly in dry oven air at 60 °C, while the degradation rate was much slower in wet compost [56]. When the transition metal contents of LDPE films was measured by using X-ray fluorescence (XRF), it could be seen that Co acetylacetonate migrated from the films during the composting while the Mn stearate remained in the films, but was somehow deactivated because the remaining catalyst no longer initiated the degradation in dry oven air. In addition to the leaching out or deactivation of the catalyst, the lower oxygen tension at the film surface can retard the oxidative degradation in water.

3.2
Biodegradation of Polyethylene

Potts and co-workers claimed that linear paraffins at $Mw=450$ were biodegraded, while branched and high molecular weight hydrocarbons (Mw higher than 450) were not [57, 58]. Haines and Alexander, however, showed that hydrocarbons containing up to 44 carbon atoms were metabolised in soil [59]. The biodegradation of commercial high molecular weight polyethylene proceeds slowly [60, 61]. By studying ^{14}C-labelled materials with the sensitive liquid scintillation (LS) technique, small changes taking place during the biodegradation of LDPE and HDPE exposed to soil, various fungi and bacteria could be followed [62, 63, 64, 65, 66]. The biodegradation was affected by irradiation with UV light [63], by the morphology and surface area of the material [62], by antioxidants [63], by additives [61, 67] and by molecular weight [63]. When the biodegradation of ^{14}C-labelled HDPE inoculated with soil or *Fusarium redolens* was compared with HDPE deprived from most of its low molecular weight material by extraction, the samples without low molecular weight fraction showed a decrease in [14]CO_2 production compared to the samples containing low molecular weight compounds [63].

When polyethylene is oxidised, the molecular weight decreases and carbonyl groups are introduced along the polyethylene chain [68]. These changes increase the hydrophilicity of the material, which in turn will increase the biodegradation rate [63]. FTIR analysis showed that the amount of carbonyl groups in the polyethylene increased with time in the abiotic environment, while the amount of carbonyl groups decreased in the biotic environment [66, 68, 69]. An increase in double bonds was also noted for samples aged in the biotic environment. The decrease in the carbonyl index during biotic ageing has later been confirmed by several studies [70, 71, 72]. In an unpublished study, NMR showed changes in the degree of short-chain branching when oxidised polyethylene was subjected to microorganisms. The studied material was LDPE mixed with 12% pro-oxidant consisting of manganese stearate, LLDPE and styrene-butadiene copolymer (SBS) (LDPE-PO). All the aged films were first subjected to 7 days at 95 °C to surpass the induction period. During the thermo-oxidation, the *Mn* and *Mw* of LDPE-PO decreased to 3200 and 22,000, respectively. The films were then aged for 67 days at 50 °C either in air, in sterile mineral medium or biotic mineral medium with mixed culture of compost microorganisms. As seen in Table 1, the amount of short-chain branching was significantly lower after ageing in biotic mineral medium compared to the ageing in sterile medium. The most significant decrease was seen in the amount of butyl branches, which were originally the most abundant branches.

Table 1 Number of methyl, ethyl, butyl and pentyl branches per 1000 C-atoms in un-aged polyethylene compared to polyethylene thermo-oxidised for 67 days at 50 °C and pre-oxidised polyethylene aged in sterile or biotic mineral medium for 67 days at 50 °C

Type of branch	Un-aged	Thermo-oxidised	Sterile medium	Biotic medium
Methyl	2.8	2.3	1.6	0.9
Ethyl	8.3	8.3	7.2	5.3
Butyl	9.4	8.8	5	3.5
Pentyl	2.2	1.4	1.3	1.2

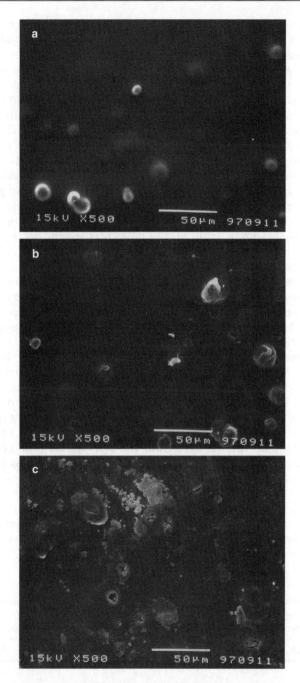

Fig. 1a–c Scanning electron micrographs showing the surface of LDPE-starch films with 7.7% starch **a** before ageing **b** after ageing in sterile mineral medium and **c** after ageing in biotic mineral medium inoculated with *A. paraffineus*

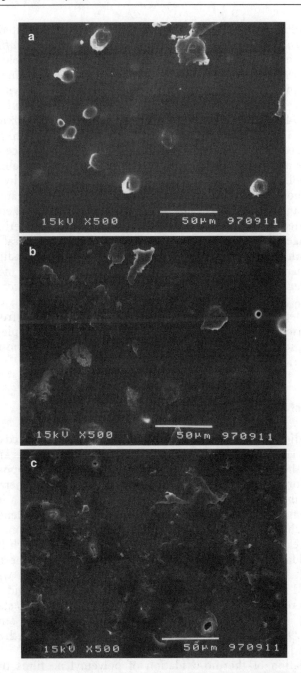

Fig. 2a–c Scanning electron micrographs showing the surface of LDPE-MB films with 7.7% starch **a** before ageing **b** after ageing in sterile mineral medium and **c** after ageing in biotic mineral medium inoculated with *A. paraffineus*

When LDPE and LDPE with 15% masterbatch were inoculated with *Fusar-ium redolens, Penicillum simplicissimum* or *Arthrobacter paraffineus*, it was noted that the inoculated samples had a slightly larger molecular weight than the sterile samples [71, 73]. Pometto et al. made the same observation when studying the degradation of various LDPE mixtures with starch and masterbatch inoculated with *Streptomyces badius, S. setonii,* or *S. viri-dosporous* [74]. Before inoculation the samples where heat-treated at 70 °C for 20 days. The authors proposed that the increase in molecular weight for biotically aged films is due to a biofilm formed on the surface of the LDPE film by the microorganisms, which limits the oxygen availability. The in-crease in molecular weight could, however, also be explained by assimilation of the low molecular weight parts of the samples by microorganisms.

Figure 1 and Fig. 2 show scanning electron micrographs taken from LDPE films modified with 7.7% starch (LDPE-starch) or 20% masterbatch (LDPE-MB) after ageing in sterile mineral medium and biotic mineral medium in-oculated with *A. paraffineus*. Before subjection to mineral medium the sam-ples were thermo-oxidised for 6 days at 100 °C to make them more suscepti-ble to biodegradation. It is clearly seen that both LDPE-starch and LDPE-MB samples are more eroded after ageing in biotic medium compared to ageing in sterile medium. Increasing the amount of starch from 3.85% to 7.7% increased the microbial attack on the surface. The samples containing starch and pro-oxidant are susceptible to both thermo-oxidation and micro-bial attack and were most severely affected.

3.3
Degradation of Polyethylene in Natural Environments

Several studies have followed the degradation of starch-filled and starch-based materials during composting and soil burial, in landfills and in seawa-ter. Griffin showed that the degradation of LDPE film in compost is acceler-ated by absorption of unsaturated lipids, resulting in the generation of per-oxides and thus enhancing autoxidation [16, 75]. There was a loss of 0–6% of starch content during 60 days of composting of polyethylene containing 0–20% starch [76]. Loss of starch was accompanied by a small molecular weight decrease in polyethylene matrix [77]. Polyolefin/starch blends with-out pro-oxidants did not show any signs of oxidation in the polyethylene matrix [78]. In addition, not more than 30% of the starch was removed. Starch-based materials (52% and 67%) demonstrated rapid and almost com-plete starch removal during 40 days of soil burial, but at 29% starch content only starch from the surface of the material had been removed after 8 months [79]. At the same time, no oxidation of the polyethylene matrix was observed.

The initiation of thermo-oxidation of polyethylene films by transition metal catalysts during composting proceeded slowly compared to oxidation at the same temperature in an oven (60–70 °C) [56]. The starch-filled poly-ethylene bags exposed on the surface of the compost broke down into small pieces while the buried bags remained intact after 49 days of exposure [55,

80]. Several authors have also compared natural and artificial weathering of polyethylene [81, 82, 83, 84]. Starch-polyethylene films deteriorated rapidly in the strawline of a marsh, while less degradation was observed in the films placed in compost, landfill, soil and seawater [85]. Polyethylene films on the surface of water showed more rapid loss of properties than those samples partially or completely submerged [86]. This can be attributed to decreased light intensity and the lack of heat build up. The degradation of LDPE placed in water was slower than when the films were kept dry. In polyethylene containing 0–30% starch and metal catalysts, very little degradation (as measured by molecular weight and tensile strength) was observed during 12 weeks in sea water [87]. Without metal catalysts no changes were observed. From 8–15% of the starch content was lost after 13 months in marine environment [88].

Photo-oxidised polyethylene with molecular weight larger than 10,000 has been shown to be rather rapidly biodegraded [89]. When polyethylene treated with hot nitric acid was the sole carbon source, polyethylene with a molecular weight higher than 100,000 was degraded to lower molecular weights by hyphae of the fungus [90]. When biodegradation of photo-degraded PS, LDPE, PP was studied, PE and PP fragments were biologically oxidised both in natural soil and in aqueous sewage sludge, while fragments from photo-degraded polystyrene were more resistant to biodegradation [91, 92]. When LDPE, PS, PVC and urea formaldehyde resin (UF) were buried in bioactive soil for more than 32 years, LDPE films were the only samples that showed any signs of biodegradation [93, 94]. The surfaces of the LDPE films, which were in contact with the soil, were severely degraded. The molecular weight of the surface was reduced to almost one-half of the molecular weight of the inner part.

Addition of starch is not enough to make a rapidly biodegradable or environmentally degradable material. The addition of pro-oxidants, photo-initiators, UV-absorbing groups or oxidative treatment before exposure to natural environments are necessary for the degradation to take place during a reasonable time frame. Even after modification, favourable environment (e.g. sunlight, moisture, presence of microorganisms, temperature, pH) will have a crucial role in whether or not the material is degraded. The way to make a rapidly biodegradable polyethylene for applications where rapid biodegradation is an advantage, could be through incorporation of ester groups into the polyethylene chain, which could render the degradation mechanism towards that of the rapidly degrading hydrolysable polymers.

4
Degradation Product Patterns in Abiotic and Biotic Environments

Since the environmental degradation of polyethylene starts with abiotic oxidation, the determination of abiotic oxidation products is an important step towards establishing the environmental degradation mechanisms and environmental impact of the material. In a secondary process, microorganisms may utilise these abiotic degradation products and the low molecular weight polymer in anabolic and catabolic cycles [60]. This assimilation leads to formation of biomass and/or formation of CO_2 and water.

4.1
Abiotic Degradation Products

We have identified more than 200 different degradation products after photo- and thermo-oxidation of polyethylene (Table 2) [37, 38, 95, 96, 97, 98, 99]. The studied polyethylenes included enhanced photo-degradable polyethylenes with Scott-Gilead photo-sensitisers and enhanced biodegradable polyethylene modified with starch (LDPE-starch) or a combination of starch and pro-oxidant in a masterbatch (LDPE-MB). The identified degradation products included alkanes, alkenes, ketones, aldehydes, alcohols, mono- and dicarboxylic acids, lactones, keto-acids and esters. Different materials degraded at different rates but not many differences were seen in the product patterns when the same extent of degradation was reached. LDPE-MB was the most susceptible material to low temperature thermo-oxidation [38], while the enhanced photo-degradable polyethylene containing iron dimethyldithiocarbamate (SG1) was the material most susceptible to photo-oxidation [95]. This sensitivity to oxidation was seen as a rapid decrease in molecular weight and large increase in the carbonyl index measured by FTIR. It was also clearly seen in the degradation product patterns.

Carboxylic and dicarboxylic acids were generally the most dominant product classes formed in all the materials, during both photo-oxidation and thermal ageing in air and water [37, 38, 95]. Carboxylic acids are the most stable oxygen-containing groups. They do not absorb UV light above 300 nm and they are not particularly prone to auto-oxidation. They accumulate during prolonged ageing when other products like aldehydes, ketones and alcohols are further oxidised to carboxylic acids. Similar products were formed during photo-oxidation and thermo-oxidation at 80 °C, but the product patterns differed somewhat during thermal degradation in water or during degradation above melting point. Larger amounts of alcohols and aldehydes were generally only detected after thermo-oxidation above melting point [37]. Alcohols and aldehydes may be formed as a result of hydroperoxide decomposition followed by hydrogen abstraction or β-scission, respectively. They are also formed when two primary or secondary alkoxy radicals react with each other. At low oxidation temperatures the concentration of free radicals is low and reactions involving two radicals are not probable. If aldehydes are formed at 80 °C their further oxidation to carboxylic acids is expected to be a much faster reaction than the formation of new aldehydes [100, 101].

Large number of dicarboxylic acids, keto-acids and/or lactones indicated severe degradation of the polyethylene matrix [37, 38, 95]. Dicarboxylic acids were the most abundant products formed during photo-oxidation and their amount increased especially after long irradiation times, i.e. in severely degraded samples [96]. The relationship between the degree of oxidation/degradation in the polymer matrix and the amount of dicarboxylic acids

Table 2 Low molecular weight degradation products identified in thermo- and photo-oxidised polyethylene

Alkanes	Ketones	Carboxylic acids	Lactones
Heptane	Acetone	Formic acid	Butyrolactone
Octane	2-Butanone	Acetic acid	5-Methyldihydro-2(3H)-furanone
Nonane	2-Pentanone	Propanoic acid	5-Ethyldihydro-2(3H)-furanone
Decane	2-Hexanone	Butanoic acid	5-Propyldihydro-2(3H)-furanone
Undecane	2-Heptanone	Pentanoic acid	5-Butyldihydro-2(3H)-furanone
Dodecane	2-Octanone	Hexanoic acid	5-Pentyldihydro-2(3H)-furanone
Tridecane	2-Nonanone	Heptanoic acid	5-Hexyldihydro-2(3H)-furanone
Tetradecane	2-Decanone	Octanoic acid	5-Heptyldihydro-2(3H)-furanone
Pentadecane	2-Undecanone	Nonanoic acid	5-Octyldihydro-2(3H)-furanone
Hexadecane	2-Dodecanone	Decanoic acid	5-Nonyldihydro-2(3H)-furanone
Heptadecane	2-Tridecanone	Undecanoic acid	5-Decyldihydro-2(3H)-furanone
Octadecane	2-Tetradecanone	Dodecanoic acid	*Alcohols*
Nonadecane	2-Pentadecanone	Tridecanoic acid	Ethanol
Eicosane	2-Hexadecanone	Tetradecanoic acid	1-Propanol
Heneicosane	2-Heptadecanone	Pentadecanoic acid	1-Butanol
Docosane	2-Octadecanone	Hexadecanoic acid	1-Pentanol
Tricosane	2-Nonadecanone	Heptadecanoic acid	1-Hexanol
Tetracosane	2-Eicosanone	Octadecanoic acid	1-Heptanol
Penatcosane	2-Heneicosanone	Nonadecanoic acid	1-Octanol
Hexacosane	2-Docosanone	Eicosanoic acid	1-Nonanol
Heptacosane	2-Tricosanone	*Dicarboxylic acids*	1-Decanol
Octacosane	2-Tetracosanone	Butanedioic acid	1-Undecanol
Nonacosane	2-Pentacosanone	Pentanedioic acid	1-Dodecanol
Triacontane	2-Hexacosanone	Hexanedioic acid	1-Tridecanol
Heneitriacontane	3-Hexanone	Heptanedioic acid	1-Tetradecanol
Dotriacontane	5-Methyl-3-hexanone	Octanedioic acid	1-Pentadecanol
3-Methyloctane	3-Heptanone	Nonanedioic acid	1-Hexadecanol
3-Methylundecane	2-Methyl-4-heptanone	Decanedioic acid	1-Heptadecanol
3-Methylpentadecane	5-Methyl-3-heptanone	Tridecanedioic acid	1-Octadecanol
3-Methylhexadecane	3-Octanone	Tetradecanedioic acid	1-Nonadecanol
3-Methylheptadecane	2-Methyl-4-octanone	Pentadecanedioic acid	1-Eicosanol
Alkenes	5-Nonanone	Hexadecanedioic acid	1-Heneicosanol
1-Octene	3-Nonanone	Heptadecanedioic acid	1-Docosanol
1-Nonene	5-Decanone	Octadecanedioic acid	2-Hexanol
1-Decene	5-Undecanone	Nonadecanedioic acid	3-Hexanol
1-Undecene	3-Undecanone	*Keto-acids*	*Esters*
1-Dodecene	5-Dodecanone	4-Oxopentanoic acid	Hexanoic acid ethylester
1-Tridecene	3-Dodecanone	5-Oxohexanoic acid	Heptanoic acid ethylester
1-Tetradecene	5-Tridecanone	6-Oxoheptanoic acid	Octanoic acid ethylester
1-Pentadecene	3-Tetradecanone	7-Oxooctanoic acid	Nonanoic acid ethylester
1-Hexedecene	3-Pentadecanone	8-Oxononanoic acid	Decanoic acid ethylester
1-Heptadecene	*Aldehydes*	9-Oxodecanoic acid	Undecanoic acid ethylester
1-Octadecene	Propanal	10-Oxoundecanoic acid	Dodecanoic acid methylester
1-Nonadecene	Butanal	11-Oxododecanoic acid	Tridecanoic acid methylester
1-Eicosene	Heptanal	12-Oxotridecanoic acid	Tetradecanoic acid methylester
1-Heneicosene	Octanal	13-Oxotetradecanoic acid	Pentadecanoic acid methylester
1-Docosene	Nonanal	14-Oxopentadecanoic acid	Hexadecanoic acid methylester
1-Tricosene	Decanal	15-Oxohexadecanoic acid	Heptadecanoic acid methylester
1-Tetracosene	Undecanal	16-Oxoheptadecanoic acid	Octadecanoic acid methylester

Table 2 (continued)

Alkanes	Ketones	Carboxylic acids	Lactones
1-Pentacosene	Dodecanal	17-Oxooctadecanoic acid	Nonadecanoic acid methylester
1-Hexacosene	Tridecanal	18-Oxononadecanoic acid	Eicosanoic acid methylester
1-Heptacosene	Tetradecanal	2-Oxopentanedioic acid	Heneicosanoic acid methylester
1-Octacosene	Pentadecanal		
1-Nonacosene	Hexadecanal		
1-Triacontene	Heptadecanal		

formed was clearly seen. Only small amounts of dicarboxylic acids were formed in the most photo-stable material containing a combination of iron dimethyldithiocarbamate (photo-initiator) and nickel dibutyldithiocarbamate (photo-stabiliser) (SG3). Depending on the dicarboxylic acid, 5–12 times higher amounts were formed in the most degraded material modified with iron dimethyldithiocarbamate (SG1). Ketones were formed both during thermo-oxidation and during photo-oxidation [37, 38, 95]. More ketones were, however, formed during thermo-oxidation. During photo-oxidation larger amount of ketones and keto-acids were formed in the Scott-Gilead materials in comparison to pure LDPE, LDPE-starch and LDPE-MB, indicating somewhat different degradation mechanism [95]. Ketones absorb UV light and undergo further reactions, like conversion to carboxylic acids on prolonged radiation. This was also seen in our study as the ketones detected, in e.g. SG1 materials after 100 h of photo-oxidation, were no longer detected after prolonged oxidation. Geuskens and Kabamba have proposed a photolytically induced reaction between ketones and hydroperoxides to explain the decrease of ketones and the formation of carboxylic acids on UV radiation [102]. Hydrocarbons and ketones were not identified after thermal degradation in water [38]. If they are formed during degradation in water, they may, due to their less polar nature, remain in the polymer matrix instead of migrating out into the water phase.

Keto-acids were formed during low temperature thermal degradation in air and in water [38] as well as during the photo-oxidation [95], but they were not detected during thermo-oxidation above melting point [37]. Instead a homologous series of γ-lactones was identified after ageing at 190 °C and 230 °C. Both keto-acids and γ-lactones were identified after UV-initiated thermal degradation at 80 °C [95, 96]. The formation of γ-lactones can take place if an acyl peroxide radical abstracts γ-hydrogen intramolecularly resulting in a formation of an alkyl radical and percarboxyl group, with the γ-carbon atom containing an unpaired electron. The homolysis of the percarboxyl group followed by intramolecular recombination results in γ-lactone. [100, 103]. The reaction of carboxylic acid and hydroxyl group generated in the 1,4-positions of the polymer backbone [104] or a 1,4-dihydroxyperoxide decomposition [105] are other mechanisms giving γ-lactones. Because polyethylene hydroperoxides usually remain isolated along the polymer chain, especially at low temperatures, the reactions between two neighbouring hydroperoxides or hydroxyl and carboxylic groups are not probable at low temperatures. γ-Lactones and especially γ-butyrolactone are, thus, mainly formed under more extreme conditions and when extensive degradation had taken place [37, 106, 107].

4.2
Assimilation of Abiotic Degradation Products in Biotic Environments

It was shown rather early that oxidation products of polyethylene could be used by thermophilic fungi as a source of carbon for growth [108, 109]. The species able to grow on polyethylene oxidation products included *Aspergillus fumigatus*, widely distributed in soil and composting systems. These findings suggested that the combined photo- and biodegradation processes in which the former is completed by the latter could ultimately convert the plastics to humus, water and CO_2 [108]. Later Cornell et al. demonstrated that oligomer fractions separated from UV-irradiated polyethylene supported microbial growth whereas polymers having high molecular weights showed minimal or no growth [110]. Photo-oxidised polyethylene showed an initial rapid rate of biodegradation as measured by CO_2 formation, followed by auto-retardation to a much slower rate [64]. The initial rapid CO_2 formation is probably due to the biodegradation of photo-oxidation products and low molecular weight polymer. Indeed the extraction of low molecular weight compounds from the photo-oxidised polyethylene with hexane reduced the rate of CO_2 formation. The loss of mass was an order of magnitude higher than the mass of CO_2 evolved from photo-oxidised polyethylene in humid soil [66]. Most of the oxidation products are, thus, initially utilised in the growth of the microorganisms which, when they die, become a part of the humic structure of the soil.

In a series of studies we subjected LDPE, LDPE-starch and LDPE-MB to different abiotic and biotic aqueous environments. Degradation was initiated by pre-heating the samples for 6 days at 100 °C in air. After pre-oxidation the samples were either aged in sterile mineral medium or in mineral medium inoculated with *A. paraffineus*. After 15 months [111] or 42 months [71] the water fractions were analysed to identify the degradation products that had migrated from the different polyethylene films. After 42 months, homologous series of mono- and dicarboxylic acids and keto-acids were identified in sterile LDPE-MB samples, while a homologous series of carboxylic acids was identified in sterile LDPE [71]. LDPE-MB samples are due to the presence of pro-oxidants more susceptible to thermo-oxidation than pure LDPE and LDPE-starch. These product patterns were, thus in agreement with our earlier studies on abiotic oxidation of polyethylene that have shown that carboxylic acids are the main products during early stages of oxidation, while keto-acids and especially dicarboxylic acids increase in more severely oxidised samples. The molecular weight for LDPE-MB in sterile mineral medium had decreased from 15,200 to 5200, while the molecular weight for pure LDPE and LDPE-starch remained practically constant during ageing. There was a marked difference in the product patterns detected after ageing in biotic environment with *Arthrobacter paraffineus*. Practically all carboxylic acids, keto-acids and dicarboxylic acids had been assimilated from the biotic mineral medium. Small traces of not completely identified products, possibly alcohols, were seen in the biotic chromatograms. The molecular weight measurements of biotically and abiotically aged polyethylene support the assimilation of the low molecular weight fraction, as the molecular weight of polyethylene aged in biotic environment was generally slightly higher than the molecular weight of polyethylene aged in abiotic environment [71]. Microorganisms, thus, not only utilised the water-soluble products, but they could also remove the low molecular weight fraction of the polyethylene films.

In an unpublished study, LDPE-PO with 12% pro-oxidant consisting of LLDPE, styrene-butadiene copolymer and manganese stearate was thermo-oxidised for 7 days at 95 °C. After the pre-oxidation LDPE-PO was aged in mineral medium with compost microorganisms or in corresponding sterile mineral medium at 50 °C. Some samples were also aged in air at 50 °C. Figure 3 shows the different degradation product patterns obtained for LDPE-

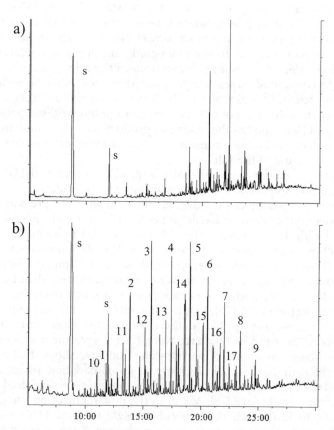

Fig. 3a, b GC-MS chromatograms showing the different degradation product patterns after ageing of pre-oxidised LDPE-PO in **a** biotic mineral medium inoculated with a mixed culture of compost microorganisms and **b** sterile mineral medium

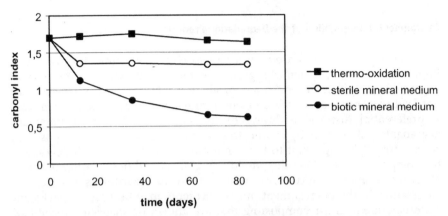

Fig. 4 Changes in carbonyl index for pre-oxidised LDPE-PO degraded at 50 °C in air, in sterile mineral medium and in biotic mineral medium with compost microorganisms

PO after ageing in sterile and biotic medium. The main products in sterile medium were different dicarboxylic acids ranging from pentanedioic acid to tridecanedioic acid (Fig. 3, peaks 1–9). In addition homologous series of keto-acids ranging from 5-oxohexanoic acid to 12-oxotridecanoic acid (Fig. 3 peaks 10–17) and small amounts of carboxylic acids were detected. As in the earlier studies, these products were absent after ageing in biotic medium. Figure 4 shows the carbonyl index for the LDPE-PO samples during ageing. During the original thermal treatment different carbonyl containing species were formed in the material and carbonyl index increased to 1.7. This value dropped during the first days in biotic and abiotic mineral medium due to the migration of low molecular weight carboxylic acids into the water phase. After the initial drop the carbonyl index of the sterile samples remained constant while the carbonyl index of biotically aged samples continued to decrease due to the ability of the microorganisms to remove and utilise carbonylated species from the polymer matrix.

The absence of carboxylic acids in the biotic medium is explained by assimilation of the low molecular weight degradation products by β-oxidation mechanism and it confirms the biodegradation mechanism of polyethylene proposed by Albertsson et al. [69]. Oxidase enzymes degrade long chain hydrocarbons and fatty acids by a β-oxidation mechanism [112]. The fatty acids are degraded by a stepwise oxidative removal of two carbon atoms as a result of oxidative attack at the β-carbon atom. The initial reaction involves activation of the fatty acid by combination with coenzyme A. The resulting fatty acetyl-CoA is then subjected to a sequence of reactions, the overall effect of which is to shorten the carbon skeleton by two carbon atoms, producing acetyl-CoA as one product, and a shortened fatty acetyl-CoA. Repeated attack on the latter will result in a complete oxidation of a long chain fatty acid to yield acetyl-CoA. The two carbon fragment acetyl-CoA enters the citric acid cycle, from which carbon dioxide and water are released.

4.3
Environmental Adaptability of the Degradation Products

Whether the degradation products are themselves environmentally safe is an important question when designing degradable polymers. The use of degradable polymers in consumer products or in agricultural waste must not lead to generation of toxic or environmentally unacceptable chemicals. A preferential disposal pathway for biodegradable and environmentally degradable polymers is by composting. For the materials to be disposed by composting it is important to know not only if the material will degrade in the compost but also how the possible fragments and degradation products affect the compost quality [113]. In order to avoid hazardous substances accumulating in the environment, no constituents may be used in packaging materials intended for composting that are known or suspected to be hazardous to the environment or could become so during composting process. In accordance, when degradable materials are used as agricultural mulch films, the polymer may be rapidly fragmented through photo-oxidation but the further degradation of the fragments usually takes longer time. The fragments and possible water-soluble degradation products should not negatively affect the soil, which may be ploughed together with the fragmented polyethylene film.

Polyethylene films containing Scott-Gilead photo-sensitisers (Plastor S-G) have been continuously used on the same fields for up to 15 years and no accumulation of plastic film or degradation products have been observed [114, 115]. No problems have been observed from the plastic residue during harvesting, tilling, sowing or transplanting [116]. The photo-sensitisers and stabilisers used in Scott-Gilead polyethylene consist of an organic part linked to a transition metal atom. There have been concerns about the use of heavy metals on the environment, on the soil and on the plants grown under these films. The transition metals used (e.g. iron, nickel) are all present in agricultural soil. Iron is essential to plant growth and is present in parts per hundred in any soil. Nickel is toxic, but it is present in low quantities in most soils. A simple calculation showed that 500 years of continuous use of plastic films would increase the nickel content of the soil by one-part per million in a soil that contained up to 300-parts per million at the outset [114]. In another study, nickel was put into the soil in greatly increased doses, simulating the use of plastic mulch on the same piece of land for up to 180 years [116]. The added nickel had no significant effect on the 'assimilable' nickel present in the soil and no changes in the constituent of the plants could be observed. Poly(vinyl alcohol) (PVA) and polyethylene have also been blended with soil and their effect on the growth behaviour of red pepper and tomato have been investigated by examining the stems, leaves and roots [117]. Biodegradable PVA significantly retarded the growth of red pepper even at concentrations as low as 0.05%. In contrast, pieces of PE film or powdery PE only negligibly influenced the growth of red pepper or tomato, even at high concentrations (up to 35 wt% in soil).

Different lactic acid based polymers including poly(lactic acid), poly(ester-urethanes) and poly (ester-amides) all degraded to over 90% of the positive control during 6 months under controlled composting conditions [118]. However, when the compost medium was used as a growth medium, the compost medium containing poly(ester-urethane) samples chain linked with 1,6-hexamethylene diisocyanate severely inhibited the growth and damaged the shoot tips of cress, radish and barley. Poly(lactic acid), poly(ester-amide) or poly(ester-urethane) chain linked with 1,4-butane diisocyanate did not cause any visible damage or growth inhibition. Degradation products can also have positive effects on plant grow e.g. it has been reported that L-lactic acid and its oligomers promote plant growth [119]. These results demonstrate that it is not enough that the material is rapidly degraded, it is equally important to establish the degradation products and how they affect the environment.

5
Outlook and Perspectives

Recent developments in catalyst systems and synthesis of functional polyolefins have created new tools to synthesise polyethylene with tailor-made structures. Well-defined functional groups can be randomly distributed along the polymer chain or at the chain ends. These functional groups introduce hydrophilicity or enable the synthesis of novel graft and blocksampolymers. These developments provide means for development of new polyethylene types specially designed for environmental degradation. The properties of the new materials can be tailored for different applications by changing the molecular weight, type and amount of functional groups, branching, morphology and so on. An advantage is that the modification will be build into the polyethylene chain and we do not need to worry about the environmental effects of the additives migrating from the material.

Nature's own waste is returned to the natural carbon cycle by biodegradation. During this process biomass is formed. If synthetic polymers are to be incorporated into this system they must be ultimately biodegradable in compost or as litter. The present generation of commodity packaging polymers are not biodegradable within a realistic time scale due to e.g. the presence of antioxidants [1]. The traditional antioxidants may in the future be replaced with new types of antioxidants that biomimic nature (e.g. vitamin E) at least in the degradable polyethylene formulations. A well-defined induction period and time-controlled degradation are important aspects for the successful development and use of degradable materials. In addition to already existing commercial controlled life-time polyethylenes, we may see the development of new degradable polyethylenes with different fragmentation times to be used in specific applications. The time to ultimate mineralisation in compost or as litter is much less important than the time taken to fragment to small particles which are subsequently biodegraded over months or years. In fact, polyolefins may have an advantage over hydrolysable polymers during composting, since once they have fragmented they mineralise slowly and increase the fertiliser value of the compost, while rapid mineralisation results in a loss of valuable nutrition [120, 121]. For applications where rapid biodegradation is advantageous, rapidly biodegradable polyethylene could be developed by copolymerisation of ester groups as a part of polyethylene chain [27, 28]. Isolated ester groups along the polyethylene chain can also be obtained by oxidising E/CO polymers by hydrogen peroxide by the Bayer-Villiger reaction [115]. We have earlier shown that biodegradation of polyethylene proceeds through intermediate formation of ester groups [69]. This modification could accelerate the biodegradation rate by directing the degradation mechanism towards rapidly hydro-biodegradable polymers.

Paralleled with the development of new materials we need to obtain better understanding of the environmental impact of degradable polymers. It is not enough that the polymer will be biodegradable or environmentally degradable it should not break down to compounds that will harm or negatively affect the environment. The ultimate fate of all individual components and the identity of intermediate and final degradation products must be well characterised in order to understand the degradation process and the environmental impact of the material. We have shown that chromatographic analysis is an effective way to study the intermediate and final degradation products formed during biotic and abiotic degradation [37, 38, 95, 96] and we have developed methods with chromatographic fingerprinting to differentiate between ageing in biotic and abiotic environments [71, 111, 122, 123]. It is really not important if the polymer is fragmented biotically or abiotically, as long as the degradation products are compatible with nature and can be utilised by microorganisms. The chemical structure of the polymer and additives determines the degradation products pattern and we have shown that, e.g. small changes in the pro-oxidant formulation can cause differences in the degradation product pattern and render the product pattern more environmentally adaptable [99]. It seems inevitable that the environmentally degradable commodity plastics will have an increasing role in the management of waste and litter in the future. However, one of the key parameters for successful use and development of environmentally degradable polymeric materials is a better understanding of the dynamic interactions between nature and degradation products.

Acknowledgements Prof. G. Scott is thanked for his comments on the manuscript.

References

1. Scott G (1999) Macromol Symp 144:113
2. Wasserbauer R, Beranová M, Vancurová D, Dolezcl (1990) Biomaterials 11:36
3. Heskins M, Guillet JE (1968) Macromolecules 1:97
4. Harlan G, Kmiec C (1995) Ethylene-carbon monoxide copolymers. In: Scott G, Gilead D (eds) Degradable polymers: principles and applications. Chapman & Hall, London, chap 8
5. Guillet JE (1973) US Patent 3 753 952

6. Guillet JE (1973) Polymers with controlled life times. In: Guillet JE (ed) Polymers and ecological problems. Plenum, New York, p 1
7. Guillet JE (1974) US Patent 3 811 931
8. Amin MU, Scott G (1974) Eur Polym J 10:1019
9. Scott G (1973) Delayed action photo-activator for the degradation of packaging polymers. In: Guillet JE (ed) Polymers and ecological problems. Plenum, New York, p 27
10. Scott G (1975) Polym Age 6:54
11. Scott G (1972) British Patent 1 436 553
12. Scott G (1978) US Patent 4 121 025
13. Gilead D, Scott G (1978) British Patent 1 586 344
14. Al-Malaika S, Marogi AM, Scott G (1986) J Appl Polym Sci 31:685
15. Griffin GJL (1974) British Patent 55195/73
16. Griffin GJL (1974) Biodegradable fillers in thermoplastics. Adv Chem Ser 134:159
17. Griffin GJL (1988) Int Patent WO 88/9354
18. Otey FH, Mark AM, Mehltretter CL, Russell CR (1974) Ind Eng Chem Prod Res Dev 13:90
19. Otey FH, Westhoff RP, Doane WM (1980) Ind Eng Chem Prod Res Dev 19:592
20. Moad G (1999) Prog Polym Sci 24:81
21. Yoo Y-D, Kim Y-W, Cho W-Y (1995) US Patent 5 461 094
22. Tudorachi N, Rusu M, Cascaval CN, Constantin L, Rugina V (2000) Cellulose Chem Technol 34:101
23. Mikulasova M, Kosiková B, Alexy P, Kacik F, Urgelová E (2001) World J Microbiol Biotechnol 17:601
24. Tilstra L, Johnsonbaugh D (1993) J Environ Polym Degrad 1:257
25. Kim J, Kim JH, Shin TK, Choi HJ, Jhon MS (2001) Eur Polym J 37:2131
26. Labuzek S, Pajak J, Nowak B, Majdiuk E, Karcz J (2002) Polimery 47:256
27. Bailey WJ, Gapud B (1985) Synthesis of biodegradable polyethylene. In: Klemchuk PP (ed) Polymer degradation and stabilization (ACS symposium series 280). ACS, Washington DC, p 423
28. Bailey WJ, Kuruganti VK (1990) Polym Mater Sci Eng 62:971
29. Austin RG (1994) US Patent 5 281 681
30. Austin RG (1994) US Patent 5 334 700
31. Holmström A, Sörvik E (1974) J Appl Polym Sci 18:3153
32. Hoff A, Jacobsson S (1981) J Appl Polym Sci 26:3409
33. Madsen J, Olsen NB, Atlung G (1985) Polym Degrad Stab 12:131
34. Bravo A, Hotchkiss JH, Acree TE (1992) J Agric Food Chem 40:1881
35. Andersson T, Wesslén B, Sandström J (2002) J Appl Polym Sci 86:1580
36. Billingham NC, Calvert PD (1982) In: Scott G (ed) Developments in polymer stabilisation -5. Applied Science, London, chap 5
37. Hakkarainen M, Albertsson A-C, Karlsson S (1996) J Chromatogr A 741:251
38. Hakkarainen M, Albertsson A-C, Karlsson S (1997) J Appl Polym Sci 66:959
39. van Sickle DE, Mill T, Mayo FR, Richardsson H, Gould CW (1973) J Org Chem 38:4435
40. Hendry DG, Gould CW, Schuetzle D, Syz MG, Mayo FR (1976) J Org Chem 41:1
41. Chien JCW (1968) J Polym Sci Part A-1 6:375
42. Iring M, Kelen T, Tüdös F (1974) Makromol Chem 175:467
43. Iring M, Kelen T, Tüdös F, Laszlo-Hedvig Zs (1976) J Polym Sci Symp 57:89
44. Gugumus F (1995) Polym Degrad Stab 49:29
45. Holmström A, Sörvik E (1978) J Polym Sci Polym Chem 16:2555
46. Gugumus F (1990) Angew Makromol Chem 182:111
47. Iring M, Tüdös F (1990) Prog Polym Sci 15:217
48. Tidjani A, Watanabe Y (1995) Polym Degrad Stab 49:299
49. Scott G (1995) Introduction to the abiotic degradation of carbon chain polymers. In: Scott G, Gilead D (eds) Degradable polymers: principles and applications. Chapman & Hall, London, chap 1
50. Carlsson DJ, Brousseau R, Zhang C, Wiles DM (1987) Polym Degrad Stab 17:303

51. Lacoste J, Carlsson DJ, Falicki S, Wiles DM (1991) Polym Degrad Stab 34:309
52. Albertsson A-C, Barenstedt C, Karlsson S (1992) Polym Degrad Stab 37:163
53. Sung W, Nikolov ZL (1992) Ind Eng Chem Res 31:1332
54. Breslin VT (1993) J Environ Polym Degrad 1:127
55. Johnson KE, Pometto III AL, Nikolov ZL (1993) Appl Environ Microbiol 59:1155
56. Weiland M, David C (1994) Polym Degrad Stab 45:371
57. Potts JE, Clendinning RA, Ackart WB, Niegisch WD (1973) The biodegradability of synthetic polymers. In: Guillet JE (ed) Polymers and ecological problems. Plenum, New York, p 61
58. Potts JE (1978) Biodegradation. In: Jellinek HHG (ed) Aspects of degradation and stabilization of polymers. Elsevier, Amsterdam, p 617
59. Haines JR, Alexander M (1974) Appl Microbiol 28:1084
60. Albertsson A-C (1978) J Appl Polym Sci 22:3419
61. Albertsson A-C, Rånby B (1979) IUPAC International symposium on long-term properties of polymer materials. Stockholm, Sweden. Appl Polym Symp 35:423
62. Albertsson A-C, Banhidi ZG, Beyer-Ericsson LL (1978) J Appl Polym Sci 22:3434
63. Albertsson A-C, Banhidi ZG (1980) J Appl Polym Sci 25:1655
64. Albertsson A-C (1980) Eur Polym J 16:623
65. Albertsson A-C, Karlsson S (1988) J Appl Polym Sci 35:1289
66. Albertsson A-C (1989) The synergism between biodegradation of polyethylenes and environmental factors. In: Patsis A (ed) Advances in stabilization and degradation of polymers. Technomic, Lancaster, p 115
67. Karlsson S, Ljungqvist O, Albertsson A-C (1988) Polym Degrad Stab 21:237
68. Albertsson A-C (1977) Studies on mineralization of^{14}C labelled polyethylenes in aerobic biodegradation and aqueous aging. PhD Thesis, Royal Institute of Technology, Stockholm, Sweden
69. Albertsson A-C, Andersson SO, Karlsson S (1987) Polym Degrad Stab 18:73
70. Weiland M, David C (1995) Polym Degrad Stab 48:275
71. Albertsson A-C, Erlandsson B, Hakkarainen M, Karlsson S (1998) J Environ Polym Degrad 6:187
72. Volke-Sepúlveda T, Favela-Torres E, Manzur-Guzmán A, Limón-González M, Trejo-Quintero G (1999) J Appl Polym Sci 73:1435
73. Erlandsson B, Albertsson A-C, Karlsson S (1998) Acta Polym 49:363
74. Pometto AL, Johnson KE, Kim M (1993) J Environ Polym Degrad 1:213
75. Griffin GJL (1976) J Polym Sci Symp 57:281
76. Chiellini E, Cioni F, Solaro R, Vallini G, Corti A, Pera A (1993) J Environ Polym Degrad 1:167
77. Vallini G, Corti A, Pera A, Solaro R, Cioni F, Chiellini E (1994) J Gen Appl Microbiol 40:445
78. Gilmore DF, Antoun S, Lenz RW, Goodwin S, Austin R, Fuller RC (1992) J Ind Microbiol 10:199
79. Goheen SM, Wool RP (1991) J Appl Polym Sci 42:2691
80. Greizerstein HB, Syracuse JA, Kostyniak PJ (1993) Polym Degrad Stab 39:251
81. David C, Trojan M, Daro A, Demarteau W (1992) Polym Degrad Stab 37:233
82. Andrady AL, Pegram JE, Song Y (1993) J Environ Polym Degrad 1:117
83. Tidjani A, Arnaud R (1993) Polym Degrad Stab 39:285
84. Hamid SH, Amin MB (1995) J Appl Polym Sci 55:1385
85. Breslin VT, Swanson RL (1993) J Air Waste Manage Assoc 43:325
86. Leonas KK, Gorden RW (1993) J Environ Polym Degrad 1:45
87. Gonsalves KE, Patel SH, Chen X (1991) J Appl Polym Sci 43:405
88. Breslin VT, Li BE (1993) J Appl Polym Sci 48:2063
89. Arnaud P, Dabin P, Lemaire J, Al-Malaika S, Chohan S, Coker M, Scott G, Fauve A, Maaroufi A (1994) Polym Degrad Stab 46:211
90. Yamada-Onodera K, Mukumoto H, Katsuyaya Y, Saiganji A, Tani Y (2001) Polym Degrad Stab 72:323

91. Spencer LM, Heskins M, Guillet JE (1976) Studies on the biodegradability of photode-graded polymers: identification of bacterial types. In: Sharpley JM, Kaplan AM (eds) Proc 3rd Int Biodegrad Symp. Applied Science, London, p 753
92. Jones PH, Prasad D, Heskins N, Morgan NH, Guillet JE (1974) Environ Sci Technol 8:919
93. Otake Y, Kobayashi T, Asabe H, Murakami N, Ono K (1995) J Appl Polym Sci 56:1789
94. Otake Y, Kobayashi T, Asabe H, Murakami N, Ono K (1998) J Appl Polym Sci 70:1643
95. Karlsson S, Hakkarainen M, Albertsson A-C (1997) Macromolecules 30:7721
96. Hakkarainen M, Albertsson A-C, Karlsson S (1997) J Environ Polym Degrad 5:67
97. Khabbaz F, Albertsson A-C, Karlsson S (1998) Polym Degrad Stab 61:329
98. Khabbaz F, Albertsson A-C, Karlsson S (1999) Polym Degrad Stab 63:127
99. Khabbaz F, Albertsson A-C (2000) Biomacromolecules 1:665
100. Iring M, Tüdös F, Fodor ZS, Kelen T (1980) Polym Degrad Stab 2:143
101. Tüdös F, Iring M (1998) Acta Polym 39:19
102. Geuskens G, Kabamba MS (1983) Polym Degrad Stab 5:399
103. Adams JH (1970) J Polym Sci A-1 8:1077
104. Lacoste J, Carlsson DJ, Falicki S, Wiles DM (1991) Polym Degrad Stab 34:309
105. Lacoste J, Arnaud R, Singh P, Lemaire J (1988) Makromol Chem 189:651
106. Albertsson A-C, Barenstedt C, Karlsson S (1994) Acta Polym 45:97
107. Albertsson A-C, Barenstedt C, Karlsson S (1995) J Chromatogr A 690:207
108. Eggins HOW, Mills J, Holt A, Scott G (1971) Biodeterioration and biodegradation of synthetic polymers. In: Sykes G, Skinner FA (eds) Microbial aspects of pollution. Academic Press, London, p 267
109. Mills J, Eggins HOW (1970) Int Biodetn Bull 6:13
110. Cornell JH, Kaplan AM, Rogers MR (1984) J Appl Polym Sci 29:2581
111. Albertsson A-C, Barenstedt C, Karlsson S, Lindberg T (1995) Polymer 36:3075
112. Schlegel HG (1979) Allgemeine Mikrobiologie. Georg Thieme, Stuttgart
113. Pagga U (1999) Appl Microbiol Biotechnol 51:125
114. Gilead D (1995) The disposal of mulching films after use. In: Scott G, Gilead D (eds) Degradable polymers: principles and applications. Chapman & Hall, London, chap 10
115. Scott G (1999) Polymers and the environment. Royal Society of Chemistry, Cambridge, chap 5
116. Fabbri A (1995) The role of degradable polymers in agricultural systems. In: Scott G, Gilead D (eds) Degradable polymers: principles and applications. Chapman & Hall, London, chap 11
117. Lee J-A, Kim M-N (2001) J Polym Environ 9:91
118. Tuominen J, Kylmä J, Kapanen A, Venelampi O, Itävaara M, Seppälä J (2002) Biomacromolecules 3:445
119. Kinnersley AM, Scott III TC, Yopp JH, Whitten GH (1990) Plant Growth Regul 9:137
120. Scott G (2000) Polym Degrad Stab 68:1
121. Scott G, Wiles DM (2001) Biomacromolecules 2:615
122. Hakkarainen M, Albertsson A-C, Karlsson S (2000) Polymer 41:2331
123. Hakkarainen M, Albertsson A-C (2002) Macromol Chem Phys 203:1357

Received: July 2003

Adv Polym Sci (2004) 169:201–229
DOI: 10.1007/b94173

Recycled Polyolefins. Material Properties and Means for Quality Determination

Sigbritt Karlsson

Department of Fibre and Polymer Science, The Royal Institute of Technology (KTH),
Teknikringen 56–58, 100 44 Stockholm, Sweden
E-mail: *sigbritt@polymer.kth.se*

Abstract Recycling of polymers has come to be a necessary part of the development of a sustainable society. Recycling of low price bulk polymers, to which group the polyolefins belong, is seen by many as a waste of time and resources when they simply may be energy recovered. Several life-cycle assessments (LCA) have, however, proven that it is also valuable to material recycle bulk polymers such as polyolefins. In many instances, polyolefins are in-plant recycled and used again in similar products, but they may also be separated and sorted from municipal solid waste. This paper will discuss recycled polyolefins, in particular their change in material properties and how to characterise these properties and show that these analyses are a basis for quality determination. In fact, an important vehicle for the success of recycled polymeric materials is to use a quality concept. Accurate determination of a series of polymeric properties will be the only way recycled polymeric materials can compete with virgin ones. Analytical methods useful in the quality concept are presented and discussed. In particular, three parameters are important for quality measurements. These are degree of degradation, polymer composition and presence of low molecular weight compounds (degradation products of polymer matrix and additives, initiator/catalysts, solvents, use-related e.g. fragrance or flavour etc.). For the future it is important to give recycled polymeric materials status as resources besides the fossil and renewable ones.

Keywords Recycling · Recycling as resources · Polyolefins · Quality · Degree of degradation · Polymer composition · Low molecular weight compounds · NIR · DSC · MAE · Chromatography · Multivariate modelling

Abbreviations

APME	Association of Plastic Producers in Europe
ATR	Attenuated total reflectance
CI	Carbonyl index
CL	Chemiluminescence
CLS	Classical least square
DSC	Differential scanning calorimetry
FDA	Food and Drug Administration
LCA	Life cycle assessment
MIR	Mid-infrared
MLR	Multiple linear regression
MSC	Multiple scattering correction

MSW	Municipal solid waste
NIR	Near infrared
OIT	Oxidation induction time
PLS	Partial least square
PCR	Post-consumer recyclate
PCW	Post-consumer waste
PCA	Principal component analysis
PCR	Principal component regression
RMSEP	Root-mean-square error of prediction

1
Introduction

Recycling of polymeric materials has emerged as a mature means to preserve material and energy. It is appropriate to introduce a concept of viewing recycled plastics as another resource, besides virgin resources (fossil and renewable). In principle, recycling can be done by mechanical, chemical and energy recovery. Mechanical recycling is performed by physical means, e.g. grinding, re-melting and processing. Chemical recycling degrades the polymers to its monomers or other low molecular weight products, which can be used in new products. Energy recovery generates heat or electricity by incineration of plastic waste.

To fully appreciate recycled polymeric materials it is central that these materials demonstrate defined properties suitable for making new products. For the future a product manufacturer shall be able to use both virgin materials but also recycled ones of various origin.

Recycling of polyolefins is perhaps the most challenging part in the area of recycling of polymeric materials, primarily from two aspects: the inherent material properties after use and reprocessing and its economic possibilities. The bulk polymers, to which polyolefins such as PE (LDPE, HDPE), PP, PS, PET and PVC belong, are available in large amounts at low prices, which makes them first choice for cheap solutions in e.g. packaging applications, building materials, etc. Polymers suffer chemical and physical changes during processing and service life. They undergo reactions with oxygen at every stage of their life cycle. New functional groups formed during the oxidation process enhance the sensitivity of the recyclate to thermal and photo-degradation. With the formation of new pro-oxidative moieties, a substantial part of the stabilisers are consumed. Consequently, the long-term stability and mechanical properties of the recycled polyolefins decrease. Furthermore, during their application and recovery plastic materials come in contact with non-polymeric impurities and with other polymers, which contaminate and affect the performance of the recycled resins. Polymers are particularly sensitive to absorption of low molecular weight compounds due to their permeable nature. The absorbed compounds may act as contaminants in a next generation use of recycled materials.

There is a demand for quality assurance of recycled polymers, not only for high-grade applications. The properties of the recyclate must be specified and guaranteed within narrow tolerances by the manufactures according to the needs of their customers.

The Association of Plastic Producer in Europe (APME) has described the potential for recycling post-use plastics waste in Western Europe: The consumption of plastics is predicted to grow to 36.9 million tons between 1995 and 2006. The amount of waste generated within this period will increase from 64 to 69% of processor total plastic consumption [1]. The packaging sector accounts for 40% of the total consumption, being in this way the major consumer of plastics. Household and industry generate large amounts of plastic waste. Over 19 million tons of post-user plastics wastes were produced in Western Europe in 1999 and more than half was produced by households [2]. In spite of the fact that plastics make up less than 1% of the total waste (2 730 million tons) and approx. 7% (20% in weight) of the MSW, plastic waste in general, and plastic packaging waste in particular, have received public attention and criticism due to its high visibility, voluminous nature, slow degradability and short use life.

Traditional landfill presents several disadvantages since the space available for landfill has become scarce. In addition, municipal waste has to be transported over increasing distances with associated wastage of energy [3]. Leachates from unprepared landfills may contain hazardous levels of substances such as ammonium salts, heavy metals and organic chemical waste that may contaminate the air, soil and ground water [4] and thus may affect crops and secondary animals and man [3, 4, 5, 6, 7].

In 1995, over one quarter of the 16 million tons of post-use plastics waste produced was recovered in Western Europe. Of this, 7.6% was mechanically recycled; feed stock recycling accounted for 0.6% and almost 17% of post-user plastic waste was subjected to energy recovery. In 2001, 9.9% of the plastic waste produced is estimated to be mechanically recycled and by 2006 this figure will to increase to 10.6 or 2.7 million tons [1, 2]. In Sweden, 37% of the plastic packaging from municipal solid waste (MSW) was recycled or re-used in 2000 [8] and 17% was recycled mechanically, which is above the minimum 15% recycling target required by the European Union's Directive on Packaging and Packaging Waste [9]. However, incineration and landfill remain the prevailing treatment of MSW disposal [10]. To maximise the potential of mechanical recycling, attention should be focused on sectors which produce rather homogeneous waste streams and are also easy to collect.

Still, the demand for recycled plastics is low, mainly due to a general uncertainty of the quality of the recycled material. The possibility of offering high quality recycled polymeric materials relays on a reliable and selective collection and sorting, an efficient recycling process and of course an available end-market. The basic technological development, thus, lies in the development of fast, reliable and relatively cheap methods for characterisation of separated and commingled plastic fractions. Recently, it was demonstrated that separation, at home, of a very small amount of products can mean a decreased use of raw oil and less emission of CO_2 to the environment [11]. If

25% of the polymeric materials used in a computer screen were made out of recycled plastic it would mean an energy gain corresponding to 5 300 m^3 raw oil and a decrease in CO_2 release corresponding to 2300 petrol-based cars.

2
Important Quality Properties

Quality assessment of recycled household and industrial plastics is of major concern since producers have to satisfy quality requirements from consumers.

To a plastic producer (i.e. processor), melt index is one property that is needed in order to evaluate whether the same process can be used irrespective of whether it uses virgin or recycled polymers. This will tell if it is possible to process the recycled polymeric materials in the same set-up as usual. Several other properties are needed in order to quality mark the materials. The melt index is related to what final tensile properties a product obtains, this in turn has an impact on the expected life-time. The purity of a recyclate stream with respect to the amount of foreign polymer in the stream has an impact on melt-index, but will also be an important factor for the final mechanical properties. Another very important property is the amount of low molecular weight compounds, which may be of vastly different types. Typically such an analysis will show the presences of additives and their degradation products, degradation products of the polymeric matrices, traces of solvents, initiators, or catalysts, compounds related to the use of the plastics and others.

The quality and properties of recyclate should be comparable to those of virgin resins. Therefore, upgrade of recycled materials might be necessary in order to improve their properties. Clearly the effect of polymeric contaminants on the end-use performance of recyclate is an important issue, and the amount of contaminants that can be tolerated depends on the final application. Polymeric contaminants do not only pose a threat for environmental impact but also impact the ultimate tensile elongation behaviour [12, 13].

It may be anticipated that the antioxidant consumption during use leads to a shorter remaining life-time and that it is necessary to add more. Recycled plastics suffer from the consequences of degradation during processing and first-life application. This leads to the introduction of new functional groups, which in particular for oxidisable polymers such as PE, enhance the sensitivity of the recyclate to thermal- and photo-degradation. With the formation of new pro-oxidative moieties, a substantial part of the stabilisers are simultaneously consumed [14].

The changes in mechanical properties due to degradation and/or blending, contamination, antioxidant consumption and general use (e.g. oxidation processes) of the polymeric material during first service life are some of the characteristics that determine if the material should be recycled mechanically or disposed of in another way (e.g. energy recovery). A successful recycling is strictly dependent on the purity and uniformity of the plastic waste.

Post-industrial waste and sorted post-consumer wastes (PCW) are suitable raw materials. However, they contain different amounts of inhomogeneities formed during the service life and non-polymeric impurities. These contaminants account for differences in mechanical properties and ageing resistance between recyclates and the respective virgin plastics. Recycled plastics are often used in low-grade applications, and it is generally believed that the mechanical performance of recyclates is lower than that of virgin materials.

The most difficult recyclates are the MSW, which usually consists of more or less dirty plastics of various types and sources. According to one report [15] the plastic portion of MSW contains typically 50% PE (mainly low density polyethylene, LDPE), 8–14% polypropylene (PP), 15% polystyrene (PS), 10% polyvinylchloride (PVC) and 5% polyethylene terephthalate (PET), and 5% other polymers, while the hard packaging fraction consists of approx. 60% of PP and HDPE [16].

3
Current European Legislation and Standardisation on Recycled Plastics and Rubbers

The main driving force towards recycling has been national and international plastic waste management policy. Within the EU, the policy is based on waste hierarchy, meaning that the first choice is waste prevention and then waste recovery (reuse, recycling and energy recovery), with preference to material recovery. Thus the type of legislation found is:

1. Directive on waste
2. Regulation on the supervision and control of shipment of waste
3. Directive on packaging and packaging waste
4. Directive on the landfill of waste
5. Directive on end-of-life vehicles

In 1999 about 1.0 million tons of plastic wastes generated in Western Europe were mechanically recycled while 4 million tons underwent energy recovery and the remaining 13 million tons were landfills. Feed stock recycling was only marginal, amounting to about 350,000 tons. The work in this area is handled by CEN, TC 249, which currently is working on a new set of standards in the area of recycling of polymers. A very important part of this work is the overview of the vocabulary used in recycling. A project recently identified contamination as one difficult word in the recycling context as this is used to denote a range of meanings from pure dirt (e.g. soil) to low molecular weight compounds from the degradation of the polymer itself [17] The standard work within CEN will develop standards for recycling of plastics (CEN/TC/WG11) and working groups deal with vocabulary, classification and marking of plastics recyclates, traceability, sampling and characterisation of plastics (polyolefins).

4
Obstacles and Possibilities for Recycling of Polyolefins

There is a general belief among the producers of polyolefins that the best choice for waste handling is incineration of waste polyolefins recovering the energy. This blocks the possibility of finding techniques and the development of separation methods that may generate pure plastic streams. Several life cycle assessments (LCA) report that material recovery is better than energy recovery, even for cheap polyolefins [11]. The main points for achieving successful material recycling of polyolefins are automatic separation into clean (no mixture of different polymers) streams with online measurements of polymeric properties. Each plastic stream must be accompanied with a quality mark specifying e.g. melt-index, degree of crystallinity, degree of degradation and level and type of low molecular weight compounds.

The objectives of this paper are to discuss and present some examples of how polymer characterisation must be used in order to determine a range of important polymeric properties in recycled plastics. These properties in turn are used to define the quality of recycled plastics.

4.1
Separation and Sorting of Mixed Plastic Waste

It is generally known that to obtain pure fractions from a mixture of different plastics is a more expensive process than separation of simple clean polymers, in plant. Therefore, it is anticipated that the separation of a municipal mixed plastic waste should be the most challenging. In such a fraction, traces of foods, labels, dirt (size mm), solvents (alcohols, petrol etc.), metals, low molecular weight products of different origins etc. could be found. Some of the sorting methods used are listed in Table 1.

Manual sorting on order basis is generally used, and the efficiency of such a sorting is often good. In some instances, it has been found that manual sorting is faster than automatic. This, however, has more to do with the still

Table 1 Some examples of sorting methods

Method	Separation property	Comments
Manual sorting	Only for large items	Very labour intensive, bad working environment
Triboelectric	Based on electrostatic charge	Only for clean, dry and non-surface-treated products
Mid infrared (MIR)	Fundamental vibrations	Surface sensitive, can measure black items, expensive
Near infrared (NIR)	Fundamental vibrations	Not applicable for dark or black products, expensive
Density sorting	Large difference in density	Fillers may alter the density. Low cost

very low technology development in the area of automatic techniques, which results in non-optimised separators, than in some sort of truth.

Online separations are often based on spectroscopic identification of polymeric type, but seldom give any indications of polymeric properties (e.g. melt index, crystallinity, mechanical properties) or of the level of low molecular weight compounds. A complicating factor when trying to define properties of recycled plastics has to do with the word 'contamination'. Depending on the context and also the person it may imply as vastly different things as dirt or low molecular weight compounds.

Electrostatic sorting is a method that charge plastics differently depending on the type of plastic. A wide variety of equipment exists. The most common means is to charge the materials by triboelectric charging, by tumbling the materials against each other. These render some materials positive and other negative. The materials are then sorted by letting them fall freely through an electric field [18]

Another common method is sorting by density in a float-sink tank or hydrocyclone. This method is, however, difficult for polyolefins as these have very similar density. Air classification uses a combination of density and shape to achieve the separation, e.g. for separating plastic films and paper residues.

A more advanced sorting is achieved by use of mid-infrared (MIR) and near-infrared (NIR) spectroscopy. These methods are used at several sites, but suffer from insensitivity when the mixed plastics are dirty and/or painted or have a paper label. In addition, for NIR, it is not possible to separate dark plastics [19, 20]

Although several very promising sorting techniques have been introduced, many recycling industries have not developed these further. This means that the items/sec to be sorted are still very low and thus it is difficult to obtain good sorting results and high volumes. The lack of investment in the area further slows down the possibility of sorting complex mixtures of dirty plastics [20].

5
Property Changes in Recycled Plastics

The material value of plastics intended for re-use is influenced by their origin and history. The post-industrial scrap is usually added to virgin material in small portions, so that the influence of the scrap on the virgin resin is negligible. Reprocessing of 100% post-industrial waste or post-industrial waste/virgin blends is often carried out in the same way and using the same technical equipment as used for virgin material. However, in this instance, upgrading by processing stabilisation is mandatory [21, 22, 23, 24]. Post-consumer plastic waste from single polymers and well-defined applications such as containers, automotive parts and electrical parts need more careful restabilisation for the second life. The commingled material is degraded to a varying extent depending on the polymer mixture, the recovery source and the material history.

5.1
Contaminations

5.1.1
Contaminations in Recycled Plastics as Defined as Structural Inhomogeneties

When compared with virgin plastics, the heterogeneity of the polymer mixture and admixed organic and mineral impurities in material made from recycled plastics results in different sensitivity to the attacking environment and consequently in its service lifetime. Some physical damage may be reversed by reprocessing but chemical changes such as crosslinking, chain scission or formation of new functional groups are irreversible. Irreversible structural changes take place at both molecular and morphological levels. The changes are induced mechano-chemically, chemically or by radiation [25, 26, 27].

Carbon-centred (alkyl) and oxygen-centred (alkylperoxyl, alkoxyl, acylperoxyl, acyl) free radicals formed due to oxidation of polymeric materials are precursors of transformation products. Oxygen-containing moieties (hydroperoxides, carbonyls, alcohols, carboxyls, lactones) and olefinically unsaturated groups (vinylic, vinylidene, allilyc) are the ultimate products [28, 29]. The complexity of transformations is increased by the formation of crosslinked structures caused by the radical recombination of low molecular fragments, by disproportionation, depolymerisation or by partial hydrolysis of some polymers containing amidic carbonate or ester groups [30].

The concentration of the new structures increases with the previous service lifetime of the polymer and the aggressiveness of the environment. Some new oxygenated structures have thermo-initiating or photo-sensitising properties. As a result they act as pro-degradants, i.e. enhance the degradation rate of polymers in which they are generated. Less degraded components of the recyclate are contaminated during reprocessing by pro-degradants, such as chromophores ($=CO$, $ROOH$) from heavily degraded components. This causes a distribution of oxidation promoters in the whole system, resulting in a material more susceptible to degradation and to diminished mechanical properties. Hence polymers that have lost their mechanical properties and have high concentrations of pro-degradants are not suitable for high quality re-use.

5.1.2
Contamination as Defined as Introduced Impurities

Impurities are mostly present in trace concentrations and accumulate in recycled plastics. Some arise from originally used additives, e.g. some phenolic antioxidants, consumed during the stabilisation process through reaction with $ROO\cdot$, generate coloured cross-conjugated dienones (cyclohexadienones, benzoquinones) [31, 32, 33]. Even though these transformation products are present in trace concentrations, some of them discolour the polymer matrix as they have high extinction coefficients in visible light. Inorganic salts are formed from heat stabilisers for PVC and all transformation products remain in the aged polymer as impurity.

Some undesirable impurities deteriorate the material properties of recycled plastics by reducing their stability. These contaminants consist not only of residues of polymerisation catalysts, but also of salts of metals introduced during polymer processing and exploitation [34, 35]. Metallic impurities arise from contaminated filler as well. Ions of copper and iron belong to the most dangerous species. They catalyse homolysis of hydroperoxides and increase the consumption of phenolic antioxidants or phenolic moieties of UV absorbers by their oxidation into dienoide compounds [36, 37, 38]. Residues of titanium and aluminium polymerisation catalysts can form coloured salts with phenolic antioxidants.

Traces of halogenated flame-retardants, printing inks, paint residues, surfactants or fatty materials are also impurities sensitising degradation of plastics. Plastic materials are particularly vulnerable to absorption of detectable amounts of contaminants because of their permeable nature. The migration of compounds out of the packaging material into the package filling should not have detrimental effects on the quality of the filling, therefore recycled materials used for food packaging require some special considerations to ensure that non-regulated chemicals or contaminants either are not present in the material or do not migrate into food. In the case of plastic packaging waste, wasted packages might have come in contact with substances other than food e.g. flavour components and other harmful contaminants which diffuse into the package material [39]. In the new life of the package, these substances may migrate from the recycled packing material into the food. This absorption/desorption process observed in plastics makes them difficult to recycle directly into new food packaging [40].

The post-consumer collected plastic may be subjected to additional reprocessing steps to reduce or remove contaminants. Some of these more rigorous processes are expensive. Therefore, it is necessary to consider alternative approaches to ensure the safe use of recycled polymers in food applications. Instead of attempting to reduce contaminant levels in recycled polymers to those found in virgin materials, an alternative may be to use diffusion theory to estimate the amount of contamination that migrates into food. In its Regulation Policy of October 1993, the Food and Drug Administration (FDA) proposed a maximum allowable migration. This threshold policy proposes a maximum safe dietary exposure of 0.5 ppb to a non-carcinogenic chemical compound [41].

One approach to reduce the contaminant levels consists in reusing the wasted plastic as the core of the new material. Residues of pesticides or harmful contaminants may limit recycling of plastics as a result of their potential toxicity. Utilisation of post-consumer plastics for pharmaceutical or food-contact applications is forbidden, and multilayer food packaging materials manufactured using functional barriers are subjected to strict regulations [9, 40, 41].

Mechanical recycling is one of the main targets of the EU Directive 94/62/EEC on packaging and packaging waste [9]. Recycled packaging materials need end-use markets and applications. However, the use of recycled materials for high-grade applications such as food contact applications is limited and has to meet the same requirements and directives regarding chemical composition and migration as applied to virgin materials. It is assumed that only pre-consumer plastic waste, referred to as plastic waste Category I, fulfils these requirements and is suitable as virgin material for food contact applications. Post-consumer plastics, so called Category II, represent a material with limited information about its purity [42].

In order to guarantee the suitability of the recyclate for direct food-contact or multilayer applications, in the presence of appropriate functional barriers, identification and quantification of low molecular weight contaminants in recycled resins becomes necessary.

6
Upgrading of Recycled Polyolefins

This section discusses the restabilisation and compatibilisation of mixed plastic recyclates. Restabilisation deals with the upgrading of recycled material by adjusted heat and light stabiliser combinations during processing. Compatibilisation is a physical and—depending on the compatibiliser—chemical method for enhancing the recovery of performance of immiscible polymer blends and especially for improving the mechanical properties.

6.1
Restabilisation by Addition of Fresh Additives

Success in the recycling of post-consumer waste (PCW) is based on preservation of the resistance of the material to deterioration by stabilisation [43]. The pro-degradant effects of structural inhomogeneities and impurities, formed or introduced during their first life time, and of loss of stabiliser have to be compensated by proper additives, which must fit the type and complexity of the recycle, the source of material and the next expected application. A constant composition of the recyclate is an important condition for upgrading. Repeated processing steps showed detrimental consequences for the polymer, such as progressive transformations at molecular level and loss of the active form of stabilisers; this loss continues during thermal ageing and weathering. Common principal transformation products of stabilisers are quinone methides generated from phenolic antioxidants [44, 45]; phosphates from organic phosphites; oxidation products of sulphides, including organic acids of sulphur [15]; and open-chain nitroso and nitro compounds from hindered amine stabilisers (HAS) [46]. Quinone methides and transformation products of sulphides contribute to resistance to degradation [47, 48].

Reprocessing of post-consumer recyclate (PCR) by exposing it to shear and heat stress produces polyolefins that have lower elongation at break, lower impact strength and less stiffness than required, due to the low remaining level of the active form of the stabiliser. Experiments indicate that even addition of phosphites alone enhances the stability during subsequent processing and heat ageing [12, 16]. Phosphites such as P-1 or P-2 reduce hydroperoxides to alcohols and hinder their thermo- and photo-initialising effect, and also lower the influence of the photosensitive carbonyl groups, but do not remove them. Additional amounts of phenols (like AO-1) as antioxidants deactivate alkylperoxy radicals (POO·) and guarantee the long-term heat stability of PCR during the next life time. Phenols and phosphites are used in synergistic combinations [16]. Up to 100% recyclate may be used for indoor purposes.

The behaviour of recyclate in outdoor exposure is more delicate because its degradation is enhanced by the photosensitive effect of chromophores, such as carbonyl groups and hydroperoxides. Blending 15% recycled HDPE with stabilised HDPE reduced the life-time of the latter in a weatherometer by about 20%. The pro-oxidative effect was progressive when increasing the content of recycled polyethylene [49]. Therefore, the restabilisation system for outdoor applications has to contain an efficient light stabiliser. The products of choice are HAS compounds or combinations of HAS with UV absorbers [50]. This can be illustrated by an HDPE recycle made from 5 year old pigmented bottle crates. The material survived only 1000 h in an accelerated ageing test until crack formation, with 40% impact retention [13, 50]. After addition of 0.1% of an HAS, namely HAS-1 (Chimassorb 944) the stability of the material reached 8000 h until crack formation, with an impact retention of 60% compared to that of the virgin material. A stabilising system consisting of 0.1% of the antioxidant mixture P-1:AO-1 (2:1) and 0.1% HAS-1 almost eliminated the sensitising effect of the aged HDPE. Addition of a UV-light absorber of the benzotriazole type LS-1 gives an additional protection against bleaching of the pigment. It has been reported that carbon black, a common light screen in virgin polyolefins, does not, alone, provide enough outdoor protection in recyclates. Restabilisation of a mixed LDPE–LLDPE, loaded with 2.5% carbon black was achieved by combining antioxidants P-1 and AO-1 with the physically persistent (non-leaching and non-volatile) HAS-3 [46]. The presence of 0.2% of a 1:2 mixture of phenol AO-1 and phosphite P-1 ensures good processing and heat stability of recycled polypropylene (PP).

6.2
Restabilisation of Mixed Plastics from Household Collection

6.2.1
Role of Compatibilisers in Mixed Plastics Recyclates

Polymers of different structures do not form homogeneous blends. Mixture of polymers that differ in structure and polarity are compatible with one another only to a very limited extent. They separate in the melt or at the latest after cooling in the solid state, as a result of weak interfaces between the components and local stress concentrations. Therefore, they cannot be mixed in extruders or kneaders homogeneously to give the molecular dispersion necessary to obtain acceptable material properties. Compatibilisation modifies polymer interfaces by reducing the interfacial tension between normally immiscible polymers in the melt during blending, leading to a fine dispersion of one phase in another, increasing the adhesion at phase boundaries and minimising phase separation in the solid state. Compatibilisers act as morphology stabilisers and prevent delamination or agglomeration by creating bridges between phases. The compatibilisers themselves are usually block or graft copolymers containing segments chemically identical to the blend components, but can also be functionalised polymers containing reactive side groups. Another possibility is the addition of a free radical initiator to the blend, which can promote in situ formation of graft copolymers during extrusion [16]. For composites, the interface was not considered to be changed during recycling, but the resulting mechanical performance of in-plant recycled fibreglass-reinforced nylon demonstrated that instead the fibre length had a strong influence on the strength [51].

6.2.2
Non-Reactive Compatibilisers

Diblock copolymers, especially those containing a block chemically identical to one of the blend components, are more effective than triblocks or graft copolymers. Thermodynamic calculations indicate that efficient compatibilisation can be achieved with multiblock copolymers [47], potentially for heterogeneous mixed blends. Miscibility of particular segments of the copolymer in one of the phases of the bend is required. Compatibilisers for blends consisting of mixtures of polyolefins are of major interest for recyclates. Random poly(ethylene-co-propylene) is an effective compatibiliser for LDPE-PP, HDPE-PP or LLDPE-PP blends. The impact performance of PE-PP was improved by the addition of very low density PE or elastomeric poly(styrene-block-(ethylene-co-butylene-1)-block styrene) triblock copolymers (SEBS) [52].

Inmiscible blends of HDPE or LDPE with PS have been compatibilised with various graft copolymers, such as PS-graft-PE, PS-graft-EPDM or block copolymers such as SBS triblocks, SEBS, PS-block-polybutadiene [53, 54]. The same block copolymers are suitable for PP-PS blends [55].

Compatibility of PE with PVC is improved by poly(ethylene-graft-vinyl chloride) or partial chlorinated PE. To compatibilise blends of PE with PET, common for the scrap of beverage bottles, EPDM or SEBS are effective additives [56].

Styrene-butadiene or styrene-(ethylene-co-propylene) block copolymers are common compatibilisers for the commingled recyclate from PCW [27]. Improvement of the mechanical properties of heterogeneous PCR (33% PE+39% PVC+28% PET) or (44% PE+1% PP+28% PET+9% PS+2% PVC+16% other plastics) was performed with EPDM or hydrogenated SBS triblocks [52].

6.2.3
Reactive Compatibilisers

Reactive compatibilisers consist of functional or reactive additives interacting in situ with components of the blend. A polymer chemically identical to one of the components is functionalised to gain a chemical reactivity with the second component. This allows the phases to be held by covalent bonds, making the blend less sensitive to physical stresses. PE, PP, EPDM or PS, grafted with maleic anhydride(MAH) or acrylic acid, poly(ethylene-co-poly-propylene), grafted with succinic anhydride, or copolymers of PP with acrylic acid, styrene with maleic anhydride are examples of effective reactive compatibilisers [57]. These grafted polymers are suitable for the compatibilisation of hydrocarbon polymers (PE, PP, PS) with polar polymers (PET, PA). The pendant carboxyl group from polyolefins grafted with acrylic acid or maleic anhydride reacts with terminal amino groups of PA 6 or 66, in shredded PP-PA industrial waste [12]. PA scraps or recycled fibres can be modified with various elastomers or high pressure ethylene copolymers. Recycled PA 6 fibres with 20–25 wt% of modifier, either EPDM grafted with MAH or ethylene copolymers containing MAH, compounded on a single screw extruder have 8–10 times higher impact strength than the non-modified fibres [58]. Approaches that use specially synthesised reactive compatibilisers and that can account for formation of compatibilising crosslinks in situ are generally very expensive for commingled PCR. Grafting of a selected monomer, in situ, to one or more components of the blend performed during processing has been successfully carried out for the reactive compatibilisation of PA 6/ABS blends. In this case, free radical copolymerisation of styrene, acrylonitrile and MAH in situ and during processing has been performed [59].

7
Polymer Characterisation for Quality Assessment of Commingled and Separated Recycled Polyethylenes

Analysis of additive levels in recycled plastics is necessary for determination of the need to add new fresh additives. Accurate measurements of additive contents are therefore very important. Traditional methods of additive analyses are generally time consuming and tedious. These methods often require additive separation by extraction with hazardous solvents and subsequent quantification by techniques like high performance liquid chromatography (HPLC), gas chromatography (GC) or thin-layer chromatography. It is worth mentioning that HPLC allows only the detection of that part of the tetrafuctional phenolic primary antioxidants (e.g. Irganox 1010) in which none of the phenolic groups has reacted. Accordingly, if even one of the four OH-groups has reacted, a stabiliser molecule is no longer detected, although the remaining groups are still active [60]. Thus, the true and effective residual concentration of the antioxidant should be somewhat higher when accounting for the undetected molecules. The thermal analysis such as differential scanning calorimetry (DSC) is widely used for polymer characterisation and can also be used for determination of antioxidant levels. The analytical methods mentioned above are destructive, time consuming and require sample preparation. In addition, these methods do not allow rapid at/in-line analysis. Larger contaminations, such as true impurities, may also be present in recycled polymeric materials. It was demonstrated that in reprocessed glass-fibre reinforced polyamide 66 a critical size–concentration zone could be determined based on changes in the tensile properties of samples containing untreated glass beads of different sizes and concentrations [61].

UV has been used as an alternative method for determining additive concentrations by direct analysis of polymer [62, 63, 64]. However, the method is constrained by excessive beam dispersion due to light scattering from the polymer crystalline regions. Additives at low concentration (0.1%) require sample thickness such that analyses must be performed in the presence of a high level of light scattering, which may change unpredictably with wave length. At lower levels of concentration and correspondingly greater sample thickness, unacceptable signal-to-noise ratios exist [65]. The fact that many additives exhibit similar or identical UV spectra makes this technique less difficult for quantitative analysis of polymers containing additive packages [66]

New methods for non-destructive quantitative analysis of additives based on MIR spectra and multivariate calibration have been presented [67, 68]. One of the limitations in the determination of additive levels by MIR spectroscopy is encountered in the detection limit of this technique, which is usually above the low concentration of additive present, due to their heavy dilution in the polymer matrix. The samples are thin polymer films with small variations in thickness (due to errors in sample preparation). The differences in thickness cause a shift in spectra and if not eliminated or reduced they may produce non-reliable results. Methods for spectral normalisation become necessary. These methods were reviewed and compared by Karstang et al. [68]. MIR is more specific than UV but the antioxidant content may be too low to give a suitable spectrum [69]. However, this difficulty can be overcome by using an additive-free polymer in the reference beam [67, 68, 69, 70]. On the other hand, UV and MIR have been successfully applied to quantify additives in polymer extracts [71, 72, 66].

Infrared spectroscopic methods are extensively used to analyse polymers due to their simplicity, rapidity, reproducibility, non-destructive character and ease of sample preparation. Degree of crystallinity [73], chain branching [74], degree of oxidation [75], density measurements [76], quantification of additives [75, 77], end-group analysis [78, 79] and other physical/chemical properties have been studied using MIR and/or NIR.

NIR in the diffuse reflectance mode is a technique that also allows non-destructive analysis of polymers with very little or no sample preparation [80, 81]. NIR is based on measurement of light reflected by the sample when exposed to electromagnetic radiation in the range from 780 nm (12820 cm^{-1}) to 2500 nm (4000 cm^{-1}). Qualitatively this is the region between the visible red and the highest frequency used in conventional mid-infrared. For NIR diffuse reflectance, samples with large effective path lengths (granular samples with large particular sizes, or poorly scattering samples) can be directly analysed without substantial preparation [81, 82]. Different multivariate calibration techniques such as classical least square (CLS), multiple linear regression (MLR), principal component regression (PCR), and partial least squares (PLS) regression can be used to extract valuable chemical information from the spectral data [83].

The degree of crystallinity of polymers (i.e. the concentration of crystallites) influences the stiffness, hardness and heat resistance. Changes in crystallinity induced by processing and autoxidation of polymers during their service life might dramatically affect the performance of next generation products. There are a number of other techniques for characterisation of crystal structure and degree of crystallinity of polymers such as infrared (MIR), dilatometric measurements, density measurements, Raman spectroscopy, nuclear magnetic resonance (NMR) spectroscopy, X-ray diffraction (SAXS, WAXS) and differential scanning calorimetry (DSC), the last two being the most used. Although these techniques yield different absolute crystallinity values due to the differences in their principles, the relative changes in crystal contents are comparable. It is important to remark that these methods often require extensive sample preparation and long analysis times. As a result they cannot be used for rapid process analysis. When rapid analysis is desired, NIR spectroscopy has shown to be a very effective technique, which allows accurate analysis on relative unprepared samples, such as meat [81], wheat [84] and bulk polymers [75, 85, 86, 87, 88, 89].

7.1
Spectroscopy for Measuring Degree of Separation

Many studies use infrared spectroscopy for quality control and quality analysis in polymer production. It is particularly used for the determination of the composition of copolymers and polymer blends and also for determination of additive and filler contents [90, 91, 92].

Generally, Swedish household hard packaging plastic waste contains approx. 33% of HDPE and 24% of PP [9]. A total separation of these two polymers is difficult due to the great similarity in their physical properties. Conventional methods of separation, e.g. float and sink method, cannot be applied. Manual sorting does not provide ideal separation due to the human factor and in some cases lack of labeling, besides this method is also cost-ineffective. More sophisticated sorting methods such as spectroscopic or electrostatic separation are still in the stage of development and do not guarantee 100% purity. Acrylonitrile-butadiene-styrene (ABS) and PP are the two major components in the automotive industry. During recycling they usually contaminate each other since a 100% separation is very difficult to achieve.

The near infrared range (12800–4000 cm^{-1}) as well as the mid-infrared range (400–4000 cm^{-1}) has been used for spectroscopic detection of physical and chemical properties of polymers. These techniques make it possible to obtain information about the chemical composition and structure of polymers. Therefore this method is widely used for compositional analysis of polymeric materials [93, 94, 95, 96, 97].

Diffuse reflectance NIR has, in competition with other spectroscopic techniques such as MIR and Raman, a high signal-to-noise (S/N) ratio and allows non-destructive sample analysis. The ease, fastness and use of fibre optics facilitate the application of NIR for in-line and on-line measurements [98, 99]. For a given molecule, many active overtone and combination bands might be present in the NIR region giving as a result a spectrum with a large number of overlapping bands, which makes spectral interpretation difficult, but this disadvantage can be overcome by using multivariate data analysis [83].

Raman spectroscopy has been widely used to study the composition and molecular structure of polymers [100, 101, 102, 103, 104]. Assessment of conformation, tacticity, orientation, chain bonds and crystallinity bands are quite well established. However, some difficulties have been found when analysing Raman data since the band intensities depend upon several factors, such as laser power and sample and instrument alignment, which are not dependent on the sample chemical properties. Raman spectra may show a non-linear base line to fluorescence (or incandescence in near infrared excited Raman spectra). Fluorescence is a strong light emission, which interferes with or totally swaps the weak Raman signal. It is therefore necessary to remove the effects of these variables. Several methods and mathematical artefacts have been used in order to remove the effects of fluorescence on the spectra [105, 106, 107].

Raman is simple to use and require virtually no sample preparation. The vibrational spectra of polymer samples in their original forms can be measured and casting or hotpressing of the material into thin films (commonly used for FTIR measurements), a very time-consuming procedure that can affect the original structure of the material, are not required. It also offers remote analysis capabilities if combined with fibre optics [108, 109]. Accurate determination of contaminant content in recycled polymeric resins have been reported in the literature [110, 111, 112].

FTIR reflectance and transmission spectroscopy is used for analysis of thin films. Nevertheless, due to the high absorptivities of mid-IR bands, the film thickness must be limited (up to 100 μm, depending on the specific bands chosen) in order to perform an accurate qualitative analysis. Other IR methods, such as attenuated total reflectance (ATR) and photoacoustic methods provide IR spectra of thick material, because they penetrate a very thin layer at the surface of a sample. However, is important to point out that the effective pathlength for the ATR and the photo-acoustic methods depends on the refractive index and thermal diffusivity, respectively. Therefore, the use of these techniques for the quantitative analysis of non-homogeneous materials can be difficult.

7.2
Chromatography for Monitoring Contamination

Low molecular weight compounds were extracted from virgin and recycled PP and PE resins [113]. Hydrocarbons, carboxylic acids, ketones, alcohols, aldehydes etc. were among the compounds identified [113, 114, 115]. The number of compounds present in the recycled material is larger than in virgin ones. Even though the HDPE samples were prepared from thin plates made by compression moulding of recycled resin pellets, they showed a very high uniformity in their composition. The contaminants and their levels were very similar between the three samples of HDPE extracted and analysed by GC-MS technique. Of 65 compounds detected, 52 of them were identified. The compounds were classified into 8 major categories as carboxylic acids, hydrocarbons, esters, fragrance and aroma compounds, ketones, alcohols, aldehydes, amines and miscellaneous. Virgin and recycled HDPE contain ethylbenzene and *o*-, *m*- and *p*-xylenes, *o*-xylene having the largest peak of the four. The presence of these aromatic hydrocarbons in the virgin resin could be explained by the fact that some additives present in both materials degrade during extraction and sample preparation procedures. Aromatic hydrocarbons without functional groups, such as ethylbenzene and xylenes, are considered highly toxic, but the height and area of their peaks suggest low concentrations. The concentration of these hazardous contaminants in the recycled resins is, however, about 5 times higher [113]. A comparison of the relative peak areas was used in order to estimate this rough relationship.

The major category of compounds identified in the virgin and recycled HDPE is comprised of aliphatic hydrocarbons, such as pentadecane, hexadecane, 1-hexadenene, branched alkanes, branched alkenes and others all oligomers of HDPE. Certain differences between virgin and recycled plastics are, however, obvious, e.g. carboxylic acids such as hexadecanoic and octadecanoic acid were found only in the recycled HDPE. Only two ketones were identified, 6-dodecanone and 2-nodadecanone, in the recycled HDPE; the former was also present in the virgin material.

In recycled HDPE, compounds widely used in cosmetics: isopropyl myristate, isopropyl palmitate, methyl stearate and methyl laurate were identified. The highest concentrations were estimated for isopropyl esters of myristic and palmitic acids. These substances are used in products for personal hygiene, cleaning agents, foods and beverages. Thus it is very possible that they are absorbed by the packaging material during storage. Alcohols, which are also used as basic compounds of cosmetic products and in manufacturing of household and industrial cleaning agents, were found in the recycled HDPE; some of these materials are 1-dodecanol,1-octadecanol, 1-pentadecanol, 1-heptadecanol and 1-nonadecanol.

Aroma and odour compounds such as limonene, 3-carene, betamyrcene and terpinolene with 3-carene having the highest estimated concentration, were found only in the recycled HDPE. The levels of these fragrance and taste materials are, however, low compared to the levels of aliphatic hydrocarbons. Fragrance and flavour compounds such as limonene, 3-carene, myrcene and terpinolene were readily extracted. Again, the highest peak levels were found for 3-carene. Only propyl myristate and isopropyl palmitate (esters used in cosmetics), were detected and identified. Alcohols were not found either in the recycled or in the virgin resin.

On the other hand, aromatic compounds without functional groups, such as benzene, toluene, xylenes and ethylbenzene have a high compatibility with PE, which slows migration out of the HDPE matrix. If this is the case, higher extraction temperatures maybe suggested.

In recycled PP, 61 compounds were detected and 35 of them were identified. Many peaks showed very low separation levels making their identification difficult. In virgin and recycled PP the following compounds were identified: Ethylbenzene and xylene were found only in the recycled resin. Present in both virgin and recycled PP were a large number of branched alkanes and n-alkanes between C_{18} and C_{25}. Octadecanoic acid, methyl ester and dibutyl palmitate, which is a typical compound used in the cosmetic industry, were found only in the recycled PP. Amines such as hexamine, 3-ethyl and $N'N''N'''$ trimethyl dipropylene triamine were identified in both PP materials. Carboxylic acids and ketones were absent in both polymeric materials and so were fragrance or flavour compounds [113].

7.3
NIR Analysis for Quantification of Antioxidants in Polyethylene

NIR of the PE powder was carried out before compounding with Irganox 1010 and Irgafos 168. It was observed that the identification and selection of specific bands or unique spectral features in the spectra is difficult. The variation in baselines is due to differences in scattering properties of the analytes. Multiplicative scattering correction (MSC) or derivation can eliminate these variations [117, 118].

A certain relationship between the samples that contain antioxidants exists, since they are gathered in two clusters, whereas the non-stabilised sample differs from the rest. By principal component analysis (PCA) the cluster on the left side is built by samples that contain a total amount of antioxidants lower or equal to 2200 ppm and the cluster on the right side is made up of samples having total antioxidant concentrations above 2500 ppm. The difference between virgin HDPE and stabilised samples may also be explained by the degradation of the virgin sample during extrusion, which has been confirmed by the presence of carbonyl groups and changes in crystallinity as measured by FTIR and DSC, respectively. The virgin sample showed a carbonyl index (CI) equal to 0.29 whereas the samples containing Irgafos 168 above 300 ppm did not show carbonyl groups at all. Samples with concentrations of Irgafos 168 bellow 300 ppm were slightly degraded, the CI was in the range 0-0.09. Small differences in the DSC crystallinity were observed among the stabilised samples, their values were in the 62-65% interval. However, a lower crystallinity value, 57% was obtained for the virgin specimen. The root-mean-square error of prediction (RMSEP) for Irganox 1010 and Irgafos 168 was 45 and 96 ppm, respectively. The models were obtained using a PLS regression with four factors over the 5000–9000 cm^{-1} spectral segment.

7.4
MIR and NIR for Simultaneous Determination of Molecular Weight and Crystallinity of Recycled HDPE

Recycled HDPE items (blow-moulding grade) produced from HDPE made by the Phillips process were analysed by IR spectroscopy. The absorption bands at 888 and 965 cm^{-1}, corresponding to unsaturations of vinylidene and vinylene type, respectively, are common in PE produced by either the Ziegler process or metallocenes [104]. The absence of these bands in the IR spectrum of the different samples confirmed that the resins had been produced using a Cr-type catalyst [118].

Another characteristic feature of Philips-manufactured HDPE is its high weight-average molecular weight, high polydispersity and two different chain ends, where one of them is a vinyl group and the other is a methyl group. It is known that the concentration of end vinyl groups can be estimated by measuring the intensity of the spectral band either at 910 or 990 cm^{-1}. The concentration of this end-group is related to the molecular weight. In addition, two spectral regions are usable to detect the presence of CH_2 and CH_3 absorbances, namely the CH bending region (1350–1500 cm^{-1}) and the stretching region (2800–3000 cm^{-1}). This region has been used by some authors to determine CH_2 (2926 cm^{-1}) and CH_3 (2962 cm^{-1}) content in fractionated polyethylene [119, 120].

The model was also validated by full cross-validation. Four PCs were necessary to explain the most variation in the spectra (99.9%), which best described the molecular weight. The root mean square error of prediction (RMSEP) is 360.

A linear correlation was observed between crystallinity determined by DSC and MIR for all the studied HDPE samples. As expected, the absolute values are not the same, but a linear correlation can be observed.

7.5
IR and Multivariate Calibration for Compositional Analysis of Commingled Recycled PE/PP

The feasibility of diffuse reflectance NIR, Fourier transform mid-IR and FT-Raman spectroscopy in combination with multivariate data analysis for in/on-line compositional analysis of binary polymer blends found in household and industrial recyclates has been reported [121, 122]. In addition, a thorough chemometric analysis of the Raman spectral data was performed.

PP shows characteristic absorption peaks at 8385 cm^{-1} and 6094 cm^{-1}, which correspond to the second and first overtone of the C–H methyl stretching. As the content of HDPE increases, the intensity of these two peaks decreases and a new C–H methylene peak at 6480 cm^{-1} appears. It can also be observed that the intensity of a peak at 8216 cm^{-1} and another at 5774 cm^{-1} increases with the increase of HDPE. These two peaks are due to the second and first overtones of C–H methylene stretching in PE [122]. Due to the high overlapping of the spectral bands an univariate calibration is difficult to perform. Therefore, a PLS calibration has been used to analyse the spectral data. The biggest advantage of this procedure is that it allows extraction of the most information from the acquired data. A PCA was also performed on the data in order to detect possible outliers and find trends in the data. The PLS calibration was preceded by a MSC.

Calibration was carried out for 0–100%wt PP content region. Two PCs fully described the model with a root square standard error of prediction (RMSEP) equal to 0.91%wt. Since the method is intended to be used for determination of very small amounts of PP in recycled HDPE, which has been carefully separated from PP and other household plastics, a PLS in a narrower interval, 0–15%wt PP, was designed. The model was validated by either cross validation or a test set consisting of three samples with 0.5, 7.0 and 12.0%wt PP content. The RMSEP was equal to 0.21%wt obtained from the calibration curves.

7.6
Fourier Transform Raman Spectroscopy for Prediction of PP Content in PP/HDPE Blends

The Raman spectra of several PP/HDPE samples in the 1500–100 cm^{-1} Raman shift region after MSC pre-treatment showed that the relative intensities of the bands at 1454, 1310, 1295, 1130 and 1059 cm^{-1} decrease with increase of PP content in the blends. An RMSEP of 0.99%wt and a correlation coefficient equal to 0.999660 were obtained. Since the PLS models are intended to be used for purity analysis, a calibration model in a narrower interval, namely 0–15%wt, was performed. A RMSEP of0.46%wt and a correlation coefficient of 0.99985 were obtained. An MLR model was also built using selected bands for PP that do not interfere with the spectral bands of HDPE. The following bands were used: 1220, 974, 841 and 399 cm^{-1}. The MRSEP was 1.04%wt in the 1–100%wt region and 0.52%wt in the 0–15%wt interval. The correlation coefficients were 0.999672 and 0.998756, respectively. In the plot, the bands corresponding to PP and HDPE show positive and negative values, respectively. This confirms that the regression coefficients for the PLS model express the relative content of PP and HDPE in the blends.

7.7
DSC for Analysing Thermal Stability

Isothermal and dynamic DSC were demonstrated to be useful tools for following the oxidative stability as a function of repeated injection moulding of glass fibre reinforced nylon [121]. Similarly DSC and OIT showed that a common feature for PP and HDPE is that the OIT sharply decreases after the two first extrusion steps due to the large amount of antioxidant initially consumed to prevent the degradation of the resins. The formation of antioxidant degradation products, which may have a stabilising effect, could be the reason why the differences in OIT are small when the materials are processed further. The shorter induction times for PP are a result of the lower resistance of the material to oxidation. The decrease in OIT in both resins is determined by the residual concentration of efficient antioxidant. Since the blends are made up of 80%wt HDPE, the OIT of the reprocessed blends is much longer than that of the extruded PP. The OIT of the blends also decreased with the number of extrusion cycles.

The DSC traces of HDPE and PE remained practically unaltered after the first and second extrusion pass, whereas further reprocessing induced changes in the peak shapes, probably owe to chain scission in PE and PP. For instance, the thermogram of HDPE after the second extrusion pass exhibited a bimodal melting peak that might be attributed to the presence of species with lower molecular weight than the original material, formed as a result of severe chain scission and not able to co-crystallise. Even though bimodality was not observed in the re-processed PP an increase in the low molar mass tail was seen after each extrusion pass. The thermograms of the multi-extruded samples showed a very small increase in the crystallinity of the HDPE fraction of the blend extruded once. However, the crystalline content in the further reprocessed blends was almost constant. The increase in crystallinity may be a result of the rearrangement of the shorter polymer chains formed as a consequence of the chain scission that probably occurred during processing. The crystals become thinner after each extrusion step, which was confirmed by the lower melting temperature. The shape of the PE peak was practically unchanged and no bimodal behaviour was observed. The PP fraction exhibits only very small changes in crystallinity and the melting point does not follow any specific trend. Consequently, certain synergism between PP and HDPE at this particular concentration might be assumed. The results presented above demonstrate that PE, PP and the blend under the used extrusion conditions undergo chain scission, which is consistent with published results [123, 124, 125].

It is important to mention that the OIT, and the Tox reported here, depend only on the total concentration of the active antioxidants, but provide no information about the oxidation process itself. Thus, it is difficult if not impossible to know whether PE and PP in the blend start degrading at the same time or any of them accelerates or delays the oxidation of the other.

7.8
Chemiluminiscence to Analyse Oxidative Stability in Multiple Recycled Polyolefins

The OIT for PP, PE and their blend was determined after each extrusion pass using chemiluminiscence (CL). The CL results showed good correlation between these techniques regarding the onset time of the oxidation phenomenon. Similar results on multi-extruded PP have been recently presented [126]. DSC and CL produced reliable and rapid OIT results when the experiments are conducted at temperatures well above the melting point. Thus, the

effectiveness of the residual stabilising package can be identically determined by either of these techniques.

The PP exhibits a sharp peak at the maximum on CL intensity whereas
the HDPE curve shows a broad bimodal behavior that has been thoroughly
described elsewhere [127]. In the CL curves of the blend all these features
were observed, which may be a strong indication of the existence of a two-
phase system in the molten state [128], although based on peroxide treatment of PP/PE blend melts. It appears that PP oxidises first and the oxidation sites created during this process accelerate, to some extent, the oxidation of the PE phase. The overlap between the PP and PE traces in the blend
can be interpreted as the interface of these two phases where the PE starts
oxidising. In addition, the shape of the curves confirms that the oxidation
mechanisms of the resins are different and that this difference remains during the oxidation of the blend in the molten state.

8
Influence of Various Processes in the Recycling Plant on the Polymeric Properties

The separation, granulation and subsequent reprocessing do also affect the
properties of the recycled plastics. During the processing of incoming recycled material, which is influence by its previous history, a further change in
the polymeric properties may occur. Some of these changes actually improve
the next use of the material as resources, such as removal of contaminations.
Figure 1 shows the melting behaviour of LDPE at three different points in a
recycling process. The material is used for the manufacturing of garbage
bags and this material was analysed with respect to thermal properties, functional groups and low molecular weight compounds [116, 117] (see Table 2).
It was demonstrated that during the course of the process in the recycling
plant the carbonyl index also increased, which is indicative of increased degradation similar to the thermograms obtained by DSC.

LDPE from three different points in recycling plant

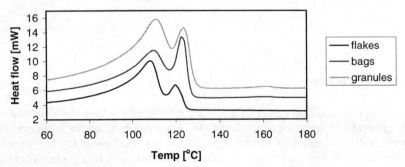

Fig. 1 DSC thermogram of LDPE directly (flakes), after washing and melting (granu-
lates) and processing (bags)

Table 2 Example of low molecular weight compounds identified in recycled LDPE at three different points in the recycling plant process

Compound	Flake	Granulate	Bag
Tridecane		x	x
Tetradecane		x	x
Nonadecane, 9-methyl-			x
Pentadecane	x	x	
Hexadecane			x
Tridecane, 5-propyl	x	x	
9-Nonadecane	x		
Cyclopentane, undecyl		x	
Heptadecane	x	x	x
Heptadecane, 3-methyl-	x		
Tetracosane		x	x
Heneicosane	x		
Dodecanoic acid, 1-methyl ester	x	x	x
1-Octadecanol		x	x
Butylated Hydroxytoluene	x		x
Octacosane		x	
Pentacosane	x		
Hexatriacontane			x
5-Heptadecene, 1-bromo-	x		
Oleyl Alcohol			x
7-Methoxy-2,2,4,8-tetramethyltricyclo[5.3.1.0(4,11)]undecane		x	x
Diisopropylnapthalene	x	x	x
2,6-Diisopropylnathalene	x	x	x
Pentanedioic acid, (2,4-di-t-buthylphenyl) ester	x	x	
Phenol, 2,4-bis(1,1-dimethylethyl)-			x
3,5,3',5'-Tetramethylbiphenyl	x	x	
Diethyl phthalate			x
3,5-di-tert-Butyl-4-hydroxybenzaldehyde	x	x	x
1,2-Benzenedicarboxylic acid, bis(2-methylpropyl) ester	x	x	x
Phenol, nonyl-	x	x	x
Dibutyl phthalate		x	x
Phenol, 2,6-di-t-butyl-4-nitro-		x	
4-Nonylphenol	x	x	x
Phenol, 2-methyl-4-(1,1,3,3-tetramethylbutyl)-	x		
9H-Flourene, 2-methoxy-		x	
Acenaphtalene, 5-acetyl-	x		x
1,2-Benzenediol, 3,5-bis(1,1-dimethylethyl)-	x	x	x

9
Future Perspective

Recycling is a necessary part of a sustainable society where the needs of future generations must be considered. We are experiencing an increased use of materials in a series of products, e.g. in biomedical products for improving health, in electronic items for industry and fun, and in simple packaging for the fast-food market. At the same time, some of the natural resources are starting to deplete, or will in the near future. This means that recycling (preservation of already made materials) is necessary. The key area to work in for the future is the technical development of automatic separation coupled with on-line characterisation. Manual separation is not the solution for the future. There is a lack of investment in this area and many recycling industries are happy with the present situation. Automatic separation that allows in situ characterisation of a series of polymeric properties will also enable separation of dirty, highly mixed plastic waste material. On-line characterisation will allow direct estimation of whether it is worthwhile to, e.g., further process the mixed waste or if part or whole of a stream should go directly for energy recovery. With such a solution, increased volumes will be possible and the process will be more economic if only the material that demonstrates good properties or contains no hazardous compounds is recycled mechanically. A problem today is also the fluctuation in availability of recycled polymeric materials. Thus, many manufacturers find it difficult to rely on the availability of the materials of their choice. For large volume producers of furniture or packaging it is necessary that the stream of recycled plastics is constant and high. This is another problem that could be solved by research and technical development in the separation and on-line characterisation of plastic waste.

Another problem that needs a good solution is the logistics of collection of non-industrial polymeric waste materials. Many collections in practise today contribute to more traffic and also to collection sites that are overfull and attract rats and birds etc. Quick and badly thought out collections lead to a negative impression of recycling and less willingness to accept it as a solution for the future.

Three different schemes for treatment of MSW (landfill, incineration and recycling) were evaluated by LCA methodology. It was shown that according to the Ecopoints method, incineration and the recycling scheme using electrostatic sorting (EU-project Tri-Repack BRE2-CT94-1010) save material and energetic resources while landfill lost them. The recycling method was, however, more environmental beneficial than incineration [129].

10
Summary

Thermal, spectroscopic and chromatographic analyses are useful for monitoring changes in the properties of recycled polymers. The use of multivariate computer modelling allows very subtle changes to be analysed. More research is needed in order to combine an automatic sorting process using spectroscopy (e.g. NIR, MIR etc.) with on-line property monitoring.

Acknowledgement Several European projects have financed different parts of the studies presented and discussed here. These are 'Electrostatic recovery of paper and plastic packaging waste' (ELREC), BRPR-CT96-0247; 'Development of multipurpose industrial units for recycling of plastic wastes by on-line pattern recognition of polymers features' (SURE-PLAST), BRPR-CT-98-0783 and 'Industrial production of high performance ecological polymeric composites based on residual/renewable cellulose fibres and post consumer thermoplastics' (ECOSITES), G5RD-CT-2000-00337. Dr. Walker Camacho, former PhD. student, and Frida Stangenberg are gratefully acknowledged for contributions to various parts of this work.

References

1. APME (1998) Assessing the potential for post-use plastics waste recycling—predicting recovery in 2001 and 2006. APME Summary Report, Reference 8023
2. APME (2001) Plastics—An analysis of plastics consumption and recovery in Western Europe. APME Report, Reference 2002
3. Scott G (1999) Polymers and the environment. The Royal Society of Chemistry, Cambridge
4. Wogrooly E (1996) In: Brandrup J (ed) Recycling and recovery of plastics. Hanser, New York
5. Rånby B, Albertsson A-C (1978) AMBIO 7(4):172
6. Haider N (2000) Consumption and loss of commercial stabilisers in polyethylene exposed to different natural environments. PhD Thesis, Royal Institute of Technology, Stockholm, Sweden
7. Council Directive 1999/31/EC on the landfill of waste (1999) Off J Eur Communities L 182:1–19
8. Förpackningsinsamligens INFO (2001) 2
9. European Parliament and Council Directive 94/62/EC of 20 December 1994 on packaging and packaging waste (1994), Off J Eur Communities L 365:10–23
10. Swedish Waste Management 2001 (2002) Renhållningverksföreningen RVF, Sweden
11. Swedish Recycling Industries (SRI), Autumn report 2002
12. Scobbo JJ (1992) Macromol Symp 57:331
13. Pospisil J, Nespurek S, Zweifel H (1997) Trends Polym Sci 5(9):294
14. Pospisil J, Sitek FA, Pfaender R (1995) Polym Degrad Stab, 48:351
15. Hope PS, Bonner JG, Milles (1994) Plast Rubber, Compos Process Appl 22 (3):147
16. Tall S (2000) Recycling of mixed plastic waste—Is separation worthwhile? PhD Thesis, Royal Institute of Technology, Stockholm, Sweden
17. Stangenberg F, Ågren S, Karlsson S (2003) Chromatographia (in press)
18. Electrostatic recovery of paper and plastic packaging waste (ELREC), BRPR-CT96-0247 (European Project)
19. Shiers J (1998) Polymer recycling. Wiley, Chichester, UK, p 1
20. Development of multipurpose industrial units for recycling of plastic wastes by on-line pattern recognition of polymers features (SURE-PLAST) (European Project) BRPR-CT-98-0783
21. Pospisil J, Nespurek S, Zweifel H (1997) Trends Polym Sci 5:294
22. Tall S, Albertsson A-C, Karlsson S (1998) J Appl Polym Sci 70(12):2981
23. Allen NS (1994) Trends Polym Sci 2:366
24. Boldizar A, Jansson A, Gevert T, Möller K (2000) Polym Degrad Stab 68:317
25. Hinsken H, Moss S, Zweifel H (1991) Polym Degrad Stab 34:279
26. Allen NS, Edge M; Mohammadian M, Jones K (1993) Polym Degrad Stab 41:191
27. Lemaire J, Arnaud R, Lacoste J (1988) Photochim Mol Macromol, Acta Polym, 39:27
28. Hoff A, Jacobsson S (1984) J Appl Polym Sci 29:465
29. Tidjani A (2000) Polym Degrad Stab 68:465
30. Denisov, ET (1989) Polym Degrad Stab 25:209
31. Klemchuk PP, Gande ME (1988) Polym Degrad Stab 22:241

32. Klemchuk PP, Gande ME (1990) Polym Degrad Stab 27:65
33. Osawa Z (1984) Polym Degrad Stab 7:193
34. Osawa Z (1988) Polym Degrad Stab 20: 203
35. Zweifel H, (1997) Stabilization of polymeric materials. Springer, Berlin Heidelberg New York
36. Allen N, Edge M (1992) Fundamentals of polymer degradation and stabilisation. Elsevier Applied Science, London
37. Scott G (1990) Polym Degrad Stab 27:183
38. Klemchuk PP, Horng PL (1991) Polym Degrad Stab 34:333
39. Hubber F, Piringer OG (1994) Food Addit Contam 11:479
40. FDA (1992) Points to consider for the use of recycled plastics in food packaging: Chemistry considerations. Chemistry Review Branch, US Food and Drug Administration, Washington DC
41. FDA (1993) US FDA FED Reg 58(195):52719–52729
42. Helberg J (1996) In: Brandrup J (ed) Recycling and recovery of plastics. Hanser, New York
43. Kartalis CN (1999) J Appl Polym Sci 73:1775
44. Pospisil J, Nespurek S, Zweifel H (1996) Polym Degrad Stab 54:7
45. Pospisil J, Nespurek S, Zweifel H (1996) Polym Degrad Stab 54:15
46. Pospisil J (1995) Adv Polym Sci 124:87
47. Pospisil J, Nespurek S (1997) Macrom Symp 115:143
48. Pospisil J (1991) Polym Degrad Stab 34:85
49. Pfnaeder R, Herbst H, Hoffmann K (1995) Angew Makromol Chem 232:193
50. Pfnaeder R, Herbst H, Hoffmann K (1995) In: Proceedings of Davos Recycle '95, Switzerland, 22–26 March, 8/2-1-8/2-14
51. Eriksson PA, Albertsson A-C, Boydel P, Prautzsch G, Manson JAE (1996) Polym Compos 17:830
52. Ha CS, Park HD, Kim Y (1996) Polym Adv Technol 7:483
53. Li T, Topolkareev VA, Hiltner A, Raer E (1995) J Polym Sci Polym Phys 33:667
54. Welander M, Rigdahl M (1989) Polymer 30:207
55. Saleem M, Baker WE (1990) J Polym Appl Sci 39:655
56. Ajji A (1995) Polym Eng Sci 35:64
57. Koning C, Vand Duin M, Jerome R (1998) Prog Polym Sci 23:707
58. Fuzessery S (1989) In: Proceedings of Davos Recycle '89, Switzerland, 10–13 April, 271
59. Hamid HS, Atiqullah M (1995) Rev Macromol Chem Phys C35 (3): 495
60. Dörner G, Lang RW (1998) Polym Degrad Stab 62:431
61. Eriksson PA, Albertsson A-C, Boydell P, Manson JAE (1998) Polym Eng Sci 38:749
62. Ruddle LH, Wilson JR (1969) Analyst,94:105
63. Burcfield HP, July JN (1960) Anal Chem 19:383
64. Klemchuk P, Horg P (1985) Plast Compounds 8:42
65. Crompto TR (1993) Practical polymer analysis. Plenum, New York
66. Haider N, Karlsson S (1999) The Analyst 124:797
67. Viregust B, Kolset K, Nordenson S, Henriksen A, Kleveland K (1991) Appl Spectrosc 2:173
68. Karstang TV, Henriksen A (1992) Chem Intell Lab Syst 14:331
69. Luongo JP (1965) Appl Spectrosc 19:117
70. Miller RGJ, Willis HA (1959) Spectrochim Acta 14:119
71. Freitag W (1983) Fresenius' Z Anal Chem 365:495
72. Rao PVC, Prasad JV, Koshi VJ (1995) Anal Chim 85:171
73. Huang JB, Hong JW, Urban MW (1992) Polymer 33:5173
74. Baker C, Maddams WF (1976) Makromol Chem 177:437
75. Camacho W, Karlsson S (2001) Int J Polym Anal Charact 41:1626
76. Shimonaya M (1998) Near Infrared Spectrosc 6:317
77. Karstang TV, Henriksen A (1992) Chemometr Intell Lab Syst 14:331
78. Kosky PG, McDonald RS Guggeheim EA (1985) Polym Eng Sci 25:389

79. Honings DE, Hirscheld TB, Hiefte M, (1985) Anal Chem 57:443
80. Siesler HW, Holland-Moritz K (1980) Infrared and Raman spectroscopy of polymers. Decker, New York
81. Burns DA, Ciurcziak EW (1992) Handbook of near infrared analysis. Decker, New York
82. Miller CE (1991) Appl Spectrosc Rev 26:277
83. Martens H, Naes T (1989) Multivariate calibration. Wiley, Chichester
84. Cowe JW, McNicol C (1985) Appl Spectrosc 39:257
85. Miller CE (1991) Appl Spectrosc Rev 26:277
86. Miller CE (1993) Appl Spectrosc 47:222
87. Miller CE (1990) Appl Spectrosc 44:496
88. Miller CE, Svendsen SA, Naes T (1993) Appl Spectrosc 47 (3):436
89. Gartner C, Sierra JD, Avakian R (19) ANTEC 98:2012
90. Prasad A (1988) Polym Eng Sci 38:1716
91. Chundury D, Scheibelhoffer A, Bauer V (1988) Society of Plastic Engineers—Annual Tech Conf 1243
92. Paroli RM, Lara J, Hechler JJ, Cole KC, Butler IS (1987) Appl Spectrosc 41:319
93. Miller CE, Eichinger BE (1990) Appl Spectrosc 44(3):496
94. Zhu C, Hieftje G (1992) Appl Spectrosc 46:69
95. Miller CE (1993) Appl Spectrosc 47(2):222
96. Lachenal G (1995) Vibrational Spectrosc 9:93
97. Lachemal G (1998) J Near Infrared Spectrosc 6:299
98. Rohe T, Becker W, Kölle S, Eisenreich N, Eyerer P (1999) Talanta 50:283
99. Michaeli W, Plessmann KW, Andrassy B, Breyer K, Laufens P (1997/8) Poly Recyc 3:287
100. Natarajan S, Michielsen S (1999) J Appl Polym Sci 73:943
101. Williams KP, Everall J (1995) J Raman Spectrosc 26:427
102. Sano K, Shimoyama M, Ohgane KM, Higashiyama H, Watari M, Tomo M, Ninomiya T, Ozaki Y (1999) J Appl Spectrosc 53:551
103. Everall N, Tayler P, Chalmers JM, MacKerron D, Ferwerda R, van der Maas J H (1994) Polymer 35:3184
104. Ewen JC, Kellar A, Galiotis C, Andrews EH (1996) Macromolecules 29:3515
105. Tomba PJ, de la Puente E, Pastor JM (2000) J Polym Sci B: Polym Phys 38:1013
106. Shimoyama M, Hayano S, Matsukawa K, Inoue H, Ninomiya T, Ozaki Y (1998) J Polym Sci B: Polym Phys 36:1529
107. Ewen JC, Kellar A, Evans M, Knowles J, Galiotis C, Andrews EH (1997) Macromolecules 30:2400
108. Everall N, King B (1999) Macromol Symp 141:103
109. Deshpande BJ, Dhamdhere MS, Li J, Hansen MG (1998) ANTEC '98:1672
110. Fraser GV, Hendra PJ, Chalmers JM, Cudby, MEA Willis, H A (1973) Makromol Chem 173:195
111. Everall N, Davis K, Owen H, Pelletier M J, Slater J (1996) Appl Spectrosc 50:388
112. Camacho W, Karlsson S (2001) Polymer Recycling 6:89
113. Camacho W, Karlsson S (2001) Polym Degrad Stab 71:123
114. Haider N, Karlsson S (2001) Polym Degrad Stab 74:103
115. Haider N, Karlsson S (2003) J Appl Polym Sci
116. Stangenberg F, Ågren S, Karlsson S (2003) In: Dhir RK, Newlands MD, Halliday JE (eds) Recycling and Rense of Waste Material. Thomas Telford Publishing, London, UK 2003, p 325
117. Stangenberg F, Ågren S, Karlsson S (2002) Refined routine analyses for quality assessment of recycled plastics in standardisation. KTH, Stockholm, Sweden
118. Camacho W, Karlsson S (2003) J Appl Polym Sci (in press)
119. Dhenim V, Rose (2002) Polym Preprints 41:285
120. Willis JN, Wheeler C (1995) Polym Preprints 19
121. Eriksson PA, Albertsson A-C, Boydell P. Manson JAE (1998) J Therm Anal Calorim 53:19
122. Camacho W, Karlsson S (2001) Polym Eng Sci 41:1626

123. Tall S, Karlsson S, Albertsson A-C (1997) Polym Polym Compos 5:417
124. Pospisil JJ, Horak Z, Kruls Z, Nespurek, S, Kurola S (1999) Polym Degrad Stab 65:405
125. Epacher E, Tolveth J, Stoll K, Pukanskkzly B (1999) J Appl Polym Sci 74:1596
126. Fearon P, Marshall N, Billingham NC, Bigger SW (2001) J Appl Polym Sci 79:733
127. Brosca R, Rychly J (2001) Polym Degrad Stab 72:271
128. Braun D, Ritcher S, Hellma GP, Rätzch M (1998) J Appl Polym Sci 68:2019
129. Albertsson A-C, Karlsson S (1997) Comparative LCA on three different schemes for the treatment of MSW. KTH Stockholm, Sweden

Received: September 2003

Adv Polym Sci (2004) 169:231–293
DOI: 10.1007/b13524

Surface Modification of Polyethylene

Shrojal M. Desai · R. P. Singh

Polymer Chemistry Division, National Chemical Laboratory, 411008 Pune, India
E-mail: shrojaldesai@hotmail.com
E-mail: singh@poly.ncl.res.in

Abstract Polyolefins such as polyethylene, polypropylene and their copolymers have excellent bulk physical/chemical properties, are inexpensive and easy to process. Yet they have not gained considerable importance as speciality materials due to their inert surface. Polyethylene in particular holds a unique status due to its excellent manufacturer- and user-friendly properties. Thus, special surface properties, which polyethylene does not possess, such as printability, hydrophilicity, roughness, lubricity, selective permeability and adhesion of micro-organisms, underscore the need for tailoring the surface of this valuable commodity polymer. The present article reviews some of the existing and emerging techniques of surface modification and characterisation of polyethylene.

Surface modification of polymers, polyethylene in particular, has been extensively studied for decades using conventional tools. Although some of these techniques are still in use, they suffer from distinct shortcomings. During the last two decades, different means of surface modification have been thoroughly explored. The increasing expectancy for smart materials in daily life has, of late, sharply influenced research in the area of surface modification. Technologies that involve surface engineering to convert inexpensive materials into valuable finished goods have become even more important in the present scenario. In this review article we have attempted to broadly address almost all conventional and modern techniques for the surface modification of different physical forms and chemical compositions of polyethylene. This article will hopefully stimulate further research in this area and result in the development of polyolefins with multi-functional and responsive surfaces, which would ultimately lead to the commodities of polyolefins with smart surfaces.

Keywords Polyethylene · Surface modification · Characterisation · Grafting · Radiation

Abbreviations

Å   Angstrom
AA   Acrylic acid
AFM   Atomic force microscopy

AIBN	α,α'-Azo bis(isobutyro nitrile)
AC	α chymotrypsin
ATR	Attenuated total reflectance
R	Alkyl
BPO	Benzoyl peroxide
BP	Benzophenone
θ	Contact angle
Co	Cobalt
XLPE	Cross linked polyethylene
DEM	Diethyl maleate
EDX	Energy-dispersive X-ray spectroscopy
EPDM	Ethylene propylene diene elastomer
EPR	Ethylene propylene rubber
EVA	Ethyl vinyl alcohol
FTIR	Fourier transform infrared
GHz	Giga hertz
GMA	Glycidyl methacrylate
HDPE	High density polypropylene
HEMA	2-Hydroxy ethyl methacrylate
HSPE	High strength polyethylene
keV	Kilo electron volt
LPE	Linear polyethylene
LLDPE	Linear low density polyethylene
LDPE	Low density polyethylene
MeV	Mega electron volt
MHz	Mega hertz
Mg	Magnesium
MA	Maleic anhydride
MFA	Multi functional acrylate
NR	Natural rubber
NVP	N-Vinyl pyrrolidone
PE	Polyethylene
PP	Polypropylene
PMMA	Polymethyl methacrylate
RBS	Rutherford back scattering
SEM	Scanning electron microscope
SSIMS	Static secondary ion mass spectroscopy
STM	Scanning tunnelling microscope
SFM	Scanning force microscopy
T	Temperature
ToF	Time-of-flight
TEM	Tunnelling electron microscopy
L2ITMS	Two-laser ion trap mass spectrometry
UV	Ultra violet
UHMWPE	Ultra high molecular weight polyethylene

ULDPE	Ultra low density polyethylene
Vis	Visible
W	Watt
XPS	X-ray photoelectron spectroscopy

1
Introduction

Many polymers are important commercial materials and constitute one of the fast moving frontiers of daily life. Polymers are important in a wide range of applications, conventionally in packaging, protective coatings, adhesion, friction and wear, composites and home-appliances and more recently in bio-materials, micro-electronic devices, high performance membranes etc. Polyolefins are amongst the oldest and most well-accepted synthetic polymers due to their excellent physical and chemical properties. Although, polyolefins, like polyethylene (PE) and polypropylene (PP), have excellent bulk physical/chemical properties and are inexpensive and easy to process, they have not gained considerable importance as speciality materials due to their inert surface. Thus, the special surface properties which polyolefins do not possess, such as printability, hydrophilicity, roughness, lubricity, selective permeability and adhesion of micro-organisms, are required for their success for specific applications [1, 2, 3, 4, 5].

Surface properties are of especial concern because the interaction of any polymer with its environment mainly occurs at the surface. The generation of technologically useful surfaces with desired interfaces using surface-modification techniques have become an important tool to convert these inexpensive polymers into valuable commercial products. Advances have been made in recent years to alter the chemical and morphological properties of the surface in a desired way with negligible change in the polymer bulk properties [6, 7, 8]. Many recent studies in this area have emphasised the need for material compatibility in the multiphase systems to provide new materials with improved properties [9, 10, 11, 12]. Surface modification of polymers is carried out taking into consideration the potential applications and desired surface properties [13, 14, 15, 16]. The following changes at the polymer interface have been used to enhance the surface properties:

- Generation of special functional groups at the surface, which can be used for secondary functionalisation
- Increase the surface free energy of polymers
- Increase hydrophilicity thereby improving dyeability and printability
- Improve adhesion of cells and micro-organisms to obtain bio-functional surfaces
- Improve surface electrical conductivity
- Improve the chemical and wear resistance

Polyolefins are amongst the most widely used polymers. Polyethylene in particular holds a unique status due to its excellent user-friendly properties. Tailoring of its surface properties would open an avenue to the most lucrative markets. The present article reviews some of the existing as well as emerging techniques of surface modification and characterisation of polyethylene.

2
Different Methods of Surface Modification

The necessity of surface modification is underscored by the above-mentioned pre-requisites of the polymer for speciality applications. However, it is very difficult to synthesise polymers with distinct bulk and surface properties [17]. The methods to tailor surface properties have, therefore; become an important and challenging area of research. The surface modification of the polymeric substrate can be brought about by various conventional and neoteric methods depending upon their desired morphological properties and their potential applications. An in-depth understanding and a close survey of the physical and chemical properties of polymer surfaces are required to bring about surface modification of the polymers. Polyethylene with its versatile physiochemical properties has ever remained an obvious choice of researchers working in the area of surface modification. During the last two decades, different means of surface modification have been extensively explored [18]. Since some of the techniques of surface modification are still in their infancy, they need to be thoroughly explored. For example, radiation-induced graft copolymerisation [19, 20] using different radiation sources has been widely experimented. However, this technique still suffers from some shortcomings such as damage to the substrate, non-homogeneous grafting, high degree of subsequent homopolymerisation and practical inconveniences. As a result of this, the area of surface modification still remains a matter of curiosity. Some of the existing, as well as recent, methods of surface modification will be briefly discussed in the following sections.

2.1
Flame Treatment

Flame is probably the oldest plasma known to humanity and flame treatment is one of the oldest methods used by industries for the modification of polymeric surfaces. Flame treatment is very often used to treat bulky objects. It is mainly employed to enhance the ink permeability on the polymer surface. Though a very simple set-up (comprising of a burner and a fuel tank) is required for this technique, a very high degree of craftsmanship is needed to produce consistent results. Oxidation at the polymer surface brought about by the flame treatment can be attributed to the high flame temperature range (1000–1500 °C) and its interaction with many exited species in the flame. For an efficient flame treatment, the variables like air-to-gas ratio, air and gas flow rates, distance between the tip of the flame and the object to be treated, treatment time and the nature of the gas should be taken into account [21]. Sutherland et al. [22] studied the effect of these variables on the flame treatment of polypropylene. The results of his studies reveal that the oxygen concentration on the polymer surface reaches the maximum when air to gas ratio in the burner is 11:1. Moreover, the decrease in the water contact angle is directly proportional to the increase in the surface oxygen concentration.

Briggs et al. [23] as well as Garbassi and co-workers [24] studied flame treatment of polyethylene using X-ray photoelectron spectroscopy. Functional groups viz. hydroxyl (>C–OH), carbonyl (>C=O) and carboxyl (-COOH and -COOR) were generated on the surface of the polyethylene as a result of flame treatment. Studies also reveal that for the best results of flame treatment, an oxidising flame should be used, which requires a slightly higher concentration of oxygen than that required for complete combustion. One of the surfaces of adhesive tape (comprising of a co-extruded support having one layer of PP and another of HDPE) was treated by flame to increase its surface energy [25]. The tape thus had improved unwinding properties without coating the releasing agent and improved printability and fixing of the adhesive on the support without using a primer. Metalisable multilayer films with enhanced bond strength based on oriented HDPE films were also prepared by flame treatment [26].

2.2
Metal Deposition

Metal coatings can also help to achieve many desirable surface properties such as electrical conductivity and optical characteristics. This method was developed in the early 1970s and still continues to fascinate the industries [27, 28, 29]. Typically, two major industrial techniques for depositing metal coatings on polymer surface are electro-less plating and vacuum deposition. Electro-less plating is different from electroplating in the sense that the electrons required for reduction are supplied by a chemical reducing agent present in the solution, whereas in the case of electroplating the electrons are supplied by an external source like a battery or generator. Moreover, electro-less plating can be applied to non-conductors like polymers and ceramics.

The most widely used vacuum deposition techniques are evaporation and sputtering, often employed for smaller substrates. In the evaporation process, heating the metal by an electron beam or by direct resistance produces the vapours. The system is operated at a very high vacuum (between 10^{-5} and 10^{-6} Torr) to allow a free path for the evaporant to reach the substrate. The rate of metal deposition by evaporation processes varies from 100 to 250,000 Å min^{-1}. These processes can be operated on a batch or a continuous scale. On the other hand, in the case of the sputtering technique, the reaction chamber is first evacuated to a pressure of about 10^{-5} Torr and then back-filled with an inert gas up to a pressure of 100 mTorr. A strong electric field in the chamber renders ionisation of the inert gas. These inert gas ions are then accelerated towards the target material, which is then deposited on the substrate. Vacuum metal deposition [30] of gold followed by zinc was performed on the non-porous surfaces of LDPE and HDPE for the development of latent fingerprints on these surfaces. The authors' results suggested that the difference between these plastic types caused variations in the gold film structure, which in turn dictated the nature of the zinc deposition.

2.3
Chemical Treatment

Chemical treatment is the most primitive method for surface modification of polymers. Surface modification of polymers by chemical treatment with different agents is one of the most important trends in the search for materials with new or improved mechanical properties. Such type of modification is almost obligatory for the specific practical application of the so-called commodity polymers such as polyethylene. Changes in surface chemical composition and also the morphology of polyethylene upon etching with strong acids has been studied for the past three decades [31, 32]. This method involves the use of chemical etchants to convert smooth hydrophobic polymer surfaces to rough hydrophilic surfaces by dissolution of amorphous regions and/or surface oxidation as shown in Fig. 1. It has been established that the amorphous phase on the interfacial layers are chiefly affected during these treatments [33]. Chemical treatment has been widely used to treat large objects that would be difficult to treat by any other prevalent industrial techniques. The choice of etchant solely depends upon the type of polymer. Usually strong acids are used for such treatment. Chromic acid is commonly used for selective removing of the amorphous regions from the polyethylene surface [34]. McCarthy and co-workers [35, 36] have developed modification reactions that change only the chemical structure of different polymer surfaces. Chlorosulphonic acid etching of polyethylene surface [37] was accomplished and the changes in its surface structure were determined by means

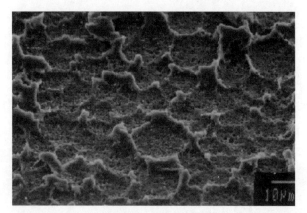

Fig. 1 Surface morphology of polysulphone after sulphuric acid treatment

of micro-hardness measurements. It was observed that the exposure of PE to chlorosulphonic acid led to amorphisation of the polymer, followed by processes of crosslinking in the amorphous layers. All these results showed that chemical and structural changes during the exposure are predominantly located in the surface layers. Chemical etching is also a potential technique for improving fibre adhesion in composite applications. The surface modification of high-strength polyethylene (HSPE) fibres [38] by chemical etching in chromic acid and investigation of the surface chemistry, functional group content, wetting ability, surface morphology and tensile strength of the modified fibres revealed that by controlling the reaction temperature and time, the tensile strength of modified fibres can be controlled and a surface with suitable wetting properties and functional groups can be achieved. However, the main shortcoming of chemical treatment is the lack of control over the process, leading to excessive bulk degradation for highly amorphous polymers.

2.4
Corona Discharge

Corona discharge is a well-accepted, relatively simple and most widely used continuous process for the surface treatment of polyolefin films [39, 40]. This technique is mainly used in the plastic industry to improve the printability [41] and adhesion [42, 43] of polyolefin films and articles. The corona treatment device is very simple and cost effective. It consists of a high voltage–high frequency generator, an electrode and a grounded metal roll covered with an insulating material as shown in Fig. 2. The whole system works as a large capacitor, with the electrode and the grounded roll as the plates of the capacitor and the roll covering and air as the dielectric. In this system, a high voltage when applied across the electrodes ionises the air producing plasma (often identified by the formation of a blue glow in the air gap). This atmospheric pressure plasma popularly known as corona discharge brings about physical and chemical changes on the polymer surface for improved

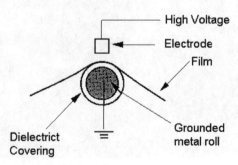

Fig. 2 Schematic diagram of corona treatment assembly

bondability and dyeability. A number of chemical reactions are bound to take place at the polymer surface as a result of corona treatment. Electrons, ions, excited neutrals and photons that are present in a discharge, react with the polymer surface to form radicals as shown in Scheme 1a.

These macro-radicals react with atmospheric oxygen [44] to give oxidation products (Scheme 1b).

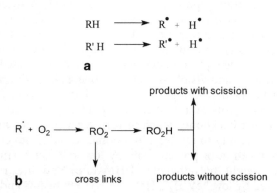

Scheme 1 a Electrons, ions, excited neutrals and photons that are present in a discharge react with the polymer surface to form radicals. **b** The macro-radicals formed on the polymer surface react with atmospheric oxygen to give oxidation products

The decomposition of these hydroperoxides produces $>$C–OH, $>$C=O, –COOH on the corona-treated polymer surface [42]. Moreover, the chain scission produces low molecular weight oxidised products [45]. Surface modification of cross-linked polyethylene [46] (XLPE) was accomplished upon exposure to partial discharges at elevated temperature. The XLPE sample surface during corona exposure was maintained at an elevated temperature ($T \leq 160$ °C). The appearance of patterns developed on the surface depended strongly on the temperature. Furthermore, the surface oxidation increased sharply with increasing temperature, an effect attributed to inward diffusion of corona-produced oxidising species. Jingxin et al. [47] employed the corona discharge technique as a means of generating chemically active sites on a LDPE film surface. The active species thus prepared at atmospheric pressure in air were exploited to subsequently induce graft copolymerisation of 2-hydroxyethyl methacrylate (HEMA) onto the LDPE film in aqueous solution. The effect of the corona discharge voltage, reaction temperature and inhibitor concentration on the grafting degree were studied in detail. Most recently, surface changes brought about by corona discharge treatment [48] of polyethylene film were used to study microbial colonisation on the modified surfaces. Though this technique has its distinct advantages it is also associated with some shortcomings such as non-consistent treatment due to the variation in ambient conditions (such as temperature and humidity), a high possibility of contamination due to the treatment carried out in air and, at times, lack of homogeneity/uniformity of surface modification.

2.5
Irradiation

Radiations can be basically differentiated into three different categories on the basis of their energy. High-energy radiation, mainly delivered by X-rays, γ-rays, electron beams from cobalt (^{60}Co) and magnesium (^{58}Mg) sources, are often known as ionising radiation. Mid-energy radiation, usually obtained from UV rays, pulsed/excimer laser [49] and plasma sources [50] (employing microwave/radio-frequency), and low-energy radiation yielded by infrared, ultrasonic, microwave and visible sources can bring about desired changes in the polymer backbone depending upon the irradiation time and energy of radiation. The primary role of any of these radiations is to activate the molecules on the polymer backbone, which in-turn reacts with the functional species present in its vicinity to render a functional surface. Toth et al. [51] studied surface modification of polyethylene by low keV ion beams. Here, ultra high molecular weight polyethylene (UHMWPE) and linear polyethylene (LPE) were treated by low keV H_2^+, He^+ and N_2^+ ion beams and the chemical changes induced in the modified surface layer were studied by FTIR and XPS. On treatment by N_2^+ ions, incorporation of N took place on the surface. The N-content of the surface layer reached the saturation value, ~11 atm% at ~10^{17} ions cm^{-2}. Ranby [52] has briefly reviewed photo-initiated surface modification of PE in the recent past. Kavc and co-workers [53] accomplished surface modification of polyethylene by photochemical introduction of sulphonic acid groups. Polyethylene samples were irradiated with UV light in a gas atmosphere containing SO_2 and air to achieve a photosulphonation of the surface. The modification, degradation and stability of different polymeric surfaces treated by reactive plasmas [54] were investigated in terms of surface morphology, etching rate and fluid holding capacity. The wetting instability reflected the hydrophobic recovery, which was attributed to the surface configurational changes. Reversible hydrophobic recovery was caused by configurational changes whereas permanent hydrophobic recovery was the result of surface cross-linking and formation of oligomers.

Radiation-induced surface grafting and surface modification are described in detail in Sect. 2.6.4.3. and Sect. 3.1, respectively.

2.6
Graft Copolymerisation

This is the most well-known and fundamental method of surface modification [55, 56]. A graft copolymer is a polymer comprising of molecules with one or more species of blocks connected as pendant chains to the backbone, having configurational features different from those in the main chain.

Graft copolymerisation can be represented by:

$\sim$ MMMMXMMMM $\sim$
  |
  G
  G
  G   (a)

Where M is the monomer unit in the backbone polymer and G is the pendant chain (graft) and X is the unit in the backbone to which the graft is attached. Graft copolymerisation can be brought about mainly through free-radical [57, 58, 59] and ionic mechanisms [60, 61]. The former mechanism includes ionisation-induced grafting, photo-grafting and plasma grafting, whereas the latter includes grafting via redox systems using various transition metal ions.

Though graft copolymerisation is a fundamental method of grafting, there are large varieties of techniques to obtain graft copolymers [62, 63]. Normally, graft polymerisation involves diffusion across a phase boundary between a monomer and the polymeric material. Depending upon the type of system, we obtain bulk grafting/surface grafting. With the advancement of technology and as an outcome of extensive research, it is possible to obtain a graft copolymer with definite chain length and lateral spacing on the polymer backbone [64, 65]. The grafting efficiency (G.E.) is usually calculated using the following equation:

$$\text{G.E.} = \frac{(\text{Mass of the monomer polymerised} - \text{mass of homopolymer})}{\text{Mass of the monomer polymerised}} \times 100$$

(1)

and the degree of grafting (%DG) using the equation:

$$\%\text{DG} = \frac{(Wgr - Wo)/Wspec}{Wo} \times 100$$

(2)

where, Wo is the initial weight of the film, Wgr is the weight of grafted film after extraction and $Wspec$ is the specific weight (per outer surface area) [66].

In the following sections we have described the different techniques of grafting.

2.6.1
Ionic Mechanism

Ionic polymerisation is a well-known technique for the preparation of graft copolymers but the fate of these reactions is determined by the reaction conditions. Since the discovery of 'living polymerisation', (anionic polymerisation) [67] it has become an excellent method for the synthesis of block and graft copolymers. In anionic polymerisation the graft copolymerisation is initiated by the anion generated by the reaction of bases with acidic protons in the polymer chain as shown in Scheme 2.

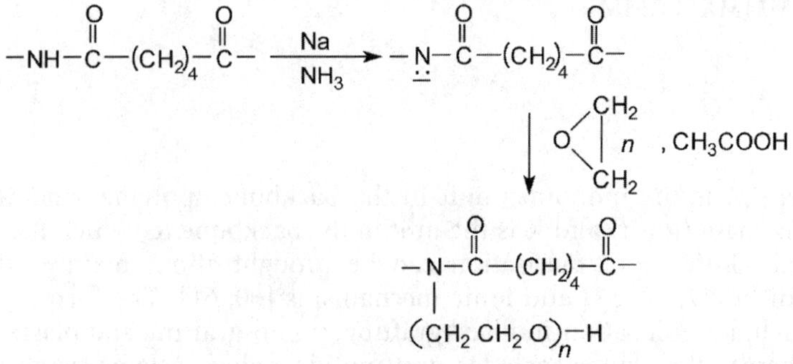

$$-NH-\overset{\overset{\displaystyle O}{\|}}{C}-(CH_2)_4-\overset{\overset{\displaystyle O}{\|}}{C}- \quad \xrightarrow[NH_3]{Na} \quad -\overset{\overset{\displaystyle O}{\|}}{N}-\overset{\overset{\displaystyle O}{\|}}{C}-(CH_2)_4-\overset{\overset{\displaystyle O}{\|}}{C}-$$

$$\downarrow \quad O\!\!\!<\!\!\begin{array}{c} CH_2 \\ | \\ CH_2 \end{array}\!\!\!>_n \;, CH_3COOH$$

$$-\overset{\overset{\displaystyle O}{\|}}{N}-\overset{\overset{\displaystyle O}{\|}}{C}-(CH_2)_4-\overset{\overset{\displaystyle O}{\|}}{C}-$$
$$|$$
$$(CH_2\, CH_2\, O)_n\!-H$$

Scheme 2 Initiation of graft copolymerisation in case of anionic polymerisation

Similarly, in the case of cationic method, the initiation reaction between labile alkyl halide [68] and Lewis acid have been utilised for cationic grafting onto halogenated polymers as shown in Scheme 3.

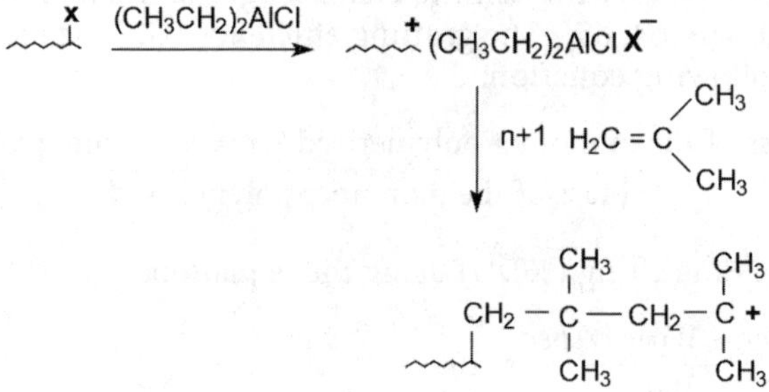

$$\overset{X}{\wedge\!\!\!\wedge\!\!\!\wedge\!\!\!\vee} \quad \xrightarrow{(CH_3CH_2)_2AlCl} \quad \overset{+}{\wedge\!\!\!\wedge\!\!\!\wedge\!\!\!\vee} (CH_3CH_2)_2AlCl\,\overset{-}{X}$$

$$\downarrow \quad n+1 \;\; H_2C = C\!\!<\!\!\begin{array}{c} CH_3 \\ CH_3 \end{array}$$

$$\wedge\!\!\!\wedge\!\!\!\wedge\!\!\!\vee\;CH_2 - \overset{\overset{\displaystyle CH_3}{|}}{\underset{\underset{\displaystyle CH_3}{|}}{C}} - CH_2 - \overset{\overset{\displaystyle CH_3}{|}}{\underset{\underset{\displaystyle CH_3}{|}}{C}} +$$

Scheme 3 Cationic grafting onto halogenated polymers can be brought about using the Lewis acids

2.6.2
Co-ordination Mechanism

Usually the Zigler-Natta co-ordination initiator system is used to graft α-olefins onto other polymers to give stereo block/graft copolymers, which contain isotactic/atactic sequences. In the Zigler-Natta co-ordination catalyst [69] system, the diethyl aluminium hydride reacts with pendant groups to form macromolecular trialkyl aluminium. The residual initiator is freed by extraction methods.

2.6.3
Coupling Mechanism

In this method the active hydrogen of the polymer is utilised to form graft copolymers. Polyamide is grafted with poly(ethylene oxide) as shown below:

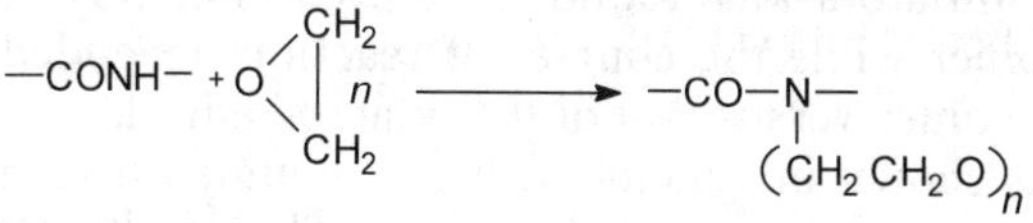

$$-CONH- + O\diagup\overset{CH_2}{\underset{CH_2}{|}}{\Big)_n} \longrightarrow -CO-N-\quad(CH_2\,CH_2\,O)_n$$

2.6.4
Free-Radical Mechanism

Free-radical mechanisms can be divided into three main categories, which are described in detail.

2.6.4.1
Chemical Grafting

This method involves graft copolymerisation using redox initiators [70] or free radical initiators [71, 72, 73] usually in the solution phase, occasionally under the influence of temperature, predominantly in the latter case. Redox systems have extensively been used to generate active sites especially on the natural polymers [74] (like cellulose). Transition metals viz. Cr^{+6}, V^{+5}, Ce^{+4},

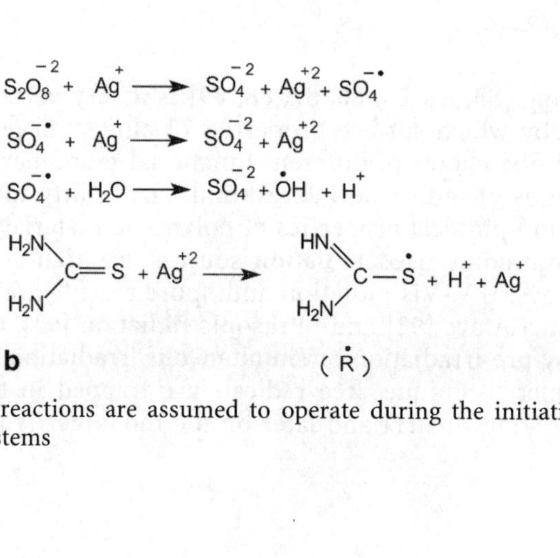

$$\underset{H_2N}{\overset{H_2N}{\diagdown}}C=S \xrightarrow{H^+} \underset{H_2N}{\overset{H_2N}{\diagdown}}\overset{+}{C}-SH \longrightarrow \underset{H_2N}{\overset{H_2N}{\diagdown}}C=SH^+$$

$$\underset{H_2N}{\overset{H_2N}{\diagdown}}\overset{+}{C}-SH + S_2O_8^{-2} \longrightarrow \underset{H_2N}{\overset{H_2N}{\diagdown}}\overset{+}{C}-\overset{\cdot}{S}$$

a $(\overset{\cdot}{R})$

$$S_2O_8^{-2} + \overset{+}{Ag} \longrightarrow SO_4^{-2} + \overset{+2}{Ag} + SO_4^{-\cdot}$$

$$SO_4^{-\cdot} + \overset{+}{Ag} \longrightarrow SO_4^{-2} + \overset{+2}{Ag}$$

$$SO_4^{-\cdot} + H_2O \longrightarrow SO_4^{-2} + \overset{\cdot}{O}H + \overset{+}{H}$$

$$\underset{H_2N}{\overset{H_2N}{\diagdown}}C=S + \overset{+2}{Ag} \longrightarrow \underset{H_2N}{\overset{HN}{\diagdown}}\overset{\cdot}{C}-S + \overset{+}{H} + \overset{+}{Ag}$$

b $(\overset{\cdot}{R})$

Scheme 4 These reactions are assumed to operate during the initiation processes involving redox systems

Co^{+3}, Mn^{+2} and Fe^{+2} have been found to be effective in producing free radical sites on the polymer backbone through the alcohol groups present on them [75]. In an alternative method, free radical initiators like BPO and AIBN are thermo-chemically activated to give rise to macro-radical sites on polymer backbone to initiate grafting of desired vinylic monomer. The efficiency of these initiators was found to be predominantly dependent on the nature of monomer while the course of reaction depended on the relative reactivity of monomer versus that of the macro-radical.

The reactions shown in Scheme 4a,b are assumed to operate during the initiation processes involving redox systems. The resultant radicals (S·, HO· and SO_4^-) were assumed to interact with the polymer, producing macro-radicals (via H abstraction) and initiating grafting in the presence of monomer [76].

2.6.4.2
Mechano-Chemical Grafting

Mechano-chemical grafting [77, 78] is carried out by a free radical mechanism, usually brought about by mastication, milling, extrusion, high-speed stirring or shaking and forcing the polymer solution through an orifice. The free radicals then generated, in the presence of monomers, initiate grafting of the latter onto the polymer surface. The advances in these techniques have brought about a new concept of grafting called 'reactive extrusion' where grafting on a polymer is brought about in the presence of a monomer, thermal initiator and compatibiliser during the process of extrusion. This is a very recent technique and possesses great potential in industrial applications [78, 79, 80, 81]. The functionalisation of ULDPE [82] with maleic anhydride (MA) and diethyl maleate (DEM) was performed by reactive extrusion in a twin-screw (Berstoff) extruder in the presence of dicumyl peroxide. This type of grafting was found to greatly influence the fracture behavior of the modified polymer.

2.6.4.3
Radiation-Induced Grafting

Radiation grafting [83, 84, 85, 86, 87, 88, 89] is a very versatile and widely used technique by which surface properties of almost all polymers can be tailored through the choice of different functional monomers. It covers potential applications of industrial interest and particularly for achieving desired chemical and physical properties of polymeric materials. In this method, the most commonly used radiation sources are high-energy electrons, γ-radiation, X-rays, U.V.-Vis radiation and, more recently, pulsed laser [90], infrared [91], microwave [92] and ultrasonic radiation [93]. Grafting is performed either by pre-irradiation or simultaneous irradiation techniques [94, 95]. In the former technique, free radicals are trapped in the inert atmosphere in the polymer matrix and later on the monomer is introduced into

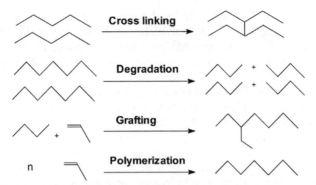

Fig. 3 Modifications produced at the polymer surface as a result of irradiation

the system to graft onto the polymer backbone. In the latter case, the polymer is irradiated in the presence of a solvent containing a monomer. In these methods, graft copolymerisation initiates from the radical sites and is generated along the polymer backbone due to the high-energy radiation. The disadvantage of this technique viz. homopolymerisation is now overcome by adding varying amount of different homopolymer inhibitors. The modifications brought about at the polymer surface as a result of irradiation are schematically presented in Fig. 3.

The grafting carried out under irradiation conditions can be broadly divided into three main categories.

2.6.4.3.1
High-energy Radiation-Induced Grafting

High-energy radiations [96, 97] viz. high-energy electron beam, X-rays and γ-rays are often known as *ionisation radiation*, since they displace electrons from the atoms and molecules, producing ions. These radiations deliver a large amount of energy to the substrate, much greater than those associated with the chemical bonds. Ionisation radiation can be classified into two categories: direct ionisation radiation, which include electrically charged α/β-particles, electrons and protons of kinetic energy sufficient enough to produce ionisation with electrons of the irradiated material through Coulombic interaction. Indirect ionisation radiation constituting X-rays and γ-rays do not produce ionisation upon interaction with the irradiated material but interact in a way that produces indirect ionising radiation. The commonly used industrial ionisation radiation sources are high-energy electrons (0.1–10 MeV) and cobalt-60 (^{60}Co) source (~1.25 MeV). Irradiation of polymers with ionisation radiation produces several chemical effects such as degradation, cross-linking as well as co-polymerisation and grafting in the presence of monomer. As previously mentioned, grafting of vinylic monomers onto a polymer can be achieved by simultaneous or pre-irradiation techniques.

Kurbanov et al. [98] studied the radiation grafting of different nitrogen-containing monomers onto the polyolefin and Lavsan articles for medical use. Here, the grafting was initiated by γ-irradiation and the effect of grafting conditions on the yield of the product and activity of various alkylating agents towards the modified surfaces were established. Yamamoto et al. [99] carried out vapour-phase mutual grafting of methyl acrylate onto polyethylene at high dose rates from an electron accelerator and found that the reaction yields the same surface graft structure as those obtained upon grafting at low dose rates from ^{60}Co sources. Yamamoto et al. [100] also studied the kinetics of the γ-ray induced surface grafting of methyl acrylate onto polyethylene using quartz helix micro-balances and found that under typical graft conditions the grafting rate increased, levelled off, and then accelerated with irradiation time. Svorcik et al. [101] carried out alanine grafting onto the ion beam-modified polyethylene surface with the aim of preparing prospective materials for biological experiments. In all the experiments, 15 μm-thick foils of PE were implanted with 156 keV Xe$^+$ ions at room temperature and subsequently reacted with a 2 wt% aqueous solution of alanine for 12 h under ambient conditions. Ion irradiation led to dehydrogenation of the PE chain and production of conjugated double bonds as an increasing function of the ion fluency, the length and concentration of which were determined by UV-Vis spectroscopy. The decrease in the double bond concentration after alanine doping, determined by the Rutherford back scattering (RBS) technique, was attributed to the alanine addition onto all the double bonds (in the form of zwitterions) and also on the macro-radicals through the -NH- group. The bond formation between alanine and surface free radicals was substantiated using EPR spectroscopy.

2.6.4.3.2
Photo-Grafting

The energy sources commonly used in radiation grafting are of three types: high-energy radiation/ionisation radiation (electron beam, X-rays and γ-rays), mid-energy (UV and plasma radiation) and low energy radiation (IR, microwave and ultrasonic radiation). The mid-energy and low energy radiation belong to photoradiation. The basic difference between ionisation radiation and photoradiation is that the energy of ionisation radiation is much higher than that delivered by the light source employed in photo-irradiation. Capital photo-grafting is usually brought about by the UV radiation obtained from various sources [102] including the most commonly used mercury vapour lamp. Although the wavelength of UV light ranges from 100–400 nm, the working range for surface photo-grafting is 200–400 nm. It is known that in a photochemical process, absorption of a photon by organic molecule (commonly called chromophore) results in excitation of the molecule from its ground state, giving rise to n-π^* and π-π^* transitions. The extra energy associated with the excited molecule is then dissipated by various processes, amongst which *energy transfer* is the most desirable process for grafting reactions. Absorption of UV light produces an excited singlet (S^{1*})

transition, which then releases a part of its absorbed energy to form an excited triplet (T^{1*}). Before the excited triplet undergoes decay, the absorbed energy is used for photochemically induced reactions [103].

The two main advantages of photo-grafting over ionisation radiation grafting are:

1. The modification produced by this technique is virtually restricted to the surface and is then helpful in generating properties like adhesion, antifogging, wear resistance, antistatic, printability, dyeability and biocompatibility at the polymer surfaces.
2. The energy associated with UV light is in competence with the chemical bond energies associated with any two atoms [104]. Hence, UV radiation often has the potential for retention of the properties of monomers and polymers while the other surface grafting techniques, which use ionisation radiation, cause damage to the substrate polymer due to excessive degradation.

Photo-grafting can be performed in two different manners: simultaneous grafting [105] and post-irradiation grafting [106]. For simultaneous grafting, the substrate, monomer, photosensitiser and solvent under inert conditions are taken in a photoreactor and irradiated with UV. In the latter case, substrate is irradiated in air to generate hydroperoxides on the surface followed by decomposition of these hydroperoxides in the presence of monomer to yield grafted surfaces.

Organic ketones, peroxides and their derivatives are well-known photo-initiators for surface photo-grafting. Benzophenone (BP) and its derivatives are the most commonly used photo-initiators. Benzophenone, when it absorbs a UV photon, is excited to the short-lived singlet state and then relaxes to a more stable triplet state, which then abstracts a hydrogen atom from the polymer backbone generating a polymer macro-radical. This then becomes the active site for the surface grafting. Initiators play a prime role in any graft copolymerisation reaction [70]. Organic ketones and peroxides are well known free radical initiators [71, 72, 73]. Under the influence of light or heat of suitable amplitude, they undergo sensitised decomposition, giving rise to a bi-radical (a pair of radicals). The general mode of their action involves abstraction of hydrogen radical from the polymer backbone generating a free radical site on the polymer backbone ($P^{.}$). This macro-radical acts as an active site for the graft copolymerisation; so called *from grafting*. Under favourable conditions, when the initiator abstracts a proton from the monomer, the later undergoes homopolymerisation, this homopolymer bearing a radical at the chain-end attacks the polymer backbone leading to a graft copolymer; so called *onto grafting*. The mechanism of surface photo-grafting is represented in Scheme 5.

Initiation:

$$I \xrightarrow{\ h\upsilon\ } {}^1I^* \longrightarrow {}^3I^*$$

$$PH + {}^3I^* \longrightarrow {}^3IH^* + P^{\cdot}$$

$$MH + {}^3I^* \longrightarrow {}^3IH^* + M^{\cdot} \quad \big[\ M/ \ MH \text{ is the monomer} \ \big]$$

$$P^{\cdot} + M \longrightarrow PM^{\cdot}$$

Propagation:

$$PM^{\cdot} + nMH \longrightarrow P(MH)_n M^{\cdot} \ [\text{grafting}]$$

$$M^{\cdot} + nMH \longrightarrow M(MH)_n M^{\cdot} \ [\text{homopolymer formation}]$$

Termination:

$$P(MH)_n^{\cdot} + R^{\cdot} \longrightarrow P(MH)_n R \ \big[\text{graft copolymer}\big]$$

$$(MH)_n M^{\cdot} + R^{\cdot} \longrightarrow M(MH)_n R \ \big[\text{homopolymer}\big]$$

Scheme 5 The mechanism of surface photo-grafting

2.6.4.3.3
Plasma Grafting

In the present scenario *plasma modification* is the most accepted technique for the surface modification of polymers on a laboratory scale [107]. It is a clean and solvent-free technique for achieving surface modification of polymers, consistently with great uniformity and close packing of functional groups. Plasma can be broadly defined as *a gas-containing charged and neutral species that includes electrons, +ve ions, −ve ions, radicals, atoms and molecules.* Plasmas are generally categorised on the basis of electron densities and electron energies. The low electron energy plasma, often called *cold plasma* is most commonly used for surface modification of polymers. The other plasmas known are interstellar plasma, alkaline vapour plasma and the hot plasma, which are often used for controlled fusion. In glow discharge plasma, the temperature of the ions and molecules are often ambient whereas that of electrons is higher by a factor almost up to 100. Thus plasmas produced by glow discharges are called *non-equilibrium* plasmas because the electron temperature is much higher than the gas temperature. A typical plasma reactor and its schematic diagram are shown in Figs. 4a and b.

The energy required to form and to sustain a plasma used for surface modification/polymerisation is obtained by an electric field which is produced either by ac or dc supply. The different ac frequencies employed for the excitation are 100 kHz, 13.56 MHz in the radio frequency range and 2.45 GHz in the microwave range. Frequencies in the range 10–20 MHz are most commonly used for industrial applications. The externally applied electric field brings about the ionisation process whereby an electron gains sufficient energy and becomes a free electron leaving behind a positive ion. An ion and two electrons thus produced can be further accelerated by the external electric field, producing further ionisation. This excited state usu-

(a)

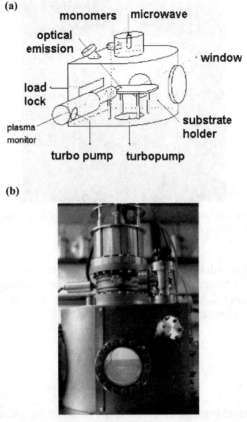

(b)

Fig. 4 a Schematic diagram of plasma reactor and **b** actual appearance of the reactor

ally stays for a very short period of time (<100 ns) and falls to a lower energy level or ground state by radiative decay. This relaxation process is known as *glow discharge* and a typical nitrogen glow in the plasma reactor is shown in the Fig. 5. The general mechanism of ionisation taking place in a plasma chamber is represented below.

The ionisation of an atom can be brought about by the influence of external energy source:

$$A + e \rightarrow A^+ + 2e \tag{3}$$

The atoms in the ground state can be promoted to exited state by electron impact:

$$A + e \rightarrow A^* + e \tag{4}$$

The exited state falls to the lower energy level by radiative decay:

$$A^* \rightarrow A + h\nu \tag{5}$$

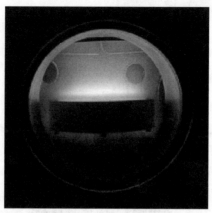

Fig. 5 A typical glow discharge produced upon applying a high voltage in a plasma reactor containing nitrogen gas

This relaxation produces a glow discharge. Upon collision of a metastable atom (energy-rich chemical reagent with a longer radiative time) with a neutral, it can cause ionisation of the later if the ionisation energy of the neutral is less than that of the exited atom:

$$A^* + B \rightarrow A + B^+ + e \,(A^* \text{ ismetastableatom}) \qquad (6)$$

or

$$A^* + B \rightarrow B^* + A \qquad (7)$$

Some other reactions like electron impact dissociation, secondary ionisation, symmetrical resonant transfer, asymmetrical charge transfer, electron–ion recombination and ion–ion recombination take place under the influence of the plasma conditions. Recently, Epaillard et al. [108] carried out functionalisation of the PE surface to generate carboxylic acid groups and tried to optimise the concentration of these functional groups by using CO_2/H_2O cold plasmas. Epaillard et al. [64] also studied the post-grafting of N-vinyl pyrrolidone (NVP) onto a plasma-modified PP surface. The equipment used to generate plasma consisted of a microwave generator (433 MHz) with a variable power output (0–250 W) which was coupled to a resonant cavity. Before each run, the system was evacuated to 10^{-6} mbar and maintained for 2 h. The plasma-treated films were dipped into a solution of NVP/distilled water/NH_4OH (60/38/2 wt%) for different time durations and the grafting yield was estimated using XPS, ATR-FTIR and contact angle measurements. The comparison of the O/C and N/C elemental ratios determined by XPS analysis showed that the increase in the carbon to oxygen and carbon to nitrogen ratio was due to the grafting of poly(vinyl pyrrolidone). The new band corresponding to the carbonyl valance vibration of poly(vinyl pyrrolidone) generated at 1683 cm^{-1} in the ATR-FTIR spectrum also confirmed that the grafting had taken place onto the PP film.

3
Surface Modification of Polyethylene

Ever since its discovery, PE has highly fascinated and greatly influenced mankind. The inertness and excellent mechanical properties of polyethylene have proved to be the greatest boon of this commodity polymer. However, researchers now are trying to harness the native properties of this polymer to develop advanced materials. With increasing modernisation and development of smart materials, demand for this wonder polymer is ever rising. Surface modification of PE has been extensively undertaken using modern and primordial methods. Today a large variety of techniques ranging from traditional to modern, and laboratory to industrial scale, are available in the literature for the surface modification of PE [109, 110, 111, 112, 113, 114, 115, 116].

3.1
Traditional and Modern Techniques

As mentioned earlier, there are a large variety of surface-modification techniques for achieving a desired surface architecture. Due to lack of sophisticated instrumental facilities, the researchers in the past had to compromise with techniques that were less surface-specific and offered little control over the process. However, with system automation, it is possible to develop surfaces with well-defined properties and morphologies, avoiding tedious processes.

3.1.1
Surface Grafting by Redox Initiators

In the early 1970s, surface modification of most polymers was achieved using redox initiators. Ce^{+4}-induced initiation was employed to achieve surface grafting of acrylamide onto LDPE film [117]. The film was first oxidised by chromic acid and then reduced with diborane to form a hydroxyl-rich surface which was then used to initiate graft polymerisation of acrylamide using Ce^{+4}/HNO_3. The mechanism of chromic-acid-facilitated surface oxidation of LDPE surface is shown in Scheme 6a and that of free-radical generation is represented in Scheme 6b.

The different functional groups generated on the surface during the course of reaction were identified using ATR-FTIR and XPS techniques.

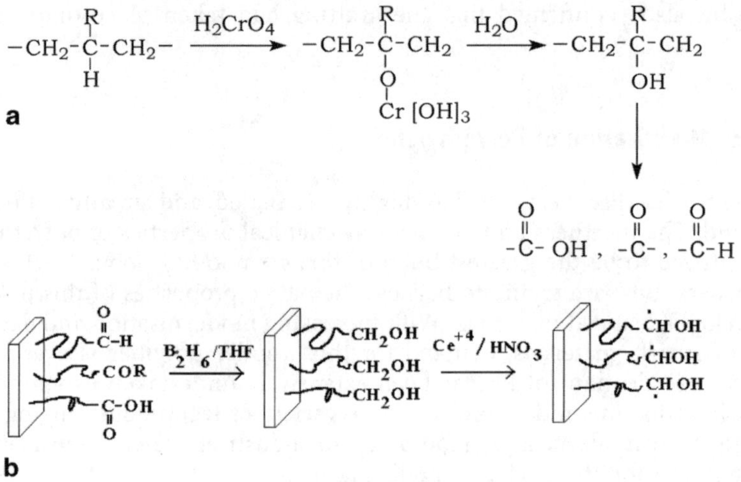

$$-CH_2-\overset{\overset{R}{|}}{\underset{\underset{H}{|}}{C}}-CH_2- \quad\xrightarrow{H_2CrO_4}\quad -CH_2-\overset{\overset{R}{|}}{\underset{\underset{O}{\|}}{C}}-CH_2 \quad\xrightarrow{H_2O}\quad -CH_2-\overset{\overset{R}{|}}{\underset{\underset{OH}{|}}{C}}-CH_2$$

a

$$Cr[OH]_3$$

$$\overset{O}{\overset{\|}{-C}}-OH,\ \overset{O}{\overset{\|}{-C}}-,\ \overset{O}{\overset{\|}{-C}}-H$$

b

(scheme b: surface oxidation structures)

C–H, COR, C–OH $\xrightarrow{B_2H_6/THF}$ CH$_2$OH, CH$_2$OH, CH$_2$OH $\xrightarrow{Ce^{+4}/HNO_3}$ CH OH, ĊHOH, CH OH

Scheme 6 a The mechanism of chromic acid facilitated surface oxidation of LDPE. **b** Free-radical generation on surface-oxidised LDPE

3.1.2
Chemical Treatment

Surface modification of LDPE film can also be brought about by chemical treatment [118] with an aqueous solution of ammoniacal ammonium persulphate in the presence of Ni^{+2} ions under variable reaction conditions. The investigation of treated surface showed the presence of polar groups (viz. carbonyl and hydroxyl) in the infrared (IR) spectroscopy, with characteristic bands at 1700, 1622 and 3450 cm^{-1}. It is known that the persulphate ion attacks the double-bond-producing epoxy or diol group. However, the destructive oxidation of saturated hydrocarbons does not occur with persulphate alone, but requires the presence of the nickel (II) ion. The authors have proposed the following mechanism of chemical treatment:

$$Ni^{+2}+S_2O_8^{-2}\rightarrow Ni^{+3}+SO_4^{-2} \tag{8}$$

Persulphate oxidises Ni^{+2} to Ni^{+3} as shown in the equation. This Ni^{+3} oxidises the carbon chain on the surface, introducing polar groups such as >C=O and -COOH. The examination of the XPS spectra of the modified and unmodified films revealed that the ratio of the area of peak O1s to C1s was less in the untreated film than that in the treated film. This difference was attributed to the greater oxygenation in the chemically treated film. The changes in the morphology of the chemically treated film can be seen in the SEM micrograph (Fig. 6). The surface modification of LDPE film by ammonium persulphate and nickel sulphate solution led to an improvement in the mechanical performance of its laminates, which is evident from the better adhesion performance with the epoxy resin. The mechanical interlocking

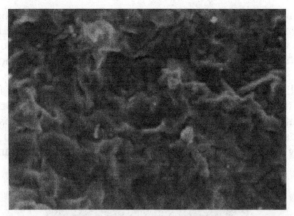

Fig. 6 SEM micrograph of the morphology of LDPE film treated with ammonium persulphate and nickel sulphate

was achieved by surface roughness and the enhanced chemical bonding with epoxy resin by the functional groups generated on the surface.

3.1.3
Grafting

Zhao et al. [119] carried out thermal and radiochemical grafting of acrylamide onto the surface of polyethylene and polystyrene. The effect of four different methods on the grafting efficiency was investigated, three of them involved two steps: first hydroxylation/hydroperoxidation of the polymer

$$S_2O_8^{-2} \rightarrow 2SO_4^{-\bullet}$$

$$SO_4^{-\bullet} + H_2O \rightarrow HSO_4^- + HO^\bullet$$

$$HO^\bullet + P\text{-}H \rightarrow H_2O + P^\bullet$$

$$HO^\bullet + P^\bullet \rightarrow P\text{---}OH$$

$$2HO^\bullet \rightarrow H_2O_2$$
a

$$P\text{---}OH + Ce^{IV} \rightarrow P\text{-}O^\bullet + H^+ + Ce^{III}$$
b

Scheme 7 a Reaction mechanism when sodium persulphate was used as hydroxylating agent for the polymer (P–H). **b** Decomposition of hydroxylated polymer P–OH at the surface, in the presence of ceric ammonium nitrate generated alkoxy radicals (P–O·) which initiated the grafting

surface, followed by grafting initiated by the thermal decomposition of hydroxyl or hydroperoxide groups formed at the surface. In the last method, the polymer was irradiated with γ-rays in the presence of an aqueous solution of monomer also containing Mohr's salt to inhibit homopolymerisation.

When sodium persulphate was used as hydroxylating agent for the polymer (P–H), the reaction underwent the mechanism shown in Scheme 7a.

After isolation of the hydroxylated polymer P–OH, decomposition of hydroxyl groups at the surface in the presence of ceric ammonium nitrate generated alkoxy radicals (P–O·) which initiated the grafting (Scheme 7b).

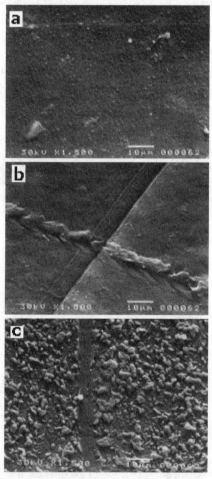

Fig. 7 γ-Radiation-induced grafting of acrylamide onto LDPE films. **a** Blank, **b** low grafting with beam damage giving rise to cracking, buckling and pock marks on the surface and **c** uniform multiphase formation with less beam damage

The γ-irradiation of the polymers in the presence of a concentrated aqueous solution of acrylamide resulted in the surface modification of PE, but not polystyrene.

Acrylamide was grafted onto low density polyethylene (LDPE) using the direct radiation grafting method [120] and the optimum conditions were determined under which homopolymerisation of acrylamide was reduced and the grafting process proceeded successfully. The study was also aimed at understanding the optical absorption spectra in solids, providing essential information about the bond structure and the energy gap in both crystalline and non-crystalline materials. Moreover, the researchers also attempted to correlate the different morphologies with the changes they induced in the fatigue cracking behavior. The samples were irradiated by γ-radiation with two ranges: low doses (0.5, 1, 1.5 Mrad) and high doses (10, 20 Mrad), using a Co^{60} source with a dose rate of 1.7 Gy s^{-1} used for all irradiations. Acrylamide was grafted onto LDPE using the direct radiation grafting method, using dioxane as solvent. The important conclusions that were drawn from the study are that the degradation induced by γ-irradiation facilitated the formation of free radicals, which were chemically active and allowed the formation of covalent bonds between different chains (cross-linking) and in turn minimised the antistropic character of the polymer, which led to a decrease in the birefringence. For blank and low grafting up to 14.6%, the beam damage was obvious as cracking, buckling and pock marking on the surface whereas for higher grafting there was more uniform multiphase formation and less beam damage as revealed by the SEM images (Fig. 7a–c).

3.1.4
Corona Treatment

The effect of corona discharge with subsequent acrylic acid grafting on LDPE film was studied in terms of surface functionality and surface energetics, to improve the dyeability [121]. The introduction of polar groups (O=C–O, C=O, and C–O) onto the corona-treated LDPE film with acrylic acid was confirmed by XPS. The analysis of corona-treated films revealed that the corona treatment of LDPE films led to an increase in its surface free energy, mainly due to the increase in its polar component. From the acid–base interaction point of view, it was found that the graft polymerisation of acrylic acid onto the corona-treated LDPE film played an important role in generating the acidic character. This is one of the specific components governing the surface free energy, resulting in an improved dyeability with basic dyeing agents.

Lei et al. [122, 123] employed the corona discharge technique as a means of generating chemically active sites on a LDPE film surface. The active species thus prepared at atmospheric pressure in air were exploited to subsequently induce graft copolymerisation of 2-hydroxyethyl methacrylate (HEMA) onto the LDPE film in aqueous solution. The effect of the corona discharge voltage, reaction time, reaction temperature and inhibitor concentration on the grafting degree are represented in Fig. 8a–d. The degree of

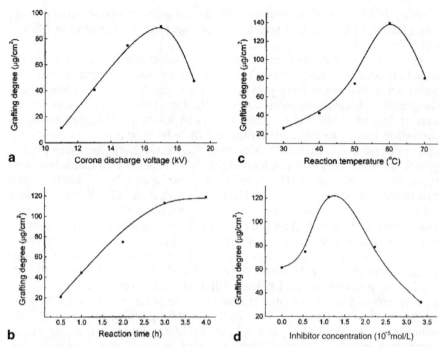

Fig. 8 a Effect of corona discharge voltage on the grafting degree of HEMA onto the LDPE films (treatment time 72 s, copolymerisation time 2 h, temperature 50 °C, HEMA concentration 5%). **b** Effect of reaction time on the grafting degree of HEMA onto LDPE films (treatment time 72 s, temperature 50 °C, HEMA concentration 5%, corona discharge voltage 15 kV). **c** Effect of reaction temperature on the grafting degree of HEMA onto LDPE films (treatment time 72 s, copolymerisation time 2 h, HEMA concentration 5%, corona discharge voltage 15 kV). **d** Effect of inhibitor concentration on the grafting degree of HEMA onto the LDPE films (treatment time 72 s, copolymerisation time 2 h, temperature 50 °C, HEMA concentration 5%)

grafting increased with an increase in the corona discharge voltage up to 17 kV for a constant treatment time of 72 s. A further increase of the discharge voltage caused the grafting degree to decrease. The active sites on the film surface increased with the corona discharge voltage, which initiated graft copolymerisation of HEMA onto the LDPE surface in the reaction solution. However, a very high corona discharge voltage caused surface degradation, forming low molecular weight oxides, which initiated homopolymerisation of HEMA in the reaction solution and thus decreased the grafting degree. The grafting degree was found to increase progressively with the reaction temperature, reaching a maximum value at 60 °C, and then decreased rapidly. This decrease in grafting yield was attributed to the homopolymerisation of HEMA in the reaction medium. The grafting degree increased with the monomer concentration. However, for the HEMA concentration (>20%) significant homopolymerisation was observed. Mohr's salt was added during

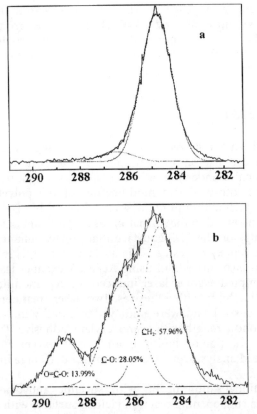

Fig. 9 XPS C1 s core-level spectra of ungrafted and HEMA-grafted LDPE films with copolymerisation times of **a** 0 h and **b** 2.0 h (72 s corona discharge treatment time, 15 kV corona discharge voltage, 50 °C copolymerisation temperature, 5% HEMA concentration)

the grafting to minimise homopolymer formation. Homopolymerisation obviously decreased with the increase in the concentration of the Mohr's salt but at higher concentration ($>10 \ \mu$mol L^{-1}) the copolymerisation was also inhibited, leading to a lower grafting yield. The presence of HEMA grafts on the LDPE film surface was confirmed from the XPS spectra of neat and modified films (Fig. 9a,b), which showed the characteristic peaks arising from the binding energies of C–O (286.6 eV) and O–C=O (289.1 eV) of HEMA. Takahashi et al. [124] studied the post grafting of HEMA onto a LDPE film treated with corona discharge. It was found that the degree of grafting increased with the increase in corona discharge time, grafting time and HEMA concentration. The characteristic peaks of HEMA grafted onto LDPE emerged at 1719 cm^{-1} on attenuated total reflectance IR spectra ($>$C=O in ester groups) and at 531 eV on electron spectroscopy for chemical analysis (XPS) spectra (O$_{1s}$). The C$_{1s}$ core level ESCA spectrum of HEMA-grafted

LDPE showed two strong peaks at ~286.6 eV (–C–O– from hydroxyl groups and ester groups) and ~289.1 eV (O=C–O– from ester groups), and the C atom ratio in the –C–O– groups and O=C–O groups was found to be 2:1. The hydrophilicity of the grafted LDPE film was remarkably improved compared to that of the ungrafted film.

3.1.5
Photochemical Treatment

Apart from the high energy radiation induced grafting, modification of the polymer surface can also be brought about by the oxidative degradation of the surface molecules under the influence of UV-Vis light irradiation. Using this technique, photo-oxidative modification of the polyethylene surface sensitised with 9,10-anthraquinone was achieved by Dalinkevich et al. [125]. This treatment led to the formation of a significant amount of carbonyl and -OH groups on the LDPE surface that greatly enhanced its adhesion strength with the epoxy adhesives. The surface etching [126] of polyolefins including PE was also carried out using Ar–F UV excimer laser (λ=193 nm). Doyle [127] attempted excimer laser induced surface modification of HDPE for enhanced bondability with adhesives. Two laser-treatment methods, viz. raw beam and focused beam, were employed to modify the surface of HDPE to enhance its bond strength with cyanoacrylate adhesive. The XPS analysis showed the incorporation of significant amount of oxygen atoms on the surface of HDPE. SEM analysis revealed no significant change in the HDPE surface morphology.

Surface functionalisation of PE with succinic anhydride followed by derivatisation via ring opening reaction yielded surfaces with different functional groups [128]. The PE films were photochemically grafted with maleic anhydride in vapour phase in a reaction involving benzophenone and acetone at 60 °C for 5 h under UV irradiation with a 400 W mercury vapour lamp. The chemical transformation of PE film bearing succinic anhydride group is schematically presented in Scheme 8.

The surface-modified PE films were characterised using XPS and ATR-FTIR spectroscopic techniques. The hydrophilicity of the modified surfaces was determined by measuring their contact angle with water.

The photochemical lamination [129] of 2-HEMA onto LDPE film was achieved by exposing the solution of 2-HEMA, acetone and benzophenone to UV radiations. The water contact angle of the PE films dropped from 97 ° for neat to 50 ° for HEMA grafted film. The minimum contact angle was achieved for LDPE film with grafting time of 2 min, but increased for longer irradiation time. The O/C ratio calculated from the XPS results indicated a rise in the oxygen species on the grafted surface. The use of the Pyrex filter prevented the homopolymer formation and resulted in stable poly(HEMA) grafts. Recently, a novel method of laminating PE surface in critical propane has also been reported [130].

Yang and Ranby [106] carried out performance evaluation of organic ketones and aldehydes as photo-initiators for surface photo-grafting of acrylic

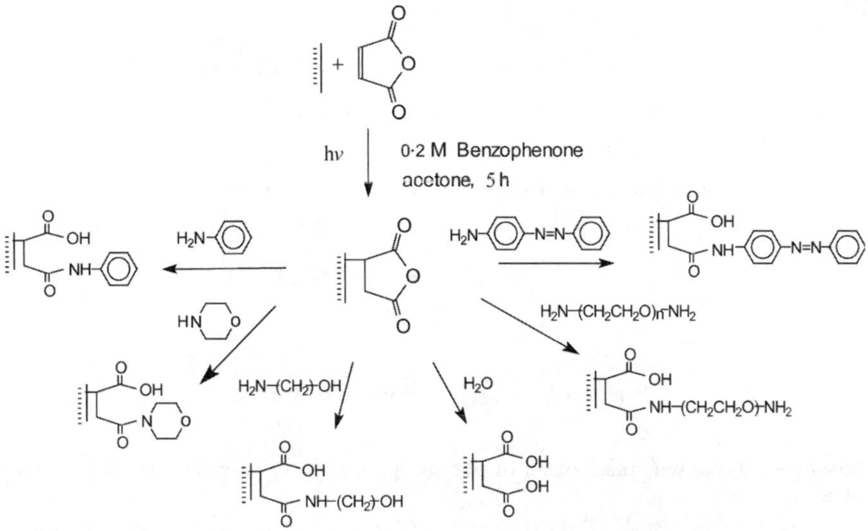

Scheme 8 Chemical transformation of PE film bearing succinic anhydride group

acid onto LDPE films. Benzophenone, xanthone, acetophenone and 9-fluo-
renone were found to be efficient photo-grafting initiators for obtaining
thick grafted layers whereas the use of anthraquinone, benzoylformic acid,
biacetyl and 4-benzylbenzophenone led to thin grafted layers. The efficiency
of the photo-initiators was determined on the basis of polymerisation reactivity, grafting efficiency and surface wetting time. The results of their study
revealed that high triplet state energy, strong UV absorption, stable molecular structure and low initiation reactivity of the ketyl free radical were the
decisive factors making the ketones efficient initiators for grafting applications.

Gaseous SO_2 absorbs in the UV region 240–320 nm with an absorbance
maximum at 285 nm. Kavc et al. [131] have recently reported surface modification of polyethylene by photochemical introduction of sulphonic acid
groups where the PE samples were irradiated with UV light in a gaseous atmosphere containing SO_2 and air for different exposure times. Irradiation
with UV light led to the excitation of ground-state SO_2 to its first singlet
state ($n-\pi^*$), followed by inter system crossing to its triplet state. The excited
SO_2 were assumed to abstract H from a hydrocarbon molecule to yield alkyl
radicals. Moreover, an energy transfer from SO_2^* to R–H via the formation
of an exciplex has also been proposed. The proposed mechanism of photo-
sulphonation of PE is shown in Scheme 9a and can be summarised as shown
in Scheme 9b.

The introduction of sulphonic acid groups (-SO_3H) onto the PE surface
was substantiated by FTIR, energy-dispersive X-ray spectroscopy (EDX) and
XPS techniques. The appearance of signal at 1038 cm^{-1} and band at
1156 cm^{-1} in the FTIR spectrum of photo-modified PE film is attributed to

$$SO_2 \xrightarrow{h\nu} SO_2{}^*$$

$$R\text{-}H \quad + \quad SO_2{}^* \quad \longrightarrow \quad SO_2 \quad + \quad R\cdot \quad + \quad H\cdot$$

$$R\cdot \quad + \quad SO_2 \quad \longrightarrow \quad R\text{-}SO_2\cdot$$

$$R\text{-}SO_2\cdot \quad + \quad O_2 \quad \longrightarrow \quad R\text{-}SO_2\text{-}O\text{-}O\cdot$$

$$R\text{-}SO_2\text{-}O\text{-}O\cdot \quad + \quad R\text{-}H \quad \longrightarrow \quad R\text{-}SO_2\text{-}O\text{-}O\text{-}H \quad + \quad R\cdot$$

$$R\text{-}SO_2\text{-}O\text{-}O\text{-}H \quad \longrightarrow \quad R\text{-}SO_2\text{-}O\cdot \quad + \quad OH\cdot$$

$$R\text{-}SO_2\text{-}O\cdot \quad + \quad R\text{-}H \quad \longrightarrow \quad R\text{-}SO_3H \quad + \quad R\cdot$$

a

$$R\text{-}H \quad + \quad SO_2 \quad + \quad \tfrac{1}{2}O_2 \quad \xrightarrow{h\nu} \quad R\text{-}SO_3H$$

b

Scheme 9 a Proposed mechanism of photosulphonation of polyethylene. **b** Summary of **a**

the symmetric and asymmetric vibrations of SO_3 units, the broad signal at 3200 cm^{-1} was assigned to O–H units of sulphonic acid. The EDX spectra of the modified PE surface showed a typical signal of sulphur at 2.3 keV. The photosulphonated PE surface showed an S $2p^{3/2}$ signal at 169.2 eV and S $2p^{1/2}$ signal at 170.6 as obvious in the XPS spectrum in Fig. 10. The influence of the gas composition on the modification reaction studied by contact angle measurements revealed that the irradiation in presence of pure SO_2 hardly induces any change in the contact angle ($\theta{\sim}90$ °). Significant hydrophilicity of the PE surface was achieved when SO_2/air composition contained slightly more SO_2 than the stoichiometric amount. The depth of photo-modification amounted to several microns. The presented method of surface modification

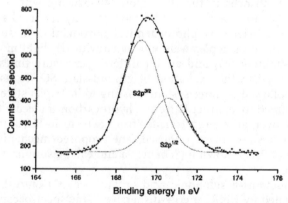

Fig. 10 X-ray photoelectron spectrum (S 2p region) of LDPE after UV irradiation in the presence of SO_2 and air (.). Experimental data (...) Gaussian fits for the S $2p^{3/2}$ and S $2p^{1/2}$ peaks with maxima at 169.2 and 170.6 eV

carried out under atmospheric pressure is considered to be an inexpensive alternative to plasma modification.

3.1.6
Halogenation

Halogenation has been one of the routes for the modification of the PE films [132]. Halogenation of PE is found to occur selectively in the amorphous regions of the PE film. Moreover, halogenated PE surfaces can be further modified to other functionalities using substitution reactions. Photobromination [133] of LDPE surface was achieved by exposing the LDPE films to the bromine vapour and irradiating them with a mercury vapour lamp for stipulated irradiation cycles (each cycle of 30 s). It was observed that irradiating the film for more than one cycle led to debromination causing surface damage. Surprisingly, the roughness of the surface after the third cycle was less than the second bromination cycle, indicating the erosion of the low molecular weight chain fragments from the topmost layer. At the end of the first bromination cycle, 4.7% bromine was found incorporated onto the LDPE surface. Moreover, the bromination process not only incorporated the bromine atoms, but also, inevitably introduced oxygen moieties

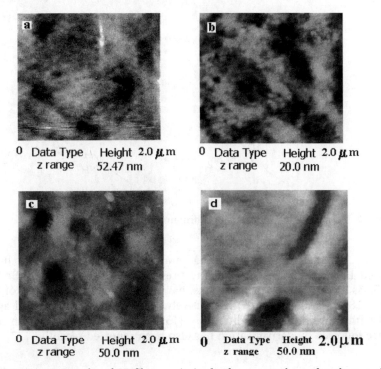

Fig. 11 AFM topography of PE films: **a** virgin, **b** after one cycle, **c** after three cycles, **d** after two bromination cycles

onto the surface. Increasing the number of bromination cycles led to the broadening of the bromine peaks in the XPS spectra, probably due to the structural and environmental effects. The excessive bromination was assumed to cause dibromination depicting an extra peak in the XPS spectra. The topographical changes in the LDPE films after each bromination cycle can be clearly seen in the AFM images (Fig. 11a–d). The bromination was followed by nucelophilic substitution of the corresponding bromine atoms with 4-aminobenzenethionate. An amine density of 3 molecules per 100 nm^2 was achieved as a result of substitution.

In a similar attempt, Vernekar et al. [134] carried out photochemical bromination of polyolefin surfaces. It was observed that the gas phase photochemical bromination occurred with high degree of regioselectivity. The bromination of LDPE surface yielded four different species (a) secondary Br, (b) allyl Br, (c) vinyl Br and (d) dibromide, in varying proportions depending upon the underlying mechanism. However, the formation of allyl Br moieties was predominant. The proposed mechanism of photochemical bromination is presented in Scheme 10.

The 1 h-brominated LDPE film revealed three bands at 550, 568 and

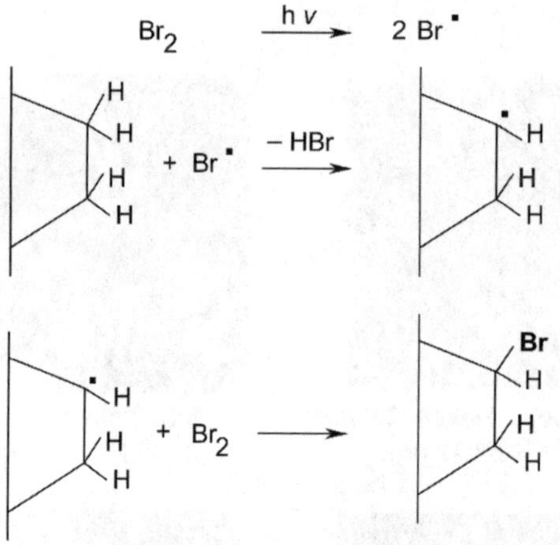

Scheme 10 Mechanism of photochemical bromination of PE

620 cm^{-1}, assigned to C-Br stretching bands. The wide scan XPS of 1 h-brominated LDPE film substantiated the presence of Br 3d, Br 3p$^{3/2}$, Br 3p$^{1/2}$ and Br 3s levels along with C1s peaks, having binding energy values of 70.2, 183.12, 189.72 and 257.04 eV, respectively. These values are in good agreement with those reported in literature [135, 136]. Photo-bromination of PE surface followed by derivatisation has also been reported recently [137].

3.1.7
Plasma Treatment

Zimmermann and co-workers [50] treated the PE films in a remote down-stream plasma reactor with oxygen, nitrogen, hydrogen and mixture of gases and investigated the surface modification using contact angle goniomctry and XPS. The hydrophilation of the PE surface occurred in decreasing order with, oxygen≅nitrogen>mixed gas>hydrogen plasma treatments. During the nitrogen plasma treatment, the trace amount of oxygen resulted in the incorporation of a relatively high amount of oxygen onto the surface. In contrast, for hydrogen plasma treatment under similar conditions, the PE surface remains hydrophobic, although the oxygen concentration should have been comparable. The most important conclusion was that the singlet molecular oxygen, which cannot be quenched by hydrogen, seems to be the major reactive species in the remote plasma reactor containing oxygen (even in trace amounts). Therefore, in hydrogen plasma, the oxygen functionalities formed by the trace oxygen are readily reduced, which does not take place in other cases. The effects of other reaction parameters on the surface modification of PE are also discussed in detail.

The effect of reactive plasma and its distance form the PE film surface has also been studied in detail [138]. The surface of polyethylene films was modified with various water-soluble polymers [(poly[2-(methacryloyloxy)ethyl phosphorylcholine] (PMPC), poly[2-(glucosyloxy)ethyl methacrylate] (PGEMA), poly(N-isopropylacrylamide) (PNIPAAm) and poly[N-(2-hydroxypropyl) methacrylamide] (PHPMA)] using Ar plasma-post polymerisation technique [139]. Here, the reactive sites were generated on the PE surface under the influence of argon plasma. These reactive sites on the surface were then utilised to covalently anchor the functional monomers as shown in Scheme 11.

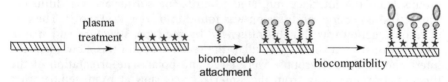

Scheme 11 Functional monomers anchored on the polymer surface utilised for achieving biocompatibility

The bio-compatibility of the grafted PE (g-PE) films was evaluated by the adsorption of serum proteins. As a conclusion, PMPC-g-PE film exhibited the highest biocompatibility among all grafted PE films because its surface adsorbed much less protein than those of the untreated PE and other g-PE films.

3.1.8
Dynamics of Oxidatively Functionalised Polyethylene Surface

The discussion on the surface modification of polyethylene would remain incomplete without referring to the surface reconstruction phenomena. The reconstruction of the interface of functionalised polyolefin surfaces has remained a matter of concern and curiosity to the researchers exploring this area [140, 141, 142]. This phenomenon of reconstruction is mainly governed by the surface thermodynamics involving the changes in the surface free-energy [143, 144]. Functionalised polyethylene surfaces are found to be highly prone to this type of phenomena. Oxidation of LDPE film with short-term treatment of corona/plasma/chromic acid results in a material having hydrophilic carboxylic acid and ketone groups in a thin oxidatively functionalised interface. These interfaces are often stable at room temperature but undergo rapid orientation with changes in temperature. Some researchers have also observed the migration of functional groups towards the sub-surface with respect to its storage time [145, 146]. Up to now there are only two techniques   that can determine the migration of the functional groups, namely XPS and contact angle measurement.

Whitesides and co-workers [147] carried out surface oxidation of PE films using aqueous chromic acid solution ($CrO_3/H_2SO_4/H_2O=29/29/42$ w/w/w) for 1 min at 72 °C to produce PE–CO_2H with 3:2 (carboxylic acid: ketone) groups present on the surface. The investigation of the surface reconstruction relied on the results of XPS and contact angle measurements. It was found that contact angle was much more sensitive than XPS towards dynamic surfaces. The pH of the sensing liquid also greatly influenced the contact angle with the surface. The contact angle at pH 13 was lower than that at pH 1 due to the conversion of interfacial carboxylic acid groups to carboxylate ions, accompanied by reconstruction of the phase. This reconstruction was assumed to be driven by solvation of carboxylate ions. On heating (35–110 °C) the oxidatively functionalised PE films under vacuum, the functional groups from the interface migrated towards the subsurface via diffusion. The activation energy of diffusion was found to be ~50 kcal mol^{-1}. The rate of reconstruction was highly influenced by the type, structure and size of the functional groups. In particular, the reconstruction was slow when the interfacial functional groups were large and polar. The orientation of the functional groups away from the interface was slow at room temperature and became rapid at temperatures close to the melting point of the polymer. The recovery of the interfacial functional groups migrated into the polymer was also investigated by heating these functionalised films in the presence of polar liquid. The results demonstrated a partial recovery, indicating the reversibility of the reconstruction phenomena.

3.2
Different Forms of Polyethylene

PE, being a commodity polymer, is used in its different physical forms viz. fibres, sheets, membranes, moulds with different backbone chemical configurations (LPE, LLDPE, LDPE, HDPE, UHMWPE, UHSPE etc). Each of these forms of PE requires surface modification at some stage of application. The surfaces of PE fibres are often modified to make them compatible in the composites, whereas PE sheets/tapes are modified to achieve adhesion. Moulds are frequently surface-modified for probability and membranes for selective permeation. In the same way, different chemical configurations of PE, by the virtue of their properties, are used for different applications after surface modification.

3.2.1
Physical Forms

Ranby et al. [148] carried out photo-initiated graft copolymerisation of acrylamide and acrylic acid onto the HDPE film tapes. The tapes were pre-soaked in the solution containing monomer and photo-initiator (benzophenone) and irradiated under nitrogen atmosphere. The pre-soaking procedure of the tape film was very important for efficient photo-grafting within short irradiation times. Figure 12 shows a sketch of the device used for continuous surface photo-grafting onto PE tapes. The effects of various reaction parameters (viz. irradiation sequence, lamp power, distance from surface of film) on the grafting efficiency were minutely investigated. The grafting rate of acrylic acid remained high even after 60 s of soaking whereas acrylamide, being solid at room temperature (in absence of a solvent),

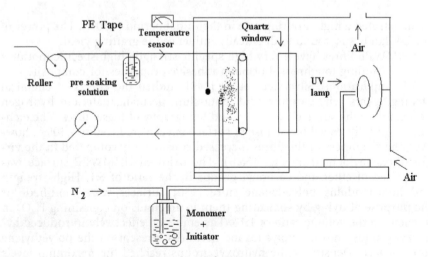

Fig. 12 Continuous photo-grafting reactor for PE tapes

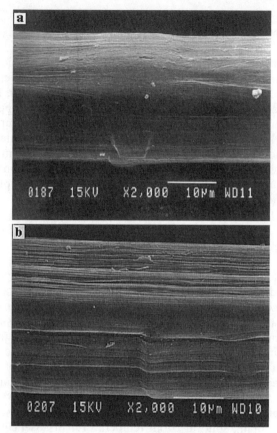

Fig. 13 a Neat PE fibres and **b** chromic acid etched PE fibres

demonstrated a high grafting rate in the initial period (<10 s). The power of the UV lamp was found to drastically influence the grafting yield.

UHMWPE fibres have a very high specific strength, high specific modulus and outstanding toughness, but poor adhesive properties for composite material applications. Silverstein et al. [149] studied the effects of chemical etching agents (viz. chromic acid, potassium permanganate and hydrogen peroxide) on the surface chemistry and topography of these fibres. The etching onto PE fibres led to the loss of surface oxygen (reduced to <10%), however, the roughness of the fibres increased considerably compared to the virgin fibres (evident from Fig. 13a,b). The oxidised UHMWPE surface was comprised of ether and carbonyl groups in the ratio of 6:1. High-strength and high-modulus polyethylene fibre [150] surfaces were modified for the purpose of dying by sonicating them in an emulsion containing H_2O_2 in o-xylene in the volume ratio of 1:10. The emulsion effectively introduced hydroxyl groups onto the polysiloxane networks, present on the polyethylene fibre surface. Density of the hydroxyl groups reached the maximum levels

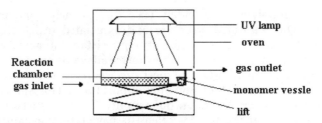

Fig. 14 Reactor used for vapour phase photo-grafting of acrylic acid onto HDPE powder

after a treatment time of 5 min at 80 °C. The dyeing of the surface-modified PE fibres with C.I. acid blue 25 resulted in a bluish fibre. Surface modification of HDPE tubes [151] for biocompatibility was accomplished by coating them with chitosan, chitosan hydrogel and heparin. Chitosan and chitosan hydrogel, immobilised on both the inside and outside surfaces of HDPE tubes, generated a pH-sensitive surface. The tube surfaces were further modified with heparin by a surface interpenetrating method to improve their blood compatibility.

Surface-modified polymer particles as performance additives represent a new dimension in material engineering. Surface modification enables formation of unique polymer–polymer composites through combination of polymers that are normally incompatible. Polymer particles (including UHMWPE, HDPE, and rubber particles) with highly polar surfaces were prepared by reactive gas processing [152]. The treatment resulted in high surface-energy and water-wettable particles. Lei et al. [153] performed photo-grafting of acrylic acid on HDPE powder in vapour phase as a means of surface functionalisation. This method of grafting curtailed the excessive homopolymerisation of solution grafting. A typical vapour phase photo-grafting reactor is shown in Fig. 14. The FTIR analysis showed an exponential rise in the $C=O/CH_2$ ratio with an increase in the reaction time, which was complementary to the XPS results. Dong and co-workers [154] performed plasma immersion ion implantation onto UHMWPE. Recently, surface modification of ultra-high strength polyethylene fibres was also carried out for enhanced adhesion to epoxy resins using an intense pulsed high-power ion beam [155]. Fuming nitric acid was used as an etchant for the treatment of UHMWPE fibres to prepare composites with epoxy resin [156].

3.2.2
Blends and Copolymers

PE is often blended and copolymerised with other thermoplastics symbiotically in order to enhance its mechanical properties and improve the processibility of the resultant polymer. Elastomers by their virtue of flexibility, hold a very special position amongst all the existing polymers and are thus employed for a wide range of applications [157]. Rubbers based on ethylene-propylene copolymer are widely used due to their good mechanical properties, highly saturated structure and associated resistance to aging and ozone deterioration. Due to their strong resistance to acids, alkalis and aqueous systems, they suffer from the shortcomings of low wettability and therefore low biocompatibility. The inert surface of ethylenic elastomers restricts them from many other potential applications. This shortcoming of the ethylenic elastomers leaves an opportunity for their surface modification. For more than three decades scientists have been constantly engaged in developing elastomers with desired surface properties using different techniques of surface modification [151, 152, 153, 154, 155, 156, 157, 158, 159, 160, 161, 162, 163, 164, 165, 166, 167]. In this section we have reviewed the surface modification and characterisation of commonly used ethylenic elastomers.

Kowalski et al. [162] performed radiation-induced surface modification of different rubbers including EPDM via sulpho-chlorination, chlorination and fluorination. The study of the reaction kinetics revealed that all these reactions followed the free-radical mechanism, with sulpho-chlorination having the highest reaction rate. The initiation step was found to depend on the irradiation intensity. Xu and co-workers [163] examined the melt grafting of glycidyl methacrylate (GMA) onto EPDM using dicumyl peroxide as radical initiator. The influence of grafting temperature and initiator (dicumyl peroxide)/monomer (GMA) concentrations on EPDM/NR blends was also investigated. Haddadi-asl and co-workers [164] studied radiation grafting of bio-monomers (HEMA, NVP, acrylamide and acrylonitrile) onto ethylene-propylene elastomers by varying molecular weight and ethylene content ranging from 40 to 70%. The grafting was carried out by the *simultaneous method* at room temperature in sealed ampoules containing monomer and rubber samples under inert conditions. The samples were irradiated at 0.4 to 7.5 kGy h^{-1} (total dose) in all the experiments. The maximum graft yield, was achieved when EPM with highest ethylene content and highest molecular weight was employed. Katbab et al. [165] reported the same trend during the grafting of functional monomers onto EPDM rubber. The authors have attributed this effect to the steric hindrance effect of side groups of pure PP in EPDM. Another simple explanation of this behaviour was attributed to the stability [tertiary>secondary] and density [secondary>tertiary] of both the secondary and tertiary macro-radicals generated due to the hydrogen abstraction from the polymeric substrate [166]. Although the macro-radicals produced by PP are more stable than those produced by PE, the higher G value of radicals produced by PE resulted in higher grafting yield on EPM (with higher ethylene content). The effects of reaction variables (oxygen effect, structure and concentration of monomer and role of homopolymer inhibitors) from the point of view of grafting yield have been thoroughly investigated. In the continuation of their work Haddadi-asl et al. [167] have extensively studied the effect of novel multifunctional acrylate (MFA) on grafting yield. The mode of action of MFAs is discussed in detail in the following section.

Bhowmick and co-workers [168] investigated the bulk and surface modification of ethylene propylene diene monomer (EPDM) rubber and fluoroelastomer by electron beam irradiation. The structure of the modified elastomers was analysed with the help of IR spectroscopy and XPS. The gel content, surface energy, friction coefficient and dynamic mechanical properties of bulk modified fluoro-elastomers and the surface-modified EPDMs were also measured. The resultant properties of the modified EPDM were correlated with the structural alterations.

4
Recent Advances in Polyethylene Surface Modification

In spite of ample literature on surface modification of polymers, there are constant efforts to develop agents and techniques with minimal shortcomings and maximum accuracy to obtain materials with desired surface properties and architectures. The expectations about smart materials are the major driving force behind this extensive research in the area of surface modification. Researchers are now trying to develop reagents with enhanced capabilities and techniques that instigate minimum changes in the native properties of the polymeric materials. As a first step towards achieving this aim, a wide range of initiators have been developed.

A variety of traditional free radical initiators and their mode of action are well documented in the literature [169]. Scientists worldwide, as a result of their extensive research, have developed a range of novel macro- and micro-initiators that outperform the conventional ones [170, 171, 172]. In order to overcome the hassles associated with solvent handling during copolymerisation and to meet the environmental norms of the present global scenario, the free radical graft copolymerisation of several functional monomers on polyolefins was studied under melt processing conditions [173] in an extrusion reactor, using organic peroxide initiators [174]. However, the use of organic peroxides led to serious discoloration [175] in LDPE. This was overcome by developing azo initiators [176] with suitable half-lives under melt graft conditions, capable of H-abstraction from the PE backbone. These free radical initiators can even be applied for other surface grafting techniques. Photo-activation of the heterocyclic azides, to generate highly reactive nitrenes, have been recently demonstrated as a novel route for the photo-modification of a wide range of surfaces, including polyolefins. The highly reactive nitrene radical formed upon UV irradiation of the heterocyclic azides (3-azidopyridine) underwent a multitude of reactions viz. insertion into C-H bonds, addition to olefins and proton abstraction to give the corresponding amine [177, 178]. The surfaces thus modified were found to display properties consistent with that of the chemically bound heterocycles.

Low energy initiation techniques [179, 180, 181] (near infrared, ultrasonic radiation and line tuneable pulse laser) have lately emerged to be better alternatives to the high-energy radiations (γ-irradiation and e-beam). Laser-induced polymerisation of monomers have attracted significant attention in recent years generating a considerable literature published on both pulsed and continuous lasers to initiate polymerisation [182, 183]. However, a little work has been done on laser-induced graft copolymerisation of monomer onto the surface of polymeric substrates. Lasers have several distinct advantages over other light sources, including narrow wavelength band, which could be tuned to the maximum absorption of the photo-initiator thus making laser treatment an energy efficient process. Moreover, high energy could be concentrated over a small area, which allows polymerisation to be initiated from a remote position within a very short time and with low beam penetration. Pulsed laser induced grafting of polyethylene has been reported [184, 185]. Pulsed lasers allow short exposure times and less thermal damage by optimising the time intervals between the pulses. A pulsed CO_2 laser has large beam sizes, large pulse energy, uses non-toxic gases, has high laser efficiency and is cheap and easy to operate.

Mirzadeh et al. [186] generated peroxides on the surface of polymers using CO_2 pulsed laser through a peroxidation mechanism, which was found to be a successful technique for initiating the graft copolymerisation of hydrophilic acrylic acid onto the surface of ethylene propylene rubber (EPR). The graft level was found to increase with the pulse and then declined above 25 pulses. The grafted poly(acrylic acid) formed a fractal type of morphology on the surface of EPR, which enabled it to impart both the hydrophilic and hydrophobic sites on the surface of EPR. Mirzadeh et al. [187] also studied the CO_2 laser induced surface grafting of HEMA, NVP and acrylamide onto the EPR surface to achieve biocompatibility. The effect of benzophenone and AIBN as photosensitisers on graft level was evaluated. Grafting was found to be dependent upon the wavelength, repetition rate and frequency of laser. The effect of multifunctional acrylate (MFA) [viz. 1,1,1-trimethylolpropane trimethacrylate (TMPTA)] as homopolymerisation inhibitor on the grafting yield was also evaluated [167].

Another excellent technique for low energy initiation is ultrasonic radiation. The efficiency of ultrasonic vibration to induce the graft copolymerisation has been known since the mid 1970s, but this versatile technique has very little been explored [188, 189]. Price and co-workers [190] have recently accomplished ultrasonically enhanced oxidation of polyethylene surface using ammonium and persulphate oxidising agents under mild conditions. Sonochemically assisted surface modification of PE using mild oxidising agents is also reported by the same authors [191].

Homopolymerisation is the most undesired phenomenon associated with surface grafting and has remained the main drawback of the radiation-induced grafting systems. During the last two and a half-decades, different homopolymerisation inhibitors [192, 193, 194] have been employed in various grafting systems and their efficiencies evaluated. Some of the latest developments in improving the grafting efficiency and reducing associated homopolymerisation are presented herewith.

Metallic ions such as Fe^{+3}, Fe^{+2} and Cu^{+2} are known to have an inhibitory effect on free-radical polymerisation and have been used for several years in the radiation grafting systems to interdict homopolymer formation. The mechanism of their action is meticulously discussed in the literature [195,

Table 1 Inhibitory effect of various additives (metal salts) on the radiation grafting of NVP, HEMA and Am onto EPM rubber

Monomer	Additive	Concentration (M)	Grafting yield
NVP 20%	No additive		0.24
	$Cu(NO_3)_2$	0.005	3.70
	$Cu(NO_3)_2$	1.000	0.0
	$FeSO_4$	0.010	0.1
	Mohr's salt	0.05	0.59
HEMA 20%	No additive		0.26
	$Cu(NO_3)_2$	0.005	0.51
	$Cu(NO_3)_2$	1.000	3.65
	$FeSO_4$	0.010	0.93
	Mohr's salt	0.05	0.40
Acrylamide 20%	No additive		0.0
	$Cu(NO_3)_2$	0.005	0.33
	$Cu(NO_3)_2$	1.000	0.53
	$FeSO_4$	0.010	0.29
	Mohr's salt	0.05	3.50

196, 197]. Table 1 displays the inhibitory effect of various metal salts on the radiation grafting of NVP, HEMA and acrylamide onto ethylene propylene rubber. Haddadi et al. [197] evaluated the effect of multifunctional acrylic additives including trimethylol propane triacrylate (TMPTA), polyethylene glycol diacrylate (PEGDA) and polyethylene glycol triacrylate (PGTA) on the γ-radiation induced grafting of hydrophilic vinyl monomers onto ethylene-propylene elastomers. Dworjanyn et al. [198] used small quantities of the crosslinking agent ethylene glycol dimethacrylate (EGDMA) to accelerate grafting of styrene onto PE, PP and cellulose substrates. Grafting was thus enhanced mainly through branching of grafted substrate chains as shown in Scheme 12.

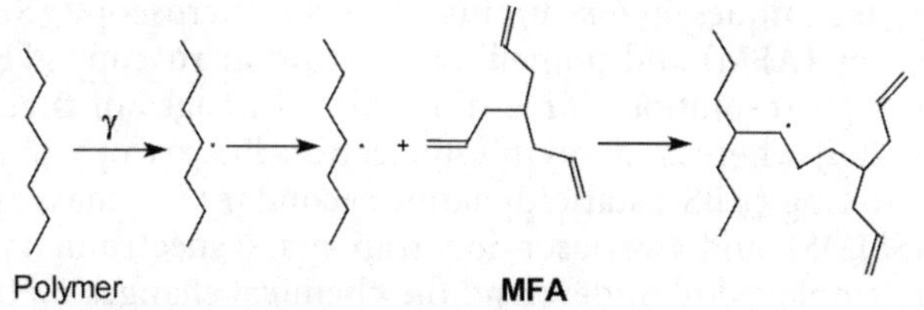

Polymer   MFA

Scheme 12 Mode of action of multi-functional acrylates in improving the grafting efficiency, through branching of grafted substrate chains

Controlled grafting via heterogeneous ATRP of polymethyl methacrylate onto poly(ethylene-co-styrene) is one of the most significant advancements in the surface modification of PE [199]. The grafting of PMMA was carried out in presence of CuBr and pentamethyldiethylenetriamine as a catalyst via the well-known ATRP mechanism, as shown in Scheme 13.

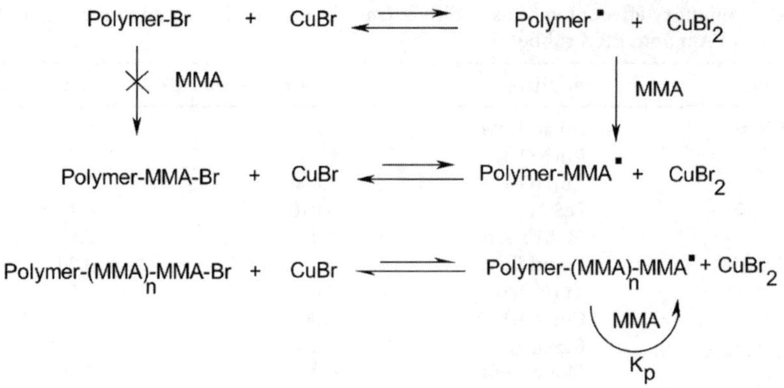

Scheme 13 Grafting of PMMA onto brominated EPDM via well-known ATRP mechanism

5
Analysis of Surface-Modified Polyethylene

The term 'surface' though it seems to be generalised, varies with the modification and characterisation techniques. The modification of a surface usually varies from a few angstroms to a couple of microns. Before modification of any surface, it is necessary to know it at a molecular level. Depending upon the type of information desired, there are a variety of analytical techniques [200, 201, 202, 203, 204] such as contact angle goniometer, X-ray photoelectron spectroscopy, attenuated total reflectance-infrared spectroscopy, surface plasmon resonance, static secondary ion mass spectroscopy, neutron refractometer etc. Therefore, the choice of the surface analysis technique must be made very judiciously, keeping in mind sample suitability, sample depth, analysis environment and the surface information desired. The information supplied by each technique is often different but complementary. The techniques like scanning electron microscopy (SEM), atomic force microscopy (AFM) and tunnelling electron microscopy (TEM) are employed when high resolution, three-dimensional images of the surfaces are desired [205, 206] whereas, X-ray photoelectron microscopy (XPS), Rutherford back scattering (RBS), static/dynamic secondary ion mass spectroscopy [207, 208] (SSIMS) and two-laser ion trap mass spectrometry [209, 210] (L2ITMS) are employed to understand the chemical changes on the substrate backbone. Many more techniques for the investigation of modified surfaces have also recently emerged [211, 212, 213]. The following section describes the uses of some of these techniques for the analysis of modified polyethylene surfaces.

5.1
Contact Angle Measurement

This technique is the fundamental, accurate and most sensitive tool for determining the surface functionality and thereby hydrophilicity. Since it is very difficult to obtain exquisite information of the outermost few angstroms of a solid surface by any other technique, solid/liquid/vapour (S/L/V) contact angle measurement emerges as one of the most surface-sensitive methods. The basis of the contact angle technique is the three-plane equilibrium, which exists at the contact point at the solid/liquid/vapour interface. This equilibrium is normally considered in terms of the surface and interfacial tension or surface and interfacial free energies. The contact is governed by the force balance at the three-phase boundary and is defined by Young's equation [214]:

$$\gamma_{LV}\mathrm{Cos}\theta = \gamma_{SV} - \gamma_{SL} \tag{9}$$

where γ_{LV} is the surface tension of the liquid in equilibrium with its saturated vapour, γ_{SV} is the surface tension of the solid in equilibrium with the saturated vapour of the liquid and $\gamma_{SL}$is the interfacial tension between the solid and liquid.

Some of the commonly used techniques for measuring contact angle [215, 216, 217] are the *sessile drop method, captive bubble method* and *Wilhelmy plate method*. These techniques have been extensively used and well documented for characterisation of modified PE surfaces [218, 219, 220, 221, 222, 223, 224, 225, 226, 227, 228, 229, 230] for various applications. Whitesides et al. [231] studied the wetting of flame-treated polyethylene film having ionisable organic acids and bases at the polymer–water interface. The effect of the size of substituted alkyl groups in amide and ester moieties on the surface hydrophilicity was also studied [232]. The biocompatibility of the polyethylene film surface modified with various water-soluble polymers was evaluated using the same technique [233]. The surface properties of hyperbranched polymers have been very recently reported [234].

5.2
Attenuated Total Reflectance-Infrared Spectroscopy

Infrared spectroscopy is a well-established and widely used technique for the identification of functional groups. Harrick and Fahrenfort [235, 236], with their pioneering efforts, developed a technique called attenuated total reflectance spectroscopy in which the IR spectra of the thin films/plates can be obtained from the near surface region (~5 μm). It is based on the principle of Maxwell's theory, which says, 'when the propagation of light takes place through an optically thin, non-absorbing medium, it forms a standing wave perpendicular to the total reflecting surface. If the sample absorbs a fraction of this radiation, the propagating wave interacts with the sample and becomes attenuated, giving rise to a reflection spectra, very similar to the absorption spectra'. An ample literature is now available on the surface characterisation of polymers [237, 238] and polyethylene [239, 240] in particular, using ATR-FTIR spectroscopy. This technique also helps in the determining the depth at which the functional groups are located. Surface analysis of electron beam irradiated thermoplastic elastomeric films of LDPE-EVA blends has been reported recently [241]. ATR-FTIR in combination with electron probe microanalysis (EPMA) was used to characterise the methyl methacrylate photo-grafted PE films [242].

5.3
Scanning Electron Microscopy

This technique is essentially employed when high resolution three-dimensional images of the surface morphology are desired [243]. When an electron beam impinges on a sample, back scattered electrons, secondary electrons and X-rays are produced. A scintillation detector detects these secondary electrons, which are emitted from a surface with low energy (50 eV). If the system is equipped with an X-ray detector (which measures either the wavelength or the energy of the X-rays), elemental analysis can also be performed [244]. The polymer samples are often coated with a metal such as gold to minimise the beam damage and to attain conductivity. Chung and co-workers [245] fabricated platelet-compatible polyethylene, where phosphorylcholine (PCe) was introduced onto the surface of PE using a novel synthetic process. The PE films were first grafted with acrylic acid and finally immobilised with PCe, after a chain of condensation reactions. These films were then taken up for in vitro platelet adhesion tests and were monitored using SEM. The SEM images (Fig. 15a–d) of the surface-modified PE revealed that the amount of platelet adhered decreased in the order P-OH>P-PTMG>P-AA>P-PCe. Moreover, it was also observed that the length of the spacer played a significant role in determining the platelet compatibility of the modified PE films. The spacer effect on the platelet adhesion is evident in Fig. 16a–d, which shows that the platelet adhesion decreased with an increase in the spacer length. Shalaby et al. [246] accomplished controlled gas phase sulphonation of LDPE films and the morphological changes were studied using SEM and EDX spectroscopic techniques.

Table 2 Semi-quantitative elemental analysis of sulphur on photo-sulphonated LDPE films obtained from the EDX spectra

Reaction time	Sulphur
min	% (mean±std. dev.)
0	0.00±0.00
5	0.16±0.03
10	0.31±0.03
30	0.47±0.03

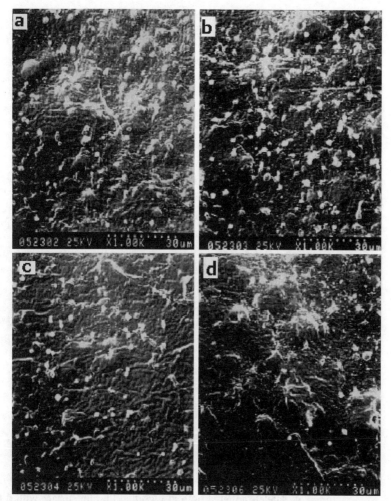

Fig. 15 SEM images of platelets adhered onto functionalised PE film surface: **a** PE-AA, **b** PE-PTMG, **c** PE-POH, **d** PE-PCe

SEM of the treated films indicated a progressive change in the surface morphology, as seen in Fig. 17. After 5 min of reaction time (Fig. 17b), the surface appeared slightly smoother than that of the control sample (Fig. 17a), upon 10 min exposure (Fig. 17c) submicron-sized blisters appeared and at 30 min (Fig. 17d), the defects covered the entire surface and were of several micron in diameter. The EDX spectra substantiated the presence of sulphur on the treated films and its absence on the control. The results of this semi-quantitative elemental analysis of sulphur on the film surface, computed from the collective EDX spectra, are shown in Table 2. The tabulated results indicate that the relative concentration of sulphur in these films increases with an increase in the reaction time.

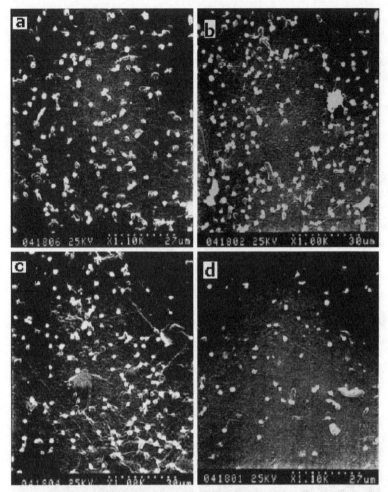

Fig. 16 SEM pictures of platelets adhered onto PCe-grafted PE film surface with different spacers of varying lengths. Spacer: **a** EG, **b** BDO, **c** PPG, **d** PTMG

5.4
Atomic Force Microscopy

Atomic force microscopy (AFM) is a novel tool, designed to achieve a high
degree of sensitivity while investigating surfaces on an atomic scale [247,
248]. AFM is based on the concept of the scanning tunnelling microscope
(STM) and is a combination of the principles of the STM and the stylus pro-
filometer. AFM incorporates a probe that imparts a lateral resolution of 30 Å
and vertical resolution of less than 1 Å. Using this technique, it is possible to
measure different types of inter-atomic forces [248, 249, 250]. AFM images
are obtained by measurement of the force on the tip of the scanner (usually

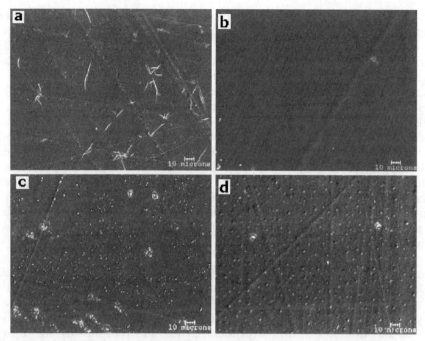

Fig. 17 SEM images of controlled gas phase sulphonation of LDPE films after different exposure times **a** control, **b** 5.0 min, **c** 10.0 min, **d** 30.0 min

diamond/silicon nitride), which is generated by its proximity to the surface of the sample. When the tip is moved horizontally, it follows the surface contours represented by trace B in Fig. 18. The experimental set-up and working of AFM are described elsewhere [251]. Recently, this technique has often been used to detect the interaction of biological systems with organic/polymeric surfaces [252].

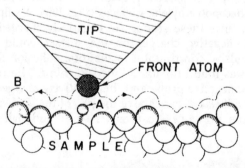

Fig. 18 Sensing tip of the AFM instrument following its path along the surface contours represented by the *trace B*

Various polymeric surfaces including PE [253] were functionalised under oxygen and dichlorosilane-RF-cold-plasma environments and were employed as substrates for further in situ derivatisation reactions and immobilisation of α-chymotrypsin (AC). The nature and morphology of the derivatised substrates and the immobilised enzymes were analysed using high-resolution XPS and AFM techniques (Fig. 19). AFM images revealed that hill-valley topographies emerged for the neat PE and AC directly connected to PE (Fig. 19a,b) whereas filigree (fibre-cluster-like) organisations were observed in the case of one-step spacer-immobilised enzyme (Fig. 19c) and functionalised PE containing spacer (Fig. 19e). The fibre-like morphologies also appeared when spacer molecules were attached to the PE surface as shown in Fig. 19d. The spacer chains offered enhanced molecular freedom for the motion of the enzymes resulting in better packing of AC molecules into specific supramolecular structures. Thus, this study helped in establishing the importance of spacer and the correlation between the spacer chain and AC packing.

Another technique is the scanning force microscopy (SFM), which measures the local changes in the surface chemical composition. Georges et al. [254] studied the time-dependent etching process of polyethylene films with chromic acid using this technique. The etching introduced alcohol, aldehyde, ketone and acid groups on the surface that increased the hydrophilicity of the surface. The time-dependent change in hydrophilicity was measured as an increase in adhesion force between the surface and an amide-functionalised SFM probe (Fig. 20). Figure 20a, was interpreted with an explanation that, during the first few minutes of the etching process, scission products (viz. aldehydes and ketone) were formed. Upon longer exposure to chromic acid, these groups were further oxidised to acid groups, which formed much stronger hydrogen bonds with the amide tip of SFM than the aldehydes and ketones formed earlier. The small increase in adhesion force in the first few minutes of the etching was due to the weak interaction with the aldehydes and ketones whereas the strong increase at longer etching times was due to the introduction of acid groups. To prove that acid groups are indeed formed, pH-dependent measurements were performed on modified and unmodified polyethylene surfaces (Fig. 20b). A sulphate SFM probe (SO_3^-) was used, since these groups are negatively charged over the whole pH range. Thus, negative charges on the surface could immediately be detected as a repulsive contribution to the total adhesion force whereas no pH-dependence was found for unmodified PE films. The repulsive interactions between the negative probe and the negatively charged surface caused a decrease in the adhesion force above the pK_a of the acid surface, which ultimately proved the presence of carboxylic acid functionalities on the modified surface.

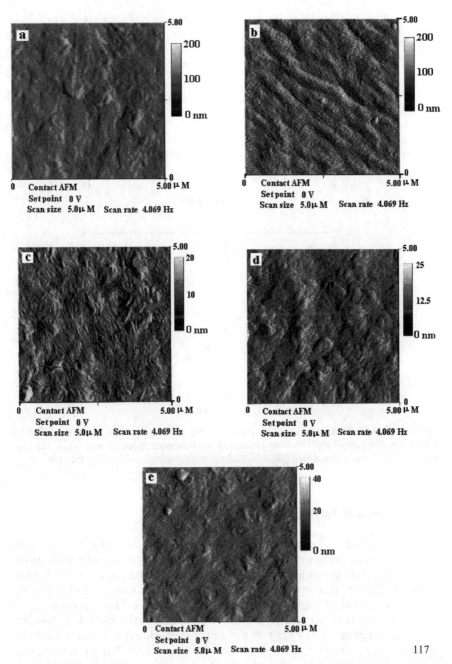

Fig. 19 AFM images of PE surface: **a** virgin PE, **b** AC directly attached to PE, **c** AC-immobilised involving one-step spacer, **d** DS functionalised PE, **e** DS-HFGA functionalised PE

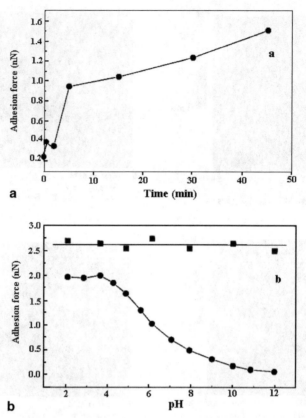

Fig. 20 a Change in adhesion forces measured by SFM using chemically modified tip during the time-dependent etching process of polyethylene films. **b** 'Adhesion' titration curves of PE films: unmodified (■) and etched with chromic acid for 45 min (●)

5.5
X-Ray Photoelectron Spectroscopy

X-ray photoelectron spectroscopy (XPS) is perhaps, one of the most substantial methods for organic surface analysis since it can identify and quantify elements/functional groups on the surface at an atomic level [255, 256]. Upon irradiation of a material with electromagnetic radiation of sufficient energy, the core electron from the atom is knocked out. The excess energy of the irradiation is converted to the kinetic energy of the ejected photoelectron. The binding energy (E_b) of the inner shell electron is determined by the difference between the X-ray photon energy and the kinetic energy, (E_k) of the ejected photoelectron:

$$E_b = h\nu - E_k \tag{10}$$

Table 3 Surface elemental analysis obtained from the XPS spectra

Element	Atomic concentration (%)			
	AA-g-HDPE	Chitosan coated HDPE	Chitosan hydrogel coated HDPE	Chitosan hydrogel coated HDPE+heparin
C	70.25	64.61	63.27	62.52
O	25.38	28.20	29.35	29.58
N	4.37	7.19	7.38	6.45

In principle, XPS analysis involves irradiation of the material surface with soft X-rays and measurement of the E_k of the photoelectrons ejected. Since the binding energy of the particular shell of an atom is unique for each element, measurement of kinetic energy and hence the binding energy allows the identification of the element [257]. Many researchers [258, 259, 260] have studied surface modification of PE exclusively using XPS.

Chitosan and chitosan hydrogel were immobilised on the surface of acrylic acid grafted high density polyethylene (HDPE) tubes [261], prepared by a two step process. The tube surfaces were further modified with heparin by surface interpenetrating method to improve blood compatibility and analysed using ATR-FTIR and XPS techniques. The surface elemental analysis obtained from the XPS spectra (Table 3) shows that the concentration of oxygen and nitrogen species on the surface increases upon grafting of AA and chitosan, respectively.

In a similar experiment, immobilisation of α-chymotrypsin (AC) on PE films [253] was studied using high resolution XPS spectroscopy. The atomic concentration changes upon functionalisation and immobilisation affirmed the successful derivatisation and immobilisation reactions.

5.6
Static Secondary Ion Mass Spectroscopy

Static secondary ion mass spectroscopy (SSIMS) is another surface selective technique for surface characterisation [262, 263, 264]. The information obtained from SSIMS is complementary to XPS because SSIMS can differentiate those polymers that give very similar XPS spectra. Moreover, it offers more surface selectivity than XPS. The typical sampling depth of SSIMS is approximately 1 nm. This method has sensitivity, sufficient to detect amounts less than a monomolecular layer, particularly when a high resolution time-of-flight (ToF) mass analyser is used [265].

In the SIMS a primary noble gas atom or ion (e.g. Ar^0, Ar^+, Xe^0, Xe^+) beam is bombarded on the sample in ultra-high vacuum, penetrating to a depth of 30–100 Å. The kinetic energy of the particle is assumed to dissipate via a collision cascade process, which causes the emission of electrons, neutral species and secondary ions, the yields of which vary with polymer surface composition and obviates the possibility of quantitative SIMS informa-

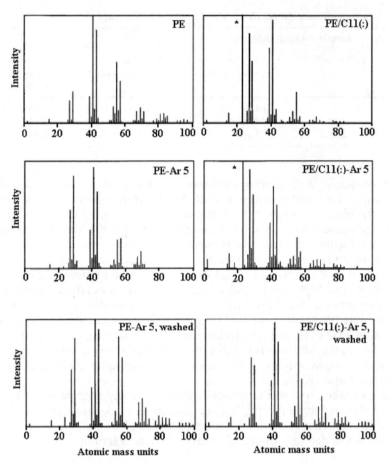

Fig. 21 Positive SSIMS spectra of PE and PE/C11(:) samples. Here, *PE-Ar 5* and *PE/C11(:)-Ar 5* indicates PE and PE/C11(:) samples treated with an argon plasma for 5 s, whereas *PE-Ar 5* and *PE/C11(:)-Ar 5, washed* indicates samples that were subsequently washed with an aqueous solution of 0.1 mM HCl

tion. The positively and negatively charged ions are analysed in terms of their mass-to-charge ratio by a quadruple or ToF mass spectrometer, yielding a positive and negative secondary ion mass spectrum.

The immobilisation of unsaturated surfactant; sodium 10-undecenoate [C11(:)] and the saturated surfactant sodium dodecanoate (C12) was carried out on PE surfaces [266] by means of an argon plasma treatment and characterised using XPS and SSIMS techniques. The typical SSIMS spectra for unmodified and modified PE films are shown in Fig. 21. The differences in the SSIMS spectra of PE and that of PE/C11(:) and PE/C11(:)-Ar 5 are merely the spectra of the surfactant which can be attributed to the presence of a layer of C11(:) on the PE sample. The SSIMS analysis showed that during the plasma treatment, surfactants were covalently coupled to the polymeric surfaces.

5.7
Confocal Laser Scanning Microscopy

Normally, the morphology of the polymer blends is studied using the prevailing techniques like optical microscopy, SEM and TEM. Although these techniques are the most sophisticated ones in their present form, still they are associated with some shortcomings such as problems in sample preparation, particularly in the case of SEM and TEM where staining methods can give rise to artefacts. In addition to this, these techniques are destructive and samples cannot be recovered back in their native form. Moreover, to obtain a three-dimensional picture of the morphology of the blends many different samples are required, which is time-consuming. With confocal laser scanning microscopy [267], which is a non-destructive method for the study of the morphology of polymer blends, it is possible to obtain three-dimensional images of the samples [268, 269, 270]. With this technique, it is also possible to obtain micrographs of the structure at different levels, starting from the surface (a few angstroms) to a maximum depth of 50 μm.

HAPEX, a bone analogue material, with similar properties to cortical bone, was added to a HDPE matrix in different volumes (20% and 40%) to produce composite materials [271]. Confocal laser scanning microscopy (CLSM) was then used to examine cell morphology on HAPEX and the surface characteristics produced by different machining protocols.

6
Applications of Surface-Modified Polyethylene

As one of the oldest and most well-accepted synthetic polymer with excellent physical and chemical properties, PE holds the largest share in the market of commodity polymers. Since surface-modified PE bears a wide range of applications, it is only possible to address a few of them and many more are included as referred literature in the following section.

6.1
Automotive Applications

High density polyethylene (HDPE) is used to make automotive fuel tanks as is an alternative material to steel. HDPE has many advantages over steel, especially the design freedom to utilise unused space and low weight. However, the shortcoming of HDPE is its permeability to gasoline and methanol so, therefore, it cannot maintain the right emission standards. Since the permeability of the liquid in the polymer depends both on the nature of the bulk as well as the surface of the polymer, the surface modification of this polymer was almost inevitable. This shortcoming is successfully overcome by fluorination of the HDPE surface using plasma/corona treatment.

6.2
Conductivity

Polymer/metal interfaces are encountered in many applications of polymers such as electronic devices, packaging applications, electrical appliances etc. Many desirable surface properties, such as electrical conductivity and optical characteristics, can be achieved with metal coating. Typically, there are two major techniques for the deposition of metal coating on polymer surfaces—sputtering and electroless plating [272]. The third and most recent approach for metallisation of PE surface is the surface grafting of conducting polymers [273] viz. polythiophene. The PE surface was modified in a three-step process to obtain a surface conductivity of 10^{-6} S cm^{-2}. A similar attempt to fabricate a conducting polymer surface was made by Neoh et al. [274].

6.3
Selective Permeability

Polymeric membranes are used in many fields to control selective flow of mixtures of gases and liquids. The transport of material through a membrane depends on the chemical and physical characteristics of both the bulk and the surface. For instance, poly trimethylsilyl propylene (PTMSP) is reported to exhibit the highest gas permeability among all existing polymers. Cation-exchange membranes [275] modified with the carboxylic acid groups for a battery separator were prepared by radiation-induced grafting of acrylic acid (AA) and methacrylic acid (MA) onto a PE film. It was found that the KOH diffusion flux of AA-grafted PE membrane and MA-grafted PE membrane increased with an increase in the degree of grafting. AA-grafted PE membrane had a higher diffusion flux than MA-grafted PE membrane. Moreover, electrical resistance of two cation-exchange membranes, modified with AA and MA, decreased rapidly with an increase in the degree of grafting. Ion-exchange membranes [276] modified with sulphonic (-SO$_3$H) and phosphonic acid (-PO$_3$H) groups were prepared by radiation-induced grafting of glycidyl methacrylate (GMA) onto PE films and subsequent sulphonation and phosphonation of poly(GMA) graft chains. The PE membrane modified with -PO$_3$H groups had a lower specific electrical resistance than that of PE membrane modified with -SO$_3$H groups.

6.4
Adhesion

Due to its inert surface, PE suffers from a large number of adhesion problems, which can be overcome by well known surface modification techniques. For example, the auto-adhesion of PE films can be improved considerably (to enhance the 'heat sealing' of PE films) by plasma, corona discharge and oxidation by inorganic oxidants. Ishihara and co-workers [277] have recently reported photo-induced graft polymerisation of a phospholipid (2-methacryloyloxyethyl phosphorylcholine) on PE membrane surface. This treatment of PE improved its resistance towards blood cell adhesion. This shows that resistance to adhesion is equally important for some specialty applications. Surface modification of HDPE by DC glow discharge significantly improved its adhesive bonding [278] to steel. Surface modification of polymer films by graft copolymerisation led to adhesive-free adhesion [279]. Excimer laser induced surface modification of ultra-high-strength PE fibres [280] enhanced their adhesion with epoxy resins. PE–PP core materials can be used for confidential postcards [281] upon surface modification with peelable laminated films. The PE–wood interfacial bonding strength was improved by using the surface-modified PE [282] in the composite material. Plasma-pretreated and graft copolymerised HDPE films [283] significantly improved their adhesion with evaporated copper.

6.5
Stabilisation

The oxidation and subsequent degradation of polymers can be partially prevented by introducing stabilisers into the polymer system. The oxidation and degradation of polymers begin from the surface and slowly proceed into the bulk. Stabilisers are therefore expected to be most effective if they are concentrated at the surface, where oxidation takes place. Singh et al. [284] have reported single- and two-step grafting procedures to covalently link photostabiliser onto the surface of polyolefins, including PE. This novel stabiliser is found to be extremely efficient in preventing photo-oxidation as compared to the conventional stabilisers.

6.6
Biomedical Applications

Together with the advances made in structural and functional substances over the last few decades, there have been an increasing number of developments in materials for their use in biomedical technology [285, 286, 287, 288, 289]. Biocompatibility and anti-thrombogeneity of several plastic materials have been improved by radiation-induced grafting [290, 291, 292, 293, 294]. A true biocompatible material should certainly posses surface merits like blood compatibility, antibacterial activity, cell/immuno-adhesivity, tissue binding etc. Among the first artificial human implants were artificial heart valves, pacemakers, vascular grafts and kidney dialysis tubes. In the following years, advances in materials engineering made possible the use of orthopaedic devices such as knee and hip joint replacements using HDPE and PP. Due to the freedom of design and their inert and non-corrosive nature, polyolefins have proved to be the most favourable alternative for bio-medical applications. Surface-modified PE balloon catheters [295] are now available for local drug delivery. Surface modification of poly(ethylene-co-vinyl alcohol) (EVA) films via carboxyl group introduction and subsequent immobilisation of collagen was carried out to generate biocompatible films [296]. Temperature-responsive controlled drug delivery [297] using biode-gradable polymers is presently under extensive study worldwide. Biocom-patible materials have been prepared by various methods since their discov-ery. PEG chains were grafted onto the surface of PE to obtain protein-resis-tant surfaces.

7
Future Trends

Looking at the present scenario, it would be right to say that polyolefins have become an irresistible element in human life. With the emerging novel applications of this commodity polymer, it has become a potent candidate in the era of specialty polymers. Though there is wide scope for the synthe-sis of new polymers, scientists today are looking forward to developing nov-el applications of this versatile polymer. In this scenario, the properties of polymer surface play a vital role in determining their importance. By tailor-ing the desired properties of the polymer surface, it would be possible to develop a wide range of exciting applications, as yet unimagined. Future research in this area would focus on the precise control of the surface prop-erties of polymeric materials by employing different strategies, such as nano-patterning, SAM, photo-lithography and coupling chemistry, to pro-vide long-term stability to these interfacial architectures. The days are not far off when polyolefins with conducting surfaces and stimuli-responsive surfaces will rule the commercial market. However, an in-depth knowledge of the polymer and its surface-associated processes is required to bring about these changes and exploit the versatility of polyolefins. Progress in the field of surface modification and characterisation shall lead to the devel-opment of next-generation elements such as micro-electronic devices and biomimetic systems with well-defined surface architectures. We hope that this review will provide up-to-date literature and a platform to the aspirants of this fascinating branch of polymer science and technology.

Acknowledgement The authors thank Dr. S. Sivaram, Director, National Chemical Laborato-ry, Pune, for his fruitful discussions and critical suggestions. Dr. S.M. Desai would like to thank C.S.I.R., India for the Senior Research Fellowship.

References

1. Wang P, Tan KL, Kang ET (2000) J Biomater Sci Polym Ed 11:169
2. Feng Z, Ranby B (1992) Die Angew Makromol Chemie 195:17
3. Kubota H, Sugiura A (1994) Polym Inter 34:313
4. Mayes AM, Irvine DJ, Griffith LG (1998) Mater Res Soc Symp Proc 530:73

5. Biesalski M, Ruehe J (1999) Macromolecules 32:2309
6. Ikada Y (1996) Macromol Symp 101:455
7. Hoffman AS (1996) Macromol Symp 101:443
8. Lucki J, Rabek JF, Ranby B (1990) Polymer 31:1772
9. Klee D, Villari RV, Hocker H, Dekker B, Mittermayer C (1994) J Mater Sci Mater Med 5:592
10. Tatiana M, Paul J (1982) Mater Plast 19:87
11. Ihara T, Yasuda HK (1990) J Appl Polym Sci Appl Polym Symp 46:511
12. Russell K (1995) J Poly Sci Part A Poly Chem 33:555
13. Hirotsu T, Asai N (1991) J Macromol Sci-Chem A 28:461
14. Pashova VS, Georgiev GS, Dakov VA (1994) J Appl Polym Sci 51:807
15. Inagaki N, Tasaka S, Inoue T (1998) J Appl Polym Sci 69:1179
16. Kang ET, Neoh KG, Shi JL, Tan KL, Liaw DJ (1999) Polym Adv Technol 10:20
17. Noda I (1991) Nature 350:143
18. Oster G, Oster GK, Moroson HJ (1959) J Polym Sci 34:671
19. Ringrose BJ, Kronfli E (2000) Eur Polym J 36:591
20. Franchina JG, Bergbreiter DE (1998) Polym Mater Sci Eng 79:17
21. Levitzsky J, Lindsey FJ, Kaghan WS (1964) SPE J 1305
22. Sutherland I, Brewis DW, Heath RJ, Sheng E (1991) Surf Interface Anal 17:507
23. Briggs D, Brewis DM, Konieczko MB (1979) J Mater Sci 14:344
24. Garbassi F, Occhiello E, Polato F (1987) J Mater Sci 22:207
25. Aldo N (1999) EP 965625
26. Bader, MJ, O'Brien JJ, Riddle KL (1996) WO 9638299
27. Mittal KL, Susko JR (eds) (1991) Metallized plastics, fundamental and applied aspects, vol 1. Plenum, New York
28. Peet RG (2000) WO 2000058088
29. Mount EM, Wagner JR (1998) WO 9832597
30. Jones N, Stoilovic M, Lennard C, Roux C (2001) Forensic Sci Int 115:73
31. Fonseca C, Pereña JM, Fatou JG, Bello A (1985) J Mater Sci 20:3283
32. Wu YL, Yu-xiang H, Fu-zeng W, Xu-qin G (1999) Gaofenzi Xuebao 12:427
33. Martínez-Salazar J, Keller A, Cagiao ME, Rueda DR, Baltá-Calleja FJ (1983) Colloid Polym Sci 261:412
34. Blaiss P, Carlsson DJ, Csullog GW, Wiles DM (1974) J Colloid Interface Sci 47:636
35. Lee KW, McCarthy TJ (1988) Macromolecules 21:2318
36. Franchina NL, McCarthy TJ (1991) Macromolecules 24:3045
37. Perena JM, Lorenzo V, Zamfirova G, Dimitrova A (1999) Polym Test 19:231
38. Yu M, Teng C, Gu L (1999) J China Text Univ 16:1
39. Sun C, Zhang D, Wadsworth LC (1999) Adv Polym Technol 18:171
40. Ogawa T, Mukai H, Osawa S (1999) J Appl Polym Sci 71:243
41. Park SJ, Jin JS (2001) J Colloid Interface Sci 236:155
42. Briggs D, Kendall CR (1982) Int J Adhes Adhes 2:13
43. Lei J, Liao X (2001) Eur Polym J 37:771
44. Strobel M, Dunatov C, Strobel JM, Lyons CS, Perron SJ, Morgen MC (1989) J Adhesion Sci Technol 3:321
45. Gerenser LJ, Elman JF, Mason MG, Wootton AB (1983) Polymer 24:47
46. Gamez-Garcia M, Bartnikas R, Wertheimer MR (1990) IEEE Trans Electr Insul 25:688
47. Lei J, Liao X (2001) J Appl Polym Sci 81:2881
48. Matsunaga M, Whitney PJ (2000) Polym Degrad Stab 70:325
49. Okoshi M, Murahara M (1994) J Photopolym Sci Technol 7:381
50. Hollaender A, Behnisch J, Zimmermann H (1993) J Appl Polym Sci 49:1857
51. Toth A, Bell T, Bertoti I, Mohai M, Zelei B (1999) Nucl Instrum Methods Phys Res Sect B 148:1131
52. Ranby B (1994) Mater Res Innovations 2:64
53. Kavc T, Kern W, Ebel MF, Svagera R, Poelt P (2000) Chem Mater 12:1053
54. Weikart CM, Yasuda HK (2000) J Polym Sci Part A Polym Chem 38:3028

55. Bamford CH, Ward JC (1961) Polymer 22:277
56. Razzak MT, Tabata Y, Otsuhata K (1993) Radiat Phys Chem 42:57
57. Lutfor M, Silong S, Yunus W, Zaki Ab Rahman M, Ahmad M, Haron M (2000) J Appl Polym Sci 77:784
58. Mishra SN, Lenka S, Nayak PL (1991) Eur Polym J 27:1319
59. Lehrle RS, Willis SL (1997) Polymer 38:5937
60. Ratner BD, Balisky T, Hoffman AS (1977) J Bioeng 1:115
61. Takayanagi M, Kayatose T (1983) J Polym Sci Polym Chem Ed 21:31
62. Yasuda H (1985) Plasma polymerization. Academic Press, New York
63. Ichijima H, Okada T, Uyama Y, Ikada Y (1991) Makromol Chem 192:1213
64. Epaillard FP, Brosse JC, Falher T (1998) Macromol Chem Phys 199:1613
65. Lee HJ, Nakayama Y, Matsuda T (1999) Macromolecules 32:6989
66. Burfield DR, Ng SC (1978) Eur Polym J 14:799
67. Mecerreyes D, Dubois P, Jenrome R, Hedrick JL (1999) Macromol Chem Phys 222:156
68. Lehman J, Dreyfuss P (1979) Adv Chem Ser 176:587
69. Zakharov VA, Bukatov GD, Chumaevskii NB, Ermakov YI (1977) Makromol Chem 178:967
70. Lenka S, Nayak PL, Mohanty IB (1985) J Appl Polym Sci 30:2711
71. Bouquet G, Kentie WC, de Theije PJG, Van Damme F (1996) Polym Prep 37:536
72. Li Y, Desimone JM, Poon CD, Samuski ET (1997) J Appl Polym Sci 64:883
73. Allen NS, Corrales T, Edge M, Peinado FCC, Pinar MB, Green A (1998) Eur Polym J 34:303
74. Mansour OY, Nagaty A (1985) Prog Polym Sci 11:91
75. Gargan K, Kronfli E, Lovell KV (1990) Radiat Phys Chem 36:757
76. Lenka S, Nayak PL, Das AP (1985) J Appl Polym Sci 30:2753
77. Rangaswamy C, Milford H (1991) Starch 43:396
78. Pan B, Moore RB (1999) Polym Prepr 40:764
79. Baker HS, Mclellan WE, Russell PJ (1997) Can Chem News 49:25
80. Ichazo MN, Rosales CM, Perera R, Sanchez AM, Rojas HA, Vivas M (1996) Lat Am Appl Res 26:77
81. Hu GH, Flat JJ, Lambla M (1997) In: Al-Malaika S (ed) Reactive modifiers for polymers. Blackie, London, UK, p 1
82. Lazzeri A, Malanima M, Pracella M (1999) J Appl Polym Sci 74:3455
83. De Los Santos Gonzalez EA, Gonzalez MJL, Gonzalez MC (1998) J Appl Polym Sci 68:45
84. Sanduja ML, Horowitz C, Thottathil P (1998) US 5763557
85. Mirzadeh H, Khorasani TM, Katbab AA, Burford R (1993) J Polym Sci Technol (Persian) 6:4
86. Hohol MD, Urban MW (1994) Polymer 35:5560
87. Minghong W, Bao B, Chen J, Xu Y, Zhou S, Ma Z (1999) Radiat Phys Chem 56:341
88. Knittel D, Schollmeyer E (1998) Polym Int 45:103
89. Ada ET, Kornienko O, Hanley L (1998) J Phys Chem B 102:3959
90. Song Q, Netravali AN (1998) J Adhes Sci Technol 12:957
91. Muratake H, Shigemitsu Y (1997) Kawamura Rikagaku Hokoku pp 67–75 (in Japanese)
92. Epaillard FP, Chevet B, Brosse JC (1990) Eur Polym J 26:333
93. Aoyama H, Kiguchi, Jpn Kokai Tokkyo Koho (1992) JP 04295818 A2
94. Yamakawa S (1976) J Appl Polym Sci 20:3057
95. Carreon MP, Aliev R, Ocampo R, Burillo G (2000) Polym Bull 44:331
96. Hebeish A, Shalaby SE, Bayazeed AM (1978) J Appl Polym Sci 22:3335
97. Hsieh YL, Shinawarta M, Castillo MD (1986) J Appl Polym Sci 31:509
98. Kurbanov SA, Musaev UN, Khakimdzhanov BS, Smurova EV, Novikova SP, Dobrova NB, Nauchn T (1976) Tashk Gos Univ (Russian) 502:129
99. Yamamoto F, Yamakawa S, Kato Y (1978) J Polym Sci Polym Chem Ed 16:1897
100. Yamamoto F, Yamakawa S (1980) J Polym Sci Polym Chem Ed 18:2257

101. Svorcik V, Proskova K, Hnatowicz V, Rybka V (2000) J Appl Polym Sci 75:1144
102. Catalog: immersion lamps for laboratory experiments. Original Hanau, Germany
103. Chan CM (1994) Surface modification and characterization of polymers. Hanser, USA
104. Ranby B, Rabek JF (1975) Photodegradation, photo-oxidation and photostabilization of polymers. Interscience, New York, p 45
105. Yang W, Rånby B (1999) Eur Polym J 35:1557
106. Uyama Y, Tadokoro H, Ikada Y (1990) J Appl Polym Sci 39:489
107. Tang KL, Woon LL, Wong HK, Kang ET, Neoh KG (1993) Macromolecules 26:2832
108. Me'dard N, Souti JC, Epaillard FP (2002) Langmuir 18:2246
109. Kim HJ, Lee KJ, Seo Y, Kwak S, Koh SK (2001) Macromolecules 34:2546
110. Novak I, Chodak I, (2001) J Macromol Sci Pure Appl Chem A 38:11
111. Wu JZ, Kang ET, Neoh KG, Wu PL, Liaw DJ (2001) J Appl Polym Sci 80:1526
112. Inagaki N, Tasaka S, Ohkubo J, Kawai H (1990) J Appl Polym Sci: Appl Polym Symp 46:399
113. Gupta B, Hilborn JG, Bisson I, Frey P (2001) J Appl Polym Sci 81:2993
114. Gupta B, Anjum N, Gupta AP (2000) J Appl Polym Sci 77:1401
115. Allmer K, Hult A, Raanby, B (1989) J Polym Sci Part A: Polym Chem 27:1641
116. Dasgupta S (1990) J Appl Polym Sci 41:233
117. Batich C, Yahiaoui A (1987) J Polym Sci Part A: Polym Chem 25:3479
118. Bandopadhay D, Panda AB, Pramanik (2001) J Appl Polym Sci 82:406
119. Zhao J, Geuskens G (1999) Eur Polym J 35:2115
120. Fayek SA, El Sayed SM, El-Arnaouty MB (2000) Polym Test 19:435
121. Park SJ, Jin JS (2001) J Colloid Interface Sci 236:155
122. Lei J, Liao X (2001) J Appl Polym Sci 81:2881
123. Lei J, Liao X (2001) Eur Polym J 37:771
124. Takahashi N, Goldman A, Goldman M, Rault J (2000) J Electrost 50:49
125. Dalinkevich AA, Kiryushkin SG, Shlyapnikov YA (1989) Vysokomol Soedin Ser A 31:1955
126. Novis Y, De Meulemeester R, Chtaib M, Pireaux JJ, Caudano R (1989) Br Polym J 21:147
127. Doyle DJ (1993) Proc SPIE-Int Soc Opt Eng, p 260
128. Sarkar N, Bhattacharjee S, Sivaram S (1997) Langmuir 13:4142
129. Edge S, Walker S, Feast J, Pacynko WF (1993) J Appl Polym Sci 47:1075
130. De Gooijer JM, Scheltus M, Koning CE (2001) SPE: Polym Eng Sci 41:86
131. Kavc T, Kern W, Ebel, Maria F, Svagera R, Poelt P (2000) Chem Mater 12:1053
132. Cross EM, McCarthy TJ (1992) Macromolecules 25:2603
133. Chanunpanich N, Ulman A, Strzhemechny YM, Schwarz SA, Janke A, Braun HG, Kratzmueller T(1999) Langmuir 15:2089
134. Balamurugan S, Mandale AB, Badrinarayanan S, Vernekar SP (2001) Polymer 42:2501
135. Balamurugan S (2000) PhD Thesis, University of Pune, India
136. Beamson G, Briggs D (1992) High resolution XPS spectra of polymers: the Scienta ESCA 300 data book. Wiley, Chichester, UK, p 274
137. Chanunpanich N, Ulman A, Malagon A, Strzhemechny YM, Schwarz SA, Janke A, Kratzmueller T, Braun HG (2000) Langmuir 16:3557
138. Foerch R, Izawa J, McIntyre NS, Hunter DH (1990) Polym Mater Sci Eng 62:428
139. Sugiyama K, Matsumoto T, Yamazaki Y (2000) Macromol Mater Eng 282:5
140. Lewis KB, Ratner BD (1993) J Colloid Interface Sci 159:77
141. Lee SH, Ruckenstein E (1987) J Colloid Interface Sci 120:529
142. Lavielle L (1988) Orientation phenomena at polymer-water interfaces. In: Andrade JD (ed) Polymer surface dynamics. Plenum, New York, p 45
143. Lavielle L, Schultz J (1985) J Colloid Interface Sci 106:446
144. Andrade JD, Gregonis DE, Smith LM (1985) Polymer surface dynamics. In: Andrade JD (ed) Surface and interfacial aspects of biomedical polymers, vol 1: surface chemistry and physics. Plenum, New York
145. Lavielle L, Schultz J (1985) J Colloid Interface Sci 106:438

146. Rasmussen JR, Bergbreiter DE, Whitsides GM (1977) J Am Chem Soc 99:4746
147. Holmes-Farley SR, Reamey RH, Nuzzo R, McCarthy TJ, Whitesides GM (1987) Langmuir 3:799
148. Yao Z, Raanby B (1990) J Appl Polym Sci 40:1647
149. Silverstein MS, Breuer O, Dodiuk H (1994) J Appl Polym Sci 52:1785
150. Fujimatsu H, Imaizumi M, Shibutani N, Usami H, Iijima T (2001) Polym J 33:509
151. Qu X, Wirsen A, Olander B, Albertsson AC (2001) Polym Bull 46:223
152. Bauman BD (1998) Addcon World '98: Addit. New Millennium, Off Book Paper, Int Plast Addit Modif Conf, Rapra Technology, Shrewsbury, UK
153. Jingxin L, Jun G, Rong Z, Baoshan Z, Jian W (2000) Polym Int 49:1492
154. Dong H, Bell T, B'lawert C, Mordike BL (2000) J Mater Sci Lett 19:1147
155. Netravali AN, Caceres JM, Thompson MO, Renk TJ (1999) J Adhes Sci Technol 13:1331
156. Taboudoucht T, Opalco R, Ishida H (1992) Polym Compos 13:81
157. Chaussy J, Escalon G, Gianese P, Roux JP (1978) Cryogenics 18:501
158. Desai SM, Singh RP (2003) Polymer (personal communication)
159. Desai SM, Bodas D, Patole M, Singh RP (2003) J Biomater Sci Polym Ed (accepted)
160. McGinniss VD, Sliemers FA (1983) WO 8303419
161. Mirzadeh H, Ekbatani AR, Katabab AA (1996) Iran Polym J 5:225
162. Kowalski A, Perkowski J, Jezierski M, Jankowski B, Wojtulewicz M (1978) Przem Chem 57:521 (in Polish)
163. Xu Q, Zhang X, Yang Y, Zhang Y (1998) Hecheng Xiangjiao Gongye 21:75
164. Haddadi-asl V, Burford RP, Garnett JL (1994) Radiat Phys Chem 44:385
165. Katbab AA, Burford RP, Garnett JL (1992) Radiat Phys Chem 39:293
166. Greco R, Maglio G, Musto PV (1987) J Appl Polym Sci 33:2513
167. Haddadi-asl V, Burford RP, Garnett JL (1995) Radiat Phys Chem 45:191
168. Bhowmick AK, Majumder PS, Banik I (1999) Macromol Symp 143:2
169. Mishra MK (1981) J Macromol Sci-Rev Makromol Chem C 20:149
170. Patil SR, Konar RS (1962) J Polym Sci 58:85
171. Nakayama Y, Matsuda T (1996) Macromolecules 29:8622
172. Anders C, Gratner R, Steinert V, Voit BI, Zschoche S (1999) J M S-Pure and Appl Chem A 36:1017
173. Gu L, Hrymak AN, Zhu S (1999) J Appl Polym Sci 76:1412
174. Naqvi MK, Reddy R (1997) Polym-Plast Technol Eng 36:585
175. Xie H, Seay M, Oliphant K, Baker WE (1993) J Appl Polym Sci 48:1199
176. Xie HQ, Baker WE (1993) In: Chung TC (ed) New advances in polyolefins. Plenum, New York
177. Harmer MA (1991) Langmiur 7:2010
178. Scriven EF (ed) (1984) Azides and nitrenes. Academic Press, New York
179. Katoot MW (1997) WO 9739838
180. Dadsetan M, Mirzadeh H, Sharifi-Sanjani N (2000) J Appl Polym Sci 76:401
181. Kondo T, Kubota H, Katakai R (1999) J Appl Polym Sci 71:251
182. Mirzadeh H, Katbab AA, Khorasani MT, Burford RP, Golastani GE (1995) Biomaterials 16:641
183. Mirzadeh H, Katbab AA, Burford RP (1995) Radiat Phys Chem 46:859
184. Brannon JH, Lankard JR (1986) J Appl Phy Lett 48:1226
185. Chtaib M, Roberfroid EM, Novis Y, Pireaux JJ, Caudano R (1990) ACS Symp Ser 440:161
186. Mirzadeh H, Ekbatani AR, Katbab AA (1996) Iran Polym J 5:4
187. Mirzadeh H, Katbab AA, Burford RP (1993) Radiat Phys Chem 41:507
188. Kondo T, Gunma-ken K (1999) Shikenjo Kenkyu Hokoku p 87
189. Kondo T, Kubota H, Katakai R (1999) J Appl Polym Sci 74:2462
190. Price G, Clifton AA, Keen F (1996) Polymer 37:5825
191. Price G, Clifton AA, Keen F (1996) Macromolecules 29:5664
192. Garnett JL, Jankiewics SV, Levot R, Sangster DF (1985) Radiat Phys Chem 25:509
193. Hsiue GH, Huang WK (1885) J Appl Polym Sci 30:1023

194. Ishigake I, Sugo T, Takayama T, Okada T, Okamoto J, Machi S (1982) J Appl Polym Sci 27:1043
195. Collinson E, Dainton F, Smith D, Trudel G, Tazuke S (1972) J Appl Polym Sci 16:921
196. Dworjanyn PA, Garnett JL (1990) In: Bellobono RI (ed) Proceedings international meeting in grafting processes on polymer films and surfaces: scientific and technological aspects. Italy, p 61
197. Haddadi-asl V, Burford RP, Garnett JL (1994) Radiat Phys Chem 44:385
198. Dworjanyn PA, Field B, Garnett JL (1989) ACS Symp Ser 38:1
199. Liu S, Sen A (2000) Poly Preprints 41:1573
200. Goss CA, Burmfield JC, Irene EA, Murray RW (1992) Langmuir 8:1459
201. Willis HA, Zichy VJI (1978) In: Clark DT, Feast WJ (eds) Polymer surfaces. Wiley, New York p 287
202. Pireaux JJ, Vermeersch M, Degosserie N, Gregoire C, Novis Y, Chtaib M, Caudano R (1989) Surf Sci 17:53
203. Sawyer LC, Grubb DT (ed) (1987) Polymer microscopy. Chapman and Hall, London
204. Ynag R, Ynag XR, Evans DF, Hendrickson WA, Baker J (1990) J Phys Chem 94:6123
205. Leung OM, Goh MC (1992) Science 255:64
206. Briggs D (1982) Surf Interface Anal 4:151
207. Briggs D (1984) Polymer 25:379
208. Batich CD, Wendt RC (1981) ACS Symp Ser 162:221
209. Park SW, Kim DH, Kim YM, Park BS, Han WS, Suh BS (1994) Anal Sci Technol 7:301
210. Kornienko O, Ada ET, Hanley L (1997) Anal Chem 69:1536
211. Koprinarov I, Lippitz A, Friedrich JF, Unger WES, Woell C (1997) Polymer 38:2005
212. Unger WES, Lippitz A, Woell C, Heckmann W (1997) J Anal Chem 358:89
213. Steiner G, Sablinskas V, Hubner A, Kuhne C, Salzer R (1999) J Mol Struct 509:265
214. Young J (1805) Phil Tran 95:82
215. Neumann AW, Good RJ (1979) Surf Colloid Sci 11:31
216. Adamson AW (1982) Physical chemistry of surfaces. Wiley, New York
217. Ambwani DS, Fort T Jr (1979) Surf Colloid Sci 11:93
218. Ko YC, Ratner BD, Hoffman AS (1981) J Colloid Interface Sci 82:25
219. Pitt WG, Young BR, Cooper SL (1987) Colloids Surf 27:345
220. Favia P, Palumbo F, D'Agostino R, Lamponi S, Magnani A, Barbucci R (1998) Plasmas Polym 3:77
221. Jaficzuk B, Bialopiotrowicz T, Zdziennicka A (1999) J Colloid Interface Sci 211:96
222. Kwon OH, Nho YC, Park KD, Kim YH (1999) J Appl Polym Sci 71:631
223. Wickson BM, Brash JL (1999) Colloids Surf A 156:201
224. Dhamodharan R, Nisha A, Pushkala K, McCarthy TJ (2001) Langmuir 17:3368
225. Praschak D, Bahners T, Schollmeyer E (2000) Appl Phys A: Mater Sci Process 71:577
226. Feiertag P, Kavc T, Meyer U, Gsoels I, Kern W, Rom I, Hofer F (2001) Synth Met 121:1371
227. Sefton MV, Sawyer A, Gorbet M, Black JP, Cheng E, Gemmell C, Pottinger-Cooper EJ (2001) Biomed Mater Res 55:447
228. Jama C, Quensierre JD, Gengembre L, Moineau V, Grimblot J, Dessaux O, Goudmand P (1999) Surf Interface Anal 27:653
229. Matsunaga M, Whitney PJ (2000) Polym Degrad Stab 70:325
230. Deng JP, Yang WT, Ranby B (2000) Polym J 32:834
231. Holmes-Farley SR, Bain CD, Whitesides GM (1988) Langmuir 4:921
232. Wilson MD, Ferguson GS, Whitesides GM (1990) J Am Chem Soc 112:1244
233. Sugiyama K, Matsumoto T, Yamazaki Y (2000) Macromol Mater Eng 282:5
234. Mackay ME, Carmezini G, Sauer BB, Kampert W (2001) Langmuir 17:1708
235. Harrick NJ (1964) J Phys Chem 64:1110
236. Fahrenfort J (1961) J Spectrochim Acta 17:698
237. Evanson KW, Urban MW (1991) J Appl Polym Sci 42:7
238. Urban MW (1996) Attenuated Total reflectance spectroscopy of polymers: polymer surfaces and interfaces series. ACS, Washington DC

239. Meichsner J (1999) Contrib Plasma Phys 39:427
240. Afanasyeva NI (1999) Macromol Symp 141:117
241. Chattopadhyay S, Ghosh RN, Chaki TK, Bhowmick AK (2001) J Adhes Sci Technol 15:303
242. Ishii T, Kuroda SI, Kubota H, Kondo T, Yoshinori I (2000) J Macromol Sci Pure Appl Chem A 37:807
243. Borgwarth K, Ricken C, Ebling DG, Heinze J (1996) J Anal Chem 356:288
244. Giurginca M, Ivan G, Herdan JM (1994) Polym Degrad Stab 44:79
245. Liu J, Jen H, Chung YC (1999) J Appl Polym Sci 74:2947
246. Jacqueline MA, Dooley RL, Shalaby SW (2000) J Appl Polym Sci 76:1865
247. Zang D, Gracias DH, Ward R, Gauckler M, Tian Y, Shen YR, Somorjai GA (1998) J Phys Chem B 102:6225
248. Binnig G, Recher H, Gerber C, Weibel E (1982) Phys Rev Let 49:57
249. Bai C, Li J, Lin Z, Tang J, Wang C (1999) Surf Interface Anal 28:44
250. Holger S, Hruska Z, Vancso GJ (2000) Macromolecules 33:4532
251. Binnig G, Quate CF, Gerber C (1986) Phys Rev Let 56:930
252. Holland NB, Marchant RE (2000) J Biomed Mater Res 51:307
253. Ganapathy R, Manolache S, Sarmadi M, Simonsick WJ Jr, Denes F (2000) J Appl Polym Sci 78:1783
254. Michel PL, Werts EW, Georges H (1997) Langmuir 13:4939
255. Gerenser LJ (1993) J Adhes Sci Technol 7:1019
256. Watts JF, Wolstenholme J (2003) An introduction to surface analysis by XPS and AES. Wiley, Europe
257. Dilks A (1981) In: Brundle CR, Baker AD (eds) Electron spectroscopy—theory, techniques and applications, vol 4. Academic Press, London
258. Klapperich CM, Komvopoulos K, Pruitt L (1999) Mater Res Soc Symp Proc 550:331
259. Jama C, Quensierre JD, Gengembre L, Moineau V, Grimblot J, Dessaux O, Goudmand P (1999) Surf Interface Anal 27:653
260. Iriyama Y (2001) J Photopolym Sci Technol 14:105
261. Qu X, Wirsén A, Olander B, Albertsson AC (2001) Polymer Bull 46:223
262. Brown A, Vickerman JC (1984) Surf Interface Anal 6:1
263. Briggs D, Hearn MJ, Ratner BD (1984) Surf Interface Anal 6:184
264. Odom RW (1993) Microbeam Anal 2:99
265. Bertrand P, Wang TI (1997) In: Brune D (ed) S mode SIMS: a review, surface characterization. Wiley-VCH, Weinheim, Germany, pp 334–353
266. Lens JP, Terlingen JGA, Engbers GHM, Feijen J (1998) J Polym Sci Part A Polym Chem 36:1829
267. Draaijer A, Houpt PM (1989) Inst of Phys Conf Ser 98, Chap 14, London
268. Semler EJ, Tjia JS, Moghe PV (1997) Biotechnol Prog 13:630
269. Verhoogt H, Van Dam JA, Posthuma de Boer A, Draaijer A, Houpt PM (1993) Polymer 34:1325
270. Bourban C, Mergaert J, Ruffieux K, Swings J, Wintermantel E (1998) ACS Symp Ser 694:194
271. Dalby MJ, Di Silvio L, Davies GW, Bonfield W (2000) J Mater Sci Mater Med 11:805
272. Charbonnier M, Romand M, Stremsdoerfer G, Fares-Karam A (1999) Rec Res Dev Macromol Res 4:27
273. Chanunpanich N, Ulman A, Strzhemechny YM, Schwarz SA, Dornicik J, Rafailovich M, Sokolov J, Janke A, Braun HG, Kratzmuller T (2000) Mater Res Soc Symp Proc 600:203
274. Liesegang J, Senn BC, Pigram PJ, Kang ET, Tan KL, Neoh KG (1999) Surf Interface Anal 28:20
275. Choi SH, Park SY, Nho YC (2000) Radiat Phys Chem 57:179
276. Choi SH, Nho YC (1999) Korean J Chem Eng 16:725
277. Ishihara K, Iwasaki Y, Ebihara S, Shindo Y, Nakabayashi N (2000) Colloids Surf B 18:325

278. Bhowmik S, Ghosh PK, Ray S, Barthwal SK (1998) J Adhes Sci Technol 12:1181
279. Kang ET, Neoh KG, Li ZF, Tan KL, Liaw DJ (1998) Polymer 39:2429
280. Song Q, Netravali AN (1998) J Adhes Sci Technol 12:983
281. Kawashima N (1998) JP 10305539
282. Razi PS, Portier R, Raman A (1999) J Compos Mater 33:1064
283. Ng CM, Oei HP, Wu SY, Zhang MC, Kang ET, Neoh KG (2000) Polym Eng Sci 40:1047
284. Desai SM, Pandey JK, Singh RP (2001) Macromol Symp Ser 169:112
285. Gottenbos B, Van der Mei HC, Busscher HJ (2000) J Biomed Mater Res 50:208
286. Haik Y, Chatterjee J, Chen CJ (2001) Polym Prepr 42:125
287. Nakaoka R, Tsuchiya T, Kato K, Ikada Y, Nakamura A (1997) J Biomed Mater Res 35:391
288. Holmberg K, Tiberg F, Malmsten M, Brink C (1997) Colloids Surf A 123–124:297
289. Udipi K, Ornberg RL, Thurmond KB II, Settle SL, Forster D, Riley D (2000) J Biomed Mater Res 51:549
290. Geckeler KE, Gebhardt R, Grunwald H (1997) Naturwissenschaften 84:150
291. Wu DY, Li S, Gutowski WS (1997) WO 9702310
292. Foerch R, Graham BA, Walzak MJ, Hunter DH (1996) CA 2177890
293. Sugawara T, Matsuda T (1997) J Polym Sci Part A Polym Chem 35:137
294. Jansen B, Steinhauser H, Prohaska (1986) Makromol Chem Macromol Symp 5:237
295. Richey T, Iwata H, Oowaki H, Uchida E, Matsuda S, Ikada Y (2000) Biomaterials 21:1057
296. Matsumura K, Hyon SH, Nakajima N, Peng C, Tsutsumi S (2000) J Biomed Mater Res 50:512
297. Ratner BD, Hoffman AS (1974) J Appl Polym Sci 18:3183

Received: July 2003

Author Index Volumes 101–169

Author Index Volumes 1-100 see Volume 100

Subject Index

Oxidative degradation   89
5-Oxohexanoic acid   193
12-Oxotridecanoic acid   193

Package material   210
Packaging waste   210
Paint residues   210
Paraffins   182
Partial least squares   216
PE *see* Polyethylene
Penicillium simplicissimum   186
Peroxidation   270
Peroxides, DMS-resistant   159
–, total luminous intensity   159
Peroxidolytic activity   132
Persulphate   252
PET   203
PGEMA   263
Phase change   106
Phenolates, metal ion   136
Phenols, hindered   128, 129
Phosphites   128, 132
Phosphonites   128
Photo-antioxidant   141
Photo-bromination   261
Photo-degradation   179
Photo-grafting   246, 258
Photo-initiators   126, 177, 180
Photo-oxidation   180
Photo-sensitisers   194
Photo-sulphonation   260
Plasma grafting   248
Plasma reactor   249
Plasma treatment   263
Plating, electro-less   236
Poly trimethylsilyl propylene   284
Poly(ethylene-*co*-propylene)   213
Poly(ethylene-*co*-vinyl alcohol)   286
Poly(4-methyl-1-ene)   142
Poly(4-methyl-1-pentene)   78
Poly(vinyl alcohol)   194
Polyamide, oxidation   163
Polyamide 6   112
Polybutene   78, 96
Polycarbonate   78
Polydispersity   8
Polyethenes, metallocene catalysis   1
Polyethylene, adjacency of chain
   folding   45
– blends, binary linear   66
–, cellulation   61, 69
–, chain folding   34
–, crystal lamellae, splaying/branching   51
–, crystal structure   31
–, crystal thickening   38

–, degradable   179
–, dielectric α process   32
–, dielectric/optical properties   32
–, grafting   231
–, high-pressure process   30
–, isothermal thickening   42
–, lenticular single crystal   49
–, linear, equilibrium melting point   44
–, liquid-like amorphous phase   48
–, low-pressure process   30
–, lozenge-shaped crystals   49
–, mass fraction, interfacial component   48
–, mechanical α process   33
–, molecular fractionation   62
– morphology   29
–, multilayer crystals   37
–, permanganic etching technique   54, 55
–, phase separation, molten state   62
–, physical forms   265
– piping   88
– processes, bimodal   21
– –, high-pressure   16
– –, low-pressure   18
–, random lamellar structure   53
–, random switchboard model   46, 48
– reactors, high-pressure   17
–, recycled   212
–, regular chain folding   45
–, ribbon   46
–, segregation   61
–, single crystal   34
–, – –, fold-chain   35
–, – –, non-planar shapes   35
–, single crystal, sectorization   35, 36
–, subsidiary lamellae   64
–, superfolding model   46
– technologies   13
–, tent-shaped   36
–, thickness of initial crystal   43
Polyethylene glycol diacrylate   271
Polymerisation catalysts, single-site,
   history   4
Polyolefins, commodity, properties   79
– oxidation   151
–, recycled   201
Post-consumer (plastic) waste   208, 211
Post-industrial waste   206
PP   125, 151, 153, 203, 223
– / HDPE   221
PP-PS   213
Principal component analysis   219
Printing inks   210
Processing temperature   130
Pro-oxidant   180
Propagation   122

Printing: Saladruck, Berlin
Binding: Stein+Lehmann, Berlin